Reviews of Environmental Contamination and Toxicology

VOLUME 258

More information about this series at https://link.springer.com/bookseries/398

Reviews of Environmental Contamination and Toxicology Volume 258

Editor
Pim de Voogt

Founding Editor
Francis A. Gunther

Volume 258

Springer

ISSN 0179-5953 ISSN 2197-6554 (electronic)
Reviews of Environmental Contamination and Toxicology
ISBN 978-3-030-88325-6 ISBN 978-3-030-88326-3 (eBook)
https://doi.org/10.1007/978-3-030-88326-3

This Springer imprint is published by the registered company Springer Nature Switzerland AG
The registered company address is: Gewerbestrasse 11, 6330 Cham, Switzerland

Foreword

International concern in scientific, industrial, and governmental communities over traces of xenobiotics in foods and in both abiotic and biotic environments has justified the present triumvirate of specialized publications in this field: comprehensive reviews, rapidly published research papers and progress reports, and archival documentations These three international publications are integrated and scheduled to provide the coherency essential for nonduplicative and current progress in a field as dynamic and complex as environmental contamination and toxicology. This series is reserved exclusively for the diversified literature on "toxic" chemicals in our food, our feeds, our homes, recreational and working surroundings, our domestic animals, our wildlife, and ourselves. Tremendous efforts worldwide have been mobilized to evaluate the nature, presence, magnitude, fate, and toxicology of the chemicals loosed upon the Earth. Among the sequelae of this broad new emphasis is an undeniable need for an articulated set of authoritative publications, where one can find the latest important world literature produced by these emerging areas of science together with documentation of pertinent ancillary legislation.

Research directors and legislative or administrative advisers do not have the time to scan the escalating number of technical publications that may contain articles important to current responsibility. Rather, these individuals need the background provided by detailed reviews and the assurance that the latest information is made available to them, all with minimal literature searching. Similarly, the scientist assigned or attracted to a new problem is required to glean all literature pertinent to the task, to publish new developments or important new experimental details quickly, to inform others of findings that might alter their own efforts, and eventually to publish all his/her supporting data and conclusions for archival purposes.

In the fields of environmental contamination and toxicology, the sum of these concerns and responsibilities is decisively addressed by the uniform, encompassing, and timely publication format of the Springer triumvirate:

Reviews of Environmental Contamination and Toxicology [Vol. 1 through 97 (1962–1986) as Residue Reviews] for detailed review articles concerned with any aspects of chemical contaminants, including pesticides, in the total environment with toxicological considerations and consequences.

Bulletin of Environmental Contamination and Toxicology (Vol. 1 in 1966) for rapid publication of short reports of significant advances and discoveries in the fields of air, soil, water, and food contamination and pollution as well as methodology and other disciplines concerned with the introduction, presence, and effects of toxicants in the total environment.

Archives of Environmental Contamination and Toxicology (Vol. 1 in 1973) for important complete articles emphasizing and describing original experimental or theoretical research work pertaining to the scientific aspects of chemical contaminants in the environment.

The individual editors of these three publications comprise the joint Coordinating Board of Editors with referral within the board of manuscripts submitted to one publication but deemed by major emphasis or length more suitable for one of the others.

Coordinating Board of Editors

Preface

The role of *Reviews* is to publish detailed scientific review articles on all aspects of environmental contamination and associated (eco)toxicological consequences. Such articles facilitate the often complex task of accessing and interpreting cogent scientific data within the confines of one or more closely related research fields.

In the 50+ years since *Reviews of Environmental Contamination and Toxicology* (formerly *Residue Reviews)* was first published, the number, scope, and complexity of environmental pollution incidents have grown unabated. During this entire period, the emphasis has been on publishing articles that address the presence and toxicity of environmental contaminants. New research is published each year on a myriad of environmental pollution issues facing people worldwide. This fact, and the routine discovery and reporting of emerging contaminants and new environmental contamination cases, creates an increasingly important function for *Reviews*. The staggering volume of scientific literature demands remedy by which data can be synthesized and made available to readers in an abridged form. *Reviews* addresses this need and provides detailed reviews worldwide to key scientists and science or policy administrators, whether employed by government, universities, nongovernmental organizations, or the private sector.

There is a panoply of environmental issues and concerns on which many scientists have focused their research in past years. The scope of this list is quite broad, encompassing environmental events globally that affect marine and terrestrial ecosystems; biotic and abiotic environments; impacts on plants, humans, and wildlife; and pollutants, both chemical and radioactive; as well as the ravages of environmental disease in virtually all environmental media (soil, water, air). New or enhanced safety and environmental concerns have emerged in the last decade to be added to incidents covered by the media, studied by scientists, and addressed by governmental and private institutions. Among these are events so striking that they are creating a paradigm shift. Two in particular are at the center of ever increasing media as well as scientific attention: bioterrorism and global warming. Unfortunately, these very worrisome issues are now superimposed on the already extensive list of ongoing environmental challenges.

The ultimate role of publishing scientific environmental research is to enhance understanding of the environment in ways that allow the public to be better informed or, in other words, to enable the public to have access to sufficient information. Because the public gets most of its information on science and technology from internet, TV news, and reports, the role for scientists as interpreters and brokers of scientific information to the public will grow rather than diminish. Environmentalism is an important global political force, resulting in the emergence of multinational consortia to control pollution and the evolution of the environmental ethic. Will the new politics of the twenty-first century involve a consortium of technologists and environmentalists, or a progressive confrontation? These matters are of genuine concern to governmental agencies and legislative bodies around the world.

For those who make the decisions about how our planet is managed, there is an ongoing need for continual surveillance and intelligent controls to avoid endangering the environment, public health, and wildlife. Ensuring safety-in-use of the many chemicals involved in our highly industrialized culture is a dynamic challenge, because the old, established materials are continually being displaced by newly developed molecules more acceptable to federal and state regulatory agencies, public health officials, and environmentalists. New legislation that will deal in an appropriate manner with this challenge is currently in the making or has been implemented recently, such as the REACH legislation in Europe. These regulations demand scientifically sound and documented dossiers on new chemicals.

Reviews publishes synoptic articles designed to treat the presence, fate, and, if possible, the safety of xenobiotics in any segment of the environment. These reviews can be either general or specific, but properly lie in the domains of analytical chemistry and its methodology, biochemistry, human and animal medicine, legislation, pharmacology, physiology, (eco)toxicology, and regulation. Certain affairs in food technology concerned specifically with pesticide and other food-additive problems may also be appropriate.

Because manuscripts are published in the order in which they are received in final form, it may seem that some important aspects have been neglected at times. However, these apparent omissions are recognized, and pertinent manuscripts are likely in preparation or planned. The field is so very large and the interests in it are so varied that the editor and the editorial board earnestly solicit authors and suggestions of underrepresented topics to make this international book series yet more useful and worthwhile.

Justification for the preparation of any review for this book series is that it deals with some aspect of the many real problems arising from the presence of anthropogenic chemicals in our surroundings. Thus, manuscripts may encompass case studies from any country. Additionally, chemical contamination in any manner of air, water, soil, or plant or animal life is within these objectives and their scope.

Manuscripts are often contributed by invitation. However, nominations for new topics or topics in areas that are rapidly advancing are welcome. Preliminary communication with the Editor-in-Chief is recommended before volunteered review manuscripts are submitted. *Reviews* is registered in WebofScience™.

Inclusion in the Science Citation Index serves to encourage scientists in academia to contribute to the series. The impact factor in recent years has increased from 2.5 in 2009 to 7.0 in 2017. The Editor-in-Chief and the Editorial Board strive for a further increase of the journal impact factor by actively inviting authors to submit manuscripts.

Amsterdam, The Netherlands
February 2020

Pim de Voogt

Contents

List of Contributors

Samir Abbès Laboratory of Genetic, Biodiversity and Bio-Resources Valorization, University of Monastir, Monastir, Tunisia

Higher Institute of Biotechnology of Béja, University of Jendouba, Jendouba, Tunisia

Jalila Ben Salah-Abbes Laboratory of Genetic, Biodiversity and Bio-Resources Valorization, University of Monastir, Monastir, Tunisia

Lu Cai Pediatric Research Institute, Department of Pediatrics, University of Louisville School of Medicine, Louisville, KY, USA

Departments of Pharmacology and Toxicology, University of Louisville School of Medicine, Louisville, KY, USA

Jonathan K. Challis Toxicology Centre, University of Saskatchewan, Saskatoon, SK, Canada

Francesco D'Agostini Department of Health Sciences (DISSAL), University of Genoa, Genoa, Italy

Herman Eijsackers Unit for Environmental Sciences and Management, North-West University, Potchefstroom, South Africa

Oluwabunmi P. Femi-Oloye Toxicology Centre, University of Saskatchewan, Saskatoon, SK, Canada

Department of Animal and Environmental Biology, Adekunle Ajasin University, Akungba-Akoko, Nigeria

John P. Giesy Toxicology Centre, University of Saskatchewan, Saskatoon, SK, Canada

Department of Biomedical Veterinary Sciences, University of Saskatchewan, Saskatoon, SK, Canada

Jianyang Guo State Key Laboratory of Environmental Geochemistry, Institute of Geochemistry, Chinese Academy of Sciences, Guiyang, China

Paul D. Jones Toxicology Centre, University of Saskatchewan, Saskatoon, SK, Canada

School of Environment and Sustainability, University of Saskatchewan, Saskatoon, SK, Canada

Sebastiano La Maestra Department of Health Sciences (DISSAL), University of Genoa, Genoa, Italy

Haiqing Liao State Key Laboratory of Environmental Criteria and Risk Assessment, Chinese Research Academy of Environmental Sciences, Beijing, China

Stella T. Lima Center for Nuclear Energy in Agriculture, University of Sao Paulo, Piracicaba, Brazil

Hong Liu Department of Emergency, Jiangxi Provincial Children's Hospital, Nanchang, Jiangxi, China

Mark Maboeta Unit for Environmental Sciences and Management, North-West University, Potchefstroom, South Africa

James S. Metcalf Brain Chemistry Labs, Jackson, WY, USA

Bahareh Nowruzi Department of Biology, Science and Research Branch, Islamic Azad University, Tehran, Iran

Femi F. Oloye Department of Chemical Sciences, Adekunle Ajasin University, Akungba-Akoko, Nigeria

Samaneh J. Porzani Department of Biology, Science and Research Branch, Islamic Azad University, Tehran, Iran

Abdelrhafour Tantaoui-Elaraki Department of Food Sciences, Hassan II Institute of Agronomy and Veterinary Medicine – Rabat, Rabat-Instituts, Témara, Morocco

Jingfu Wang State Key Laboratory of Environmental Geochemistry, Institute of Geochemistry, Chinese Academy of Sciences, Guiyang, China

Hongchen Wu State Key Laboratory of Environmental Geochemistry, Institute of Geochemistry, Chinese Academy of Sciences, Guiyang, China

University of Chinese Academy of Sciences, Beijing, China

Lijuan Xiong Department of Emergency, Jiangxi Provincial Children's Hospital, Nanchang, Jiangxi, China

Department of Pediatrics, Pediatric Research Institute, University of Louisville School of Medicine, Louisville, KY, USA

Zhiqi Zhao School of Earth Science and Resources, Chang'an University, Xi'an, China

Bin Zhou Department of Emergency, Jiangxi Provincial Children's Hospital, Nanchang, Jiangxi, China

Abdellah Zinedine Faculty of Sciences, BIOMARE Laboratory, Applied Microbiology and Biotechnologies, Chouaib Doukkali University, El Jadida, Morocco

Review on Health Impacts from Domestic Coal Burning: Emphasis on Endemic Fluorosis in Guizhou Province, Southwest China

Jianyang Guo, Hongchen Wu, Zhiqi Zhao, Jingfu Wang, and Haiqing Liao

Contents

Abstract Endemic fluorosis in Guizhou Province, Southwest China was firstly reported by Lyth in 1946 and was extensively concerned since the early 1980s. Initially, the pathological cause of endemic fluorosis in Guizhou Province was

---

J. Guo (✉) · J. Wang
State Key Laboratory of Environmental Geochemistry, Institute of Geochemistry, Chinese Academy of Sciences, Guiyang, China
e-mail: guojianyang@vip.gyig.ac.cn; wangjingfu@vip.gyig.ac.cn

H. Wu
State Key Laboratory of Environmental Geochemistry, Institute of Geochemistry, Chinese Academy of Sciences, Guiyang, China

University of Chinese Academy of Sciences, Beijing, China
e-mail: wuhongchen@vip.gyig.ac.cn

Z. Zhao
School of Earth Science and Resources, Chang'an University, Xi'an, China
e-mail: zhaozhiqi@chd.edu.cn

H. Liao (✉)
State Key Laboratory of Environmental Criteria and Risk Assessment, Chinese Research Academy of Environmental Sciences, Beijing, China
e-mail: liaohaiqing@vip.skleg.cn

P. de Voogt (ed.), *Reviews of Environmental Contamination and Toxicology Volume 258*,
Reviews of Environmental Contamination and Toxicology 258,
https://doi.org/10.1007/398_2021_71

instinctively ascribed to the drinking water. However, increasing evidences pointed that the major exposure route of fluorine for the local residents is via the roasted foodstuffs, especially the roasted pepper and corn. Source of fluorine in roasted foodstuffs was once blamed on the local coal and subsequently imputed to clay mixed in the coal. In fact, both are probably the source. Geogenic fluorine concentration in soil and clay is indeed high in Guizhou Province, but is not likely to be the direct cause for endemic fluorosis. The real culprit for endemic fluorosis in Guizhou Province is the unhealthy lifestyle of the local residents, who usually roasted their foodstuffs using local coal or briquettes (a mixture of coal and clay), resulting in the elevated fluorine in roasted foodstuffs. Nowadays, endemic fluorosis in Guizhou Province has substantially mitigated. Nevertheless, millions of confirmed cases of dental fluorosis remain left. In addition to endemic fluorosis, other health problems associated with domestic coal burning may also exist, because of the enrichment of toxic/harmful elements in the local coal. It is necessary to determine how serious the situation is and find out the possible solution. As people in other developing countries may suffer from similar health issues, same health issues around the world deserve more attention.

Keywords Epidemiology · Fluorine · Roasted foodstuffs · Source · Toxic trace elements

1 Introduction

Coal plays an important role in fueling the world industrialization and remains an important energy source, especially in the developing countries such as China and India (Finkelman et al. 2002). Taking China as an example, although proportion of coal in the energy mix has declined, consumption rate of coal has continued increasing (Dai et al. 2012; You and Xu 2010). During the formation of coal, potentially harmful or toxic elements can be incorporated into the coal, such as fluorine (F), arsenic (As), antimony (Sb), selenium (Se), mercury (Hg), chromium (Cr), and cadmium (Cd) (Dai et al. 2006a, b, 2012; Li et al. 2006). These elements can be released into the surrounding environments during the mining, storage, and combustion of coal, resulting in a variety of environmental and health problems (Finkelman et al. 1999, 2002; Tian et al. 2010). Considering the widespread utilization and the huge consumption of coal, pollution caused by the coal burning is not only a local or regional issue, but also a global issue.

China is the largest producer and consumer of coal in the world (Zhao and Luo 2018). In China, coal is extensively used for domestic purposes such as house heating and cooking, due to its cost-effectiveness and easy-accessibility. This situation is quite common in Guizhou Province, Southwest China where the winter is cold and damp. Although health problem associated with coal used for electric utility is less reported, health issues associated with domestic coal burning have been frequently reported, especially in Guizhou Province (Finkelman et al. 1999, 2002). The most prominent health issue is endemic fluorosis, with millions of cases of dental fluorosis being confirmed (Dai et al. 2004).

Fluorine (F) is the 13th most abundant element in the Earth's crust and the lightest member of the halogens. As the most electronegative and reactive member of all elements, F is naturally occurred as fluoride-bearing minerals in rocks and dissolved fluoride in water (Ali et al. 2016; Schafer et al. 2018, 2020). Similar to many other trace elements, F is beneficial to human health in trace amounts, but can be harmful in excess (Fordyce et al. 2007). Dental protection benefitted from low intake of F is well documented (Ayoob and Gupta 2006), while dental fluorosis or skeletal fluorosis caused by excessive intake of F is also found worldwide, because of the powerful calcium-seeking property of F (Fordyce et al. 2007). The narrow margin between the desired and the harmful dose of F makes it difficult to keep a balance between the dental protection and the fluorosis. This is probably the main reason for the ubiquitous incidents of fluorosis (Ali et al. 2016). Due to the double-sided nature of F, it is critical to understand the geological and chemical provenance of F in different environmental settings, from a public health perspective.

Endemic fluorosis in Guizhou Province has been extensively investigated within the last several decades. Studies conducted so far, however, were mostly confined to local or regional scale and were inclined to derive descriptive conclusions. It is desired to get a comprehensive understanding on endemic fluorosis in Guizhou Province. In addition to endemic fluorosis, other health problems have also been reported in Guizhou Province associated with the domestic coal burning. Therefore, in the present work, a full description on endemic fluorosis in Guizhou Province associated with the domestic coal burning was provided at first, including the origin of the health issue, the possible exposure routes of F for the local residents, possible sources of F in the roasted foodstuffs, and the real culprits. Secondly, other health problems hidden behind the endemic fluorosis were presented briefly. Lastly, possible research aspects associated with the domestic coal burning were proposed.

2 Endemic Fluorosis in Guizhou Province

2.1 Origin of the Health Issue

Fluorosis was firstly recognized at the beginning of the last century by McKay and Black (1916). They found that enamel developmental imperfection was prevalent in Colorado, a phenomenon confirmed to be related to elevated F^- in the local drinking water. After that, fluorosis was found in various countries/regions, especially in China and India (Sun 2017).

As to Guizhou Province, it can retrospect to 1934 when endemic fluorosis was firstly realized in Southwest Guizhou Province and Northeast Yunnan Province, an area covering approximately 2×10^4 km^2 in Southwest China (Kilborn et al. 1950). This phenomenon was firstly reported by Lyth at Kweichow, a small village in

Guizhou Province. In this work, 134 cases of dental fluorosis were investigated and four cases of skeletal fluorosis were described in detail (Lyth 1946), while the cause of this health issue was not carefully explored. Based on the high contents of F^- in a little stream running out of a coal-mine (Lyth 1946), the pathological cause of endemic fluorosis was instinctively ascribed to F^- in the local water, a viewpoint proven to be wrong subsequently.

Prior to 1979, endemic fluorosis is less concerned by the Chinese central government. Early tentative efforts on endemic fluorosis from the central government were mainly paid to the dental fluorosis in Northeast/Northwest China, where elevated F^- was usually found in the local drinking water. Meanwhile, endemic fluorosis in Guizhou Province was rarely mentioned in the documents and scientific literature. In early 1980s, endemic fluorosis in Guizhou Province was confirmed to be caused by the domestic coal burning rather than the drinking water (Guiyang Epidemic Prevention Station et al. 1981). It was subsequently named as "coal-burning type of endemic fluorosis," a special type of endemic fluorosis found to be prevalent in Southwest China, with millions of cases of dental fluorosis being confirmed (Sun 2017). With the accumulation of epidemiological data, endemic fluorosis in Guizhou Province was extensively concerned.

2.2 *Epidemiological Data*

Epidemiological survey on endemic fluorosis in Guizhou Province can track back to 1979 when a small survey was conducted at a heavily polluted village in Zhijin, western Guizhou Province, in which 192 volunteers were involved (He et al. 2007). The results indicated that 98.9% of the volunteers were the confirmed cases of dental fluorosis and 77.6% of the adult volunteers were the confirmed cases of skeletal fluorosis. Same serious situation was also found in another heavily polluted village in Jinsha, northwestern Guizhou Province, with all volunteers being the confirmed cases of dental fluorosis and 94% of adult volunteers being the confirmed cases of skeletal fluorosis (He et al. 2007). In fact, the prevalence of endemic fluorosis in Guizhou Province is far beyond a few small villages, but is widespread in the whole province (Zhang et al. 2017a, b). After that, more extensive surveys on endemic fluorosis have been conducted. The results were mostly published in Chinese and were seldom available for foreigners (An et al. 2009; Gao et al. 2015; Li et al. 2003, 2005a, b; Wang et al. 2013). Some of which have been summarized in a recent work (Zhang et al. 2017a, b) and were schematically shown in Fig. 1.

For the adolescent, the latest survey was conducted in 2014 (17,962 volunteers were involved), covering 23 administrative regions of Guizhou Province (Zhang et al. 2017a, b). The results indicated that the average prevalence rate of dental fluorosis was 32.3% (Fig. 1). According to the survey conducted at the same regions in 2007 (502,457 volunteers were involved), the confirmed cases of dental fluorosis was 229,943, with the average prevalence rate of 45.8%. As to the survey conducted in 2000 (188,642 volunteers were involved), the confirmed cases of dental fluorosis

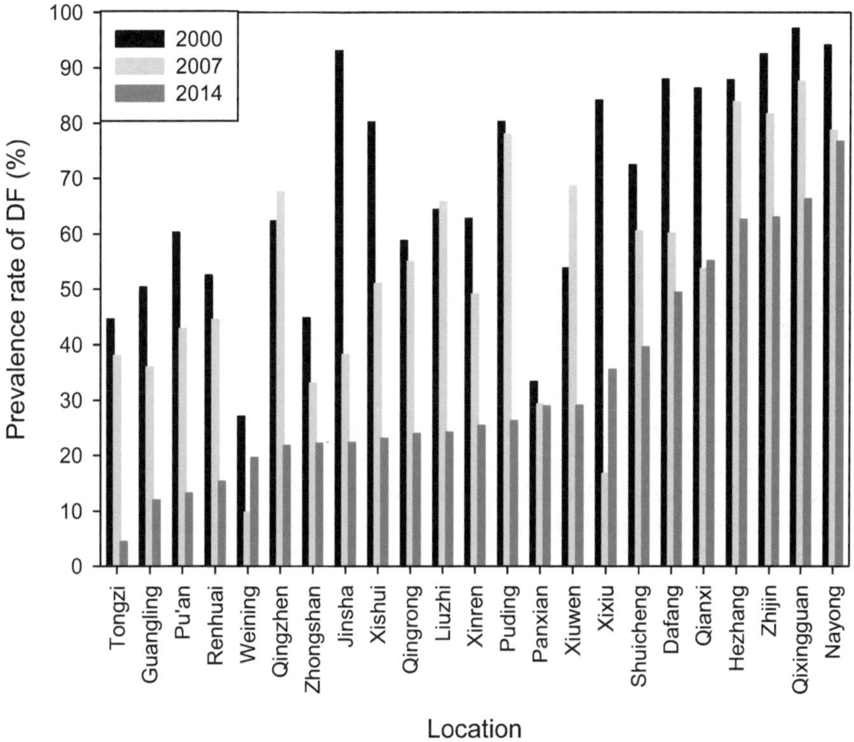

Fig. 1 Prevalence rate of dental fluorosis in 2000, 2007, and 2014 in Guizhou Province (Data source: Zhang et al. 2017a, b)

were 138,256, with the average prevalence rate of 73.3% (Zhang et al. 2017a, b). This indicated that the prevalence rate of dental fluorosis in Guizhou Province was substantially declined since 2000, owing to the great efforts from the local governments (Sun 2017). However, based on the latest statistics, there still have 8.79 million cases of dental fluorosis in Guizhou Province by the end of 2018 (Chinese Health Statistics Yearbook 2019). In addition, significantly high prevalence rates of dental fluorosis were still found in some administrative regions, such as Dafang, Qianxi, Zhijin, Qixingguan, and Nayong, according to the latest survey (Zhang et al. 2017a, b).

As to the adults, based on the survey conducted in seven administrative regions of Guizhou Province during 2001–2003 (122,275 volunteers were involved), the suspected cases of skeletal fluorosis were 33,074, with a suspected rate of 27.03% (Wang et al. 2013). Among the suspected cases, 62.54% of which was confirmed (Li et al. 2005a, b). Among different administrative regions, Qianxinan has the highest suspected rate of skeletal fluorosis (50.76%), followed by Liupanshui, Bijie, and Anshun. Combined with the confirmed rates of skeletal fluorosis, the highest prevalence rate of skeletal fluorosis was found in Liupanshui (30.7%),

Table 1 Results of epidemiological survey conducted in 2001–2003 in Guizhou Province (Data source: Wang et al. 2013)

	Volunteers	Suspected cases	Suspected rate (%)	Confirmed rate (%)	Prevalence rate of skeletal fluorosis (%)
Liupanshui	14,992	4,735	31.58	97.15	30.7
Qianxinan	5,875	2,982	50.76	52.98	26.9
Anshun	19,250	5,970	31.01	74.77	23.2
Bijie	28,304	8,845	31.25	58.12	18.2
Zunyi	21,096	4,755	22.54	43.59	9.83
Qiannan	14,392	3,104	21.57	45.33	9.78
Guiyang	18,366	2,656	14.46	52.76	7.63
Total	122,275	33,047	27.03	62.54	16.9

followed by Qianxinan, Anshun, and Bijie. In comparison, situation is much better in Guiyang, Qiannan, and Zunyi. The detailed information was shown in Table 1.

Historically, endemic fluorosis is very serious in Southwest China, with millions of cases of dental fluorosis being found (Sun et al. 2001), especially in Guizhou Province and Yunnan Province. Fortunately, the confirmed cases of skeletal fluorosis in Guizhou Province have greatly declined and only 2,592 cases were left by the end of 2018 (Chinese Health Statistics Yearbook 2019). During the control and prevention of endemic diseases in China (including the endemic fluorosis in Guizhou Province), great efforts have been done and have been summarized in a recent book (Sun 2017). With regard to endemic fluorosis in Guizhou Province, the possible exposure routes of F for the local residents, the possible sources of F in the foodstuffs and the real culprits have been carefully investigated (Dai et al. 2004, 2007; Luo et al. 2010; Finkelman et al. 1999). This makes the whole story of endemic fluorosis in Guizhou Province becoming clear.

2.3 *Possible Exposure Routes*

Generally speaking, there are three main routes for human exposure to pollutants, i.e., dietary intake, respiratory inhalation, and dermal exposure. Among them, dietary intake is the major route for most pollutants entering into the human body, although respiratory inhalation is also very important for volatile or semi-volatile organic pollutants. Drinking water and foodstuffs are the major ways of dietary intake of pollutants.

2.3.1 Drinking Water

As far as F^- is concerned, drinking water is the most common way for human exposure in most regions, including China, India, Africa, Australia, Europe, and the USA (Ali et al. 2016; Arif et al. 2012), while this is not the case at all in

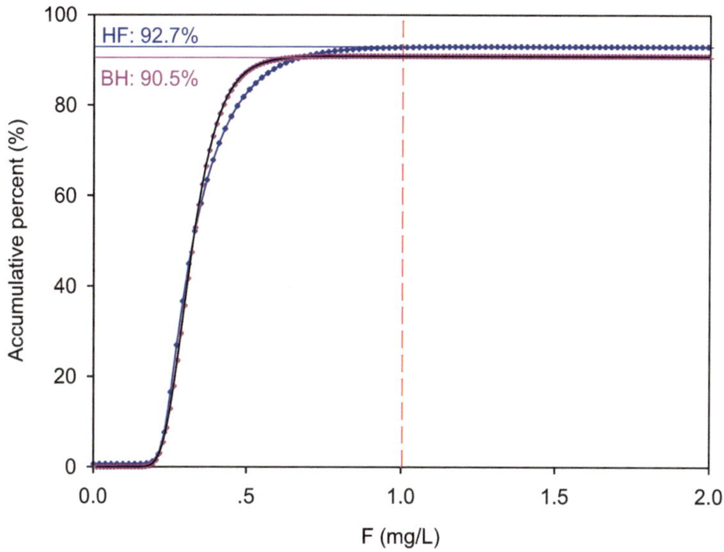

Fig. 2 Contents of F in drinking water sources (Blue line: HF; Pink line: BH) in Guizhou Province

Guizhou Province. Although Guizhou Province is one of the most serious regions suffered from endemic fluorosis, F^- in the local drinking water is seldom to be a problem. The first evidence comes from the long-term monitoring of F^- at two drinking water sources adjacent to Guiyang, central Guizhou Province (5,155 samples were analyzed). The results indicated that content of F^- in more than 90% of the samples was less than the permissible limit proposed by the WHO (1.5 mg/L) and content of F^- in more than 80% of the samples was less than the optimal value (0.5–1.0 mg/L) recommended by the WHO (Fig. 2). The second evidence comes from an extensive survey on F^- in groundwater (1,023 samples were analyzed) (Pu et al. 2013).The results indicated that in Guizhou Province, contents of F^- in the groundwater were within the range of 0.002–3.72 mg/L (mean: 0.313 mg/L), most of which were <1.0 mg/L (Table 2). If these evidences are not strong enough, data on F^- in surface water, groundwater, and drinking water from the heavily polluted areas can further support that endemic fluorosis in Guizhou Province cannot put the blame on the drinking water. As summarized in Table 2, contents of F^- in water are quite low in Guizhou Province, even in the heavily polluted regions. This suggests that F^- in drinking water is not the major cause for endemic fluorosis in Guizhou Province. In the early 1980s, local organizations have realized that endemic fluorosis in Guizhou Province has nothing to do with F^- in the drinking water (Guiyang Medical College et al. 1981). Therefore, people' concerns were shifted to the foodstuffs.

Table 2 Fluorine in groundwater, drinking water, and surface water collected from Guizhou Province

Water type	Location	No. of sample	Content of F (mg/L)	Description	Reference
Ground water	Guizhou province	1,023	96.7% <1.0	Non-specific	Pu et al. (2013)
Drinking water	Liupanshui	354	96.6% <0.5	Heavy polluted area	Li and Yan (1994)
Surface water	Guiyang	7	0.36–0.91 (0.49)	Polluted area	Yu et al. (1994)
Drinking water	Zhijin	5	0.09–0.42 (0.19)	Heavy polluted area	Dai et al. (2004)
Drinking water	Zhijin	13	0.05–0.14	Heavy polluted area	Wu et al. (2004)
Drinking water	Zhijin, Nayong, and Pingba	76	<0.02–0.44 (0.09)	Heavy polluted area	Xie et al. (2010)
Ground water	Dafang	31	0.09–0.37 (0.15)	Heavy polluted area	Zhang et al. (2016)
Drinking water	Shuicheng	10	0.03–0.14	Heavy polluted area	Xiao et al. (2016)
Surface water	Zhijin	15	0.05–0.38 (0.19)	Heavy polluted area	Li et al. (2016)
Ground water	Zhijin	20	0.01–0.20 (0.08)	Heavy polluted area	Li et al. (2016)

2.3.2 Foodstuffs

It has long been suspected that endemic fluorosis in Guizhou Province may associate with F^- in the foodstuffs. The earliest data on F^- in the local foodstuffs were reported in 1981, in which ten kinds of food items were analyzed and contents of F^- in the foodstuffs were in the range of 0.5–9.3 mg/kg (Guiyang Medical College et al. 1981). However, to date, there is no extensive survey on F^- in the foodstuffs and most reported data were associated with the roasted pepper and corn. The available data were summarized in Table 3.

In 1990s, An and his co-authors (1996) conducted a survey on F^- in foodstuffs collected from Puding, Zhijin, and Guiyang, in which seven villages were involved and 20 families were randomly selected in each village. The results indicated that extremely high level of F^- was found in roasted pepper (Table 3). In the following case studies conducted in Zhijin, western Guizhou Province, extremely high level of F^- in roasted pepper was also found at Chengguan Township, Qimo Township, and Sanjia Township (Li et al. 2004), as well as some small villages (such as Hualuo, Majiazhuang, and Pianpozhai) (Dai et al. 2007; Wu et al. 2004). Even very recently, extremely high level of F^- (574 ± 297 mg/kg) was still reported in roasted pepper collected from Nayong, western Guizhou Province (Liu et al. 2013). These data are two or three orders of magnitude higher than the permitted level proposed by the Chinese government (1.0 mg/kg, GB 2762–2005).

Table 3 Contents of fluorine in foodstuffs collected from Guizhou Province

Location	Date	Food type	No. of sample	F (mg/kg)	Reference
Liupanshui	1994	Roasted pepper	17	167–1,267	Li and Yan (1994)
Xinzhai, Zhijin	1996	Roasted pepper	20	1,126 ± 604	An et al. (1996)
Daga, Zhijin	1996	Roasted pepper	20	674 ± 513	An et al. (1996)
Hehua, Zhijin	1996	Roasted pepper	20	1,136 ± 727	An et al. (1996)
Haoyun, Puding	1996	Roasted pepper	20	349 ± 714	An et al. (1996)
Xiayan, Guiyang	1996	Roasted pepper	20	593 ± 424	An et al. (1996)
Erguai, Guiyang	1996	Roasted pepper	20	459 ± 471	An et al. (1996)
Chengguan, Zhijin	2002	Roasted pepper	5	222 ± 175	Li et al. (2004)
Qimo, Zhijin	2002	Roasted pepper	5	408 ± 305	Li et al. (2004)
Sanjia, Zhijin	2002	Roasted pepper	5	362 ± 306	Li et al. (2004)
Hualuo, Zhijin	2003	Roasted pepper	10	513 ± 389	Wu et al. (2004)
Majiazhuang, Zhijin	2003	Roasted pepper	7	343 ± 238	Wu et al. (2004)
Pianpozhai, Zhijin	2003	Roasted pepper	13	281 ± 225	Wu et al. (2004)
Zhijin	NA[a]	Roasted pepper	9	Ave: 1419	Dai et al. (2007)
Four villages, Nayong	NA	Roasted pepper	11	574 ± 297	Liu et al. (2013)
Liupanshui	1994	Roasted corn	88	4.2–300	Li and Yan (1994)
Xinzhai, Zhijin	1996	Roasted corn	20	69.8 ± 24.2	An et al. (1996)
Daga, Zhijin	1996	Roasted corn	20	34.2 ± 14.3	An et al. (1996)
Hehua, Zhijin	1996	Roasted corn	20	23.9 ± 11.9	An et al. (1996)
Chengguan, Zhijin	2002	Roasted corn	5	8.33 ± 5.33	Li et al. (2004)
Qimo, Zhijin	2002	Roasted corn	5	25.0 ± 22.8	Li et al. (2004)
Sanjia, Zhijin	2002	Roasted corn	5	18.9 ± 6.38	Li et al. (2004)
Hualuo, Zhijin	2003	Roasted corn	10	30.6 ± 7.6	Wu et al. (2004)
Majiazhuang, Zhijin	2003	Roasted corn	7	30.4 ± 13.2	Wu et al. (2004)
Pianpozhai, Zhijin	2003	Roasted corn	13	49.8 ± 29.0	Wu et al. (2004)
Zhijin	NA	Roasted corn	9	Ave: 110	Dai et al. (2007)
Weining, Bijie	2006, 2008	Roasted corn	9	9.50–28.9	Luo et al. (2011a)

(continued)

Table 3 (continued)

Location	Date	Food type	No. of sample	F (mg/kg)	Reference
Four villages, Nayong	NA	Roasted corn	11	23.2 ± 12.8	Liu et al. (2013)
Haoyun, Puding	1996	Rice	20	0.92 ± 0.30	An et al. (1996)
Xiayan, Guiyang	1996	Rice	20	4.13 ± 2.85	An et al. (1996)
Erguai, Guiyang	1996	Rice	20	5.51 ± 3.22	An et al. (1996)
Liupanshui	1994	Fresh pepper	10	17.7–18.5	Li and Yan (1994)
Liupanshui	1994	Fresh corn	10	1.2–1.6	Li and Yan (1994)
Weining, Bijie	2006, 2008	Fresh corn	9	0.24–2.21	Luo et al. (2011a)

[a]NA: Not available

In addition to roasted pepper, high content of F^- was also found in roasted corn. As displayed in Table 3, contents of F^- in roasted corn collected from Xinzhai village, Daga village, and Hehua village were 69.8 ± 24.2 mg/kg, 34.2 ± 14.3 mg/kg, and 23.9 ± 11.9 mg/kg, respectively (An et al. 1996). These data were close to the results reported at Hualuo village (30.6 ± 7.6 mg/kg), Majiazhuang village (30.4 ± 13.2 mg/kg), and Pianpozhai village (49.8 ± 29.0 mg/kg) (Wu et al. 2004), and were slightly higher than the reported data at Chengguan Township (8.33 ± 5.33 mg/kg), Qimo Township (25.0 ± 22.8 mg/kg), and Sanjia Township (18.9 ± 6.38 mg/kg) (Li et al. 2004). Although contents of F^- in roasted corn varied among different regions, most were dozens of the permitted level proposed by the Chinese government (1.5 mg/kg, GB 2762–2005). Fortunately, contents of F^- in roasted pepper and corn have substantially decreased, based on the survey conducted in 2013 in the heavily polluted area (covering 23 administrative regions of Guizhou Province) (Zhang et al. 2017a, b).

As to F^- in other foodstuffs, related data were limited and scatted in different literature. The earliest data collected from Zhijin indicated that mean contents of F^- in wheat and rice were 9.3 mg/kg and 5.4 mg/kg, while they were within the range of 0.5–5.5 mg/kg in different kinds of vegetables (Guiyang Medical College et al. 1981). Contents of F^- in rice were also reported at two villages in Guiyang, with the mean values of 4.13 ± 2.85 mg/kg and 5.51 ± 3.22 mg/kg, respectively (An et al. 1996). They were clearly less than that in roasted pepper and corn, but still several times that of the permitted level proposed by the Chinese government (GB 2762–2005).

Pepper is the favorite food item for the local residents and is usually roasted before consumption. Extremely high F^- level in roasted pepper implied that high F^- exposure risk is expected via the consumption of roasted pepper, and things may be worse if the roasted corn is the staple food for the local residents.

2.3.3 Other Exposure Routes

In addition to dietary intake, other possible routes for human exposure to F^- include respiratory inhalation and dermal exposure. Dermal exposure of F^- for the local residents is seldom mentioned, while respiratory inhalation of F^- is occasionally reported. Based on the collected data, respiratory intakes of F^- for the local residents were very limited, although they were varied among different surveys (Table 4). For example, respiratory intake of F^- estimated for the local residents in Zhijin was 0.2 mg/day/capita (Guiyang Medical College et al. 1981). This is similar to that of a recent survey conducted in Nayong, western Guizhou Province, with the respiratory intake of F^- of 0.16 mg/day/capita during the house-heating season (Liu et al. 2013). Based on the survey conducted in 2006 in Zhijin, Jinsha, and Guiyang (Li et al. 2011), the respiratory intake of F^- was even less, within the range of 0.03–0.12 mg/day/capita. Compared with the diary intake of F^-, the respiratory intake of F^- is quite low (Guiyang Medical College et al. 1981; Li et al. 2011; Liu et al. 2013). For example, respiratory intake of F^- for the local residents in Zhijin is only 0.2 mg/day/capita, accounting for 3.2–7.1% of the total intakes of F^- (Guiyang Medical College et al. 1981). While in Nayong, respiratory intake of F^- is only 1.4% of the total intakes of F^- (Liu et al. 2013). If more food items are included as estimated by Li and his co-authors (2011), the proportion of respiratory intake of F^- to the total exposure of F^- would further decrease to <1%. Therefore, the exposure of F^- for the local residents is mostly via the roasted foodstuffs (usually more than 90%), rather than the air and drinking water (Table 4).

Different organizations have set different safe doses for human exposure to F^-, most of which were within the range of 1.5–4.0 mg/day/capita. For example, the guideline for daily intake of F^- recommended by WHO is 2 mg/day/capita for children and 4 mg/day/capita for adults (WHO 2002). As to China, the recommended daily allowances (RDAs) in coal burning fluorosis area were set to be 2.0 mg/day/capita for children aged 8–15 years and 3.0 mg/day/capita for adults; While in drinking water fluorosis area, the RDAs were set to be 2.4 mg/day/capita for children aged 8–15 years and 3.5 mg/day/capita for adults (Li et al. 2015). Guizhou Province is the typical coal burning fluorosis area, and most of the actual exposure levels (as listed in Table 4) are much higher than the proposed RDAs for the local residents.

2.4 *Possible Sources of Fluorine*

Because human exposure of F^- in Guizhou Province is mainly via the food consumption, it is easy to link endemic fluorosis to F in soil. After all, high content of F^- has been found in vegetables (6.64–10.4 mg/kg) cultivated in soil with high-F background in Xuzhou, East China (Zhu et al. 2000), and elevated F^- in crops (ranging from 3.3–9.8 mg/kg) has been reported in Zhijin, western Guizhou

Table 4 Human exposure of F (mg/day/capita) for local residents in Guizhou Province

Location	Year	Pepper	Corn	Rice	Vegetables	Food[a]	Drinking water	Air	Total intake	Reference
Zhijin	2006	15.4	2.35	1.03	0.50	19.3	0.28	0.03	19.6	Li et al. (2011)
	1979	16.3	31.6	0.23	0.50	48.6	0.28	0.34	49.3	Li et al. (2011)
Jinsha	2006	13.9	16.2	0.26	0.51	30.9	0.24	0.08	31.1	Li et al. (2011)
	1986	15.0	40.2	0.23	0.50	55.9	0.24	0.45	56.6	Li et al. (2011)
Bijie	2006	24.7	7.29	1.03	0.50	33.5	0.22	0.07	33.9	Li et al. (2011)
Bijie	2006	22.0	3.51	1.03	0.50	27.0	0.14	0.06	27.2	Li et al. (2011)
Zhijin	2006	7.97	---	1.29	0.50	9.76	0.09	0.12	9.96	Li et al. (2011)
Jinsha	2006	0.32	---	1.29	0.50	2.11	0.18	0.11	2.39	Li et al. (2011)
Zhijin	NA[a]	---	---	---	---	4.3	0.20	0.37	4.87	Guiyang Medical College et al. (1981)
Zhijin	NA[a]	---	---	---	---	6.3	0.20	0.37	6.87	Guiyang Medical College et al. (1981)
Zhijin	NA[a]	---	---	---	---	8.6	0.20	0.37	9.17	Guiyang Medical College et al. (1981)
Zhijin	NA[a]	---	---	---	---	9.8	0.20	0.37	10.4	Guiyang Medical College et al. (1981)
Zhijin	NA[a]	---	---	---	---	10.9	0.20	0.37	11.5	Guiyang Medical College et al. (1981)
Nayong	NA[a]	8.60	2.30		---	10.9	---	0.16	11.1	Liu et al. (2013)

[a]Food: sum of all the food items included in the survey. *GMC* Guiyang Medical College, *NA* Not available

Province (Guiyang Medical College et al. 1981; Li and Chen 1997). Secondly, extremely high level of F^- is usually found in roasted foodstuffs, especially the roasted pepper and corn. This implied that endemic fluorosis in Guizhou Province is more likely associated with F^- in the coal, an energy source used for food roasting. In addition, during the source seeking of F^- in roasted foodstuffs, clay mixed in the coal is also the possible source.

2.4.1 Fluorine in Soil

Based on the pervious investigation (Fig. 3), geological background of F^- in Guizhou Province is highest nationwide, falling in the range of 800–899 mg/kg. This is approximately twice that of F^- in surface soil in China (440 mg/kg) and at least four times that of F^- in soil worldwide (200 mg/kg) (Wu et al. 2008). According to the survey conducted in 2007–2008 (more than 40,000 soil samples were involved), the arithmetic mean value of F^- was 929 mg/kg (Zheng et al. 2009), slightly higher than the previously reported data. If different strata were taken into account, high contents of F^- were usually found in soil developed in Triassic, Permian, Silurian, Ordovician, and Cambrian strata (Zheng et al. 2009). Recently, a survey conducted in Guiyang, central Guizhou Province indicated that contents of

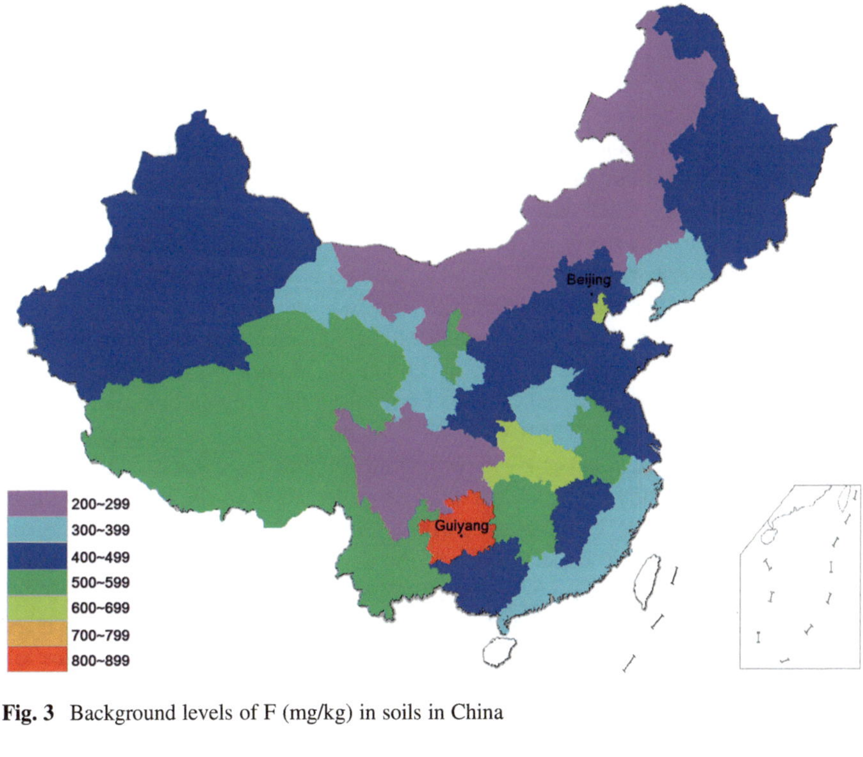

Fig. 3 Background levels of F (mg/kg) in soils in China

F^- in surface and deep soil were in the range of 274–3,663 mg/kg (mean: 1103 mg/kg) and 246–3,695 mg/kg (mean: 1382 mg/kg), respectively (Pan et al. 2018). All the results suggested that geological background of F^- is indeed high in Guizhou Province and could be the fundamental cause for endemic fluorosis.

Our concerns about the geological background of F^- stemmed from the hypothesis that high geological background of F^- may lead to elevated F^- in crops or vegetables, and further resulting in endemic fluorosis via the food consumption. Fortunately, the capability of crops or vegetables to accumulate F from soil is limited (Wu et al. 2008; Zhu et al. 2000). Although extremely high content of F^- was found in the roasted pepper and corn, content of F^- in the fresh pepper and corn is fairly low, even in samples collected from the heavily polluted areas (Table 3). For example, mean contents of F^- in fresh pepper and corn collected from Zhijin were 1.64 mg/kg and 0.79 mg/kg (Dai et al. 2007). This is close to the pervious reported data with mean value of 1.7 mg/kg in fresh pepper and 0.65 mg/kg in fresh corn (An et al. 1995). Similar results can also be found in Liupanshui and Weining, where contents of F^- in fresh corn were in the range of 1.2–1.6 mg/kg and 0.24–2.24 mg/kg, respectively (Li and Yan 1994; Luo et al. 2010). Therefore, high geological background of F^- is not necessary resulting in high content of F^- in the foodstuffs and is not necessary the direct cause for endemic fluorosis in Guizhou Province. High content of F^- in roasted foodstuffs should blame on the roasting processes, rather than the foodstuffs themselves. This understanding is helpful to find the real source of F^- in foodstuffs.

2.4.2 Fluorine in Coal

Because the foodstuffs themselves is seldom to be a problem, high content of F^- in roasted foodstuffs is more likely associated with coal used in the roasting processes. This is the main reason that endemic fluorosis in Guizhou Province is defined as the coal-burning type of fluorosis, a typical endemic fluorosis found in Southwest China.

Early on, it was assumed that the local coal could be abundant in F^-, resulting in elevated F^- in roasted foodstuffs. Because some coal samples from Guizhou Province is do abundant in F^- (Table 5), such as coal samples collected from Anlong, Southwest Guizhou Province (Liu et al. 2008a, b), domestic coal samples from Zhijin (Dai et al. 2007), and coal samples from Zhijin, Nayong and Pingba, western Guizhou Province (Xie et al. 2010). However, as the accumulation of data on F^- in coal, we realized that content of F^- in coals from Guizhou province is modest and is not so high as expected, even in coal samples collected from the heavily polluted areas (Dai et al. 2004; Hong et al. 2016; Liang et al. 2011). For example, in anthracite coal samples collected from Pu'an and Puding, western Guizhou Province, contents of F^- were in the range of 90.2–149 mg/kg (Liang et al. 2011). In 38 outcrop coal samples from western and northwestern Guizhou Province, contents of F^- were between 44 mg/kg and 382 mg/kg, with mean value of 125 mg/kg (Hong et al. 2016). While in seam channel coals from western Guizhou Province, contents of F^- were in the range of 16.6–500 mg/kg, with mean value of

Table 5 Fluorine in coal and clay of Guizhou Province

Sample	Location	No. of sample	Content of F (mg/kg)	Reference
Coal	Anlong	7	271–1,416 (925)[a]	Liu et al. (2008a, b)
Domestic coal	Zhijin	9	79.5–599 (237)	Dai et al. (2007)
Anthracite coal	Zhijin, Nayong, and Pingba	39	62–535 (202)	Xie et al. (2010)
Anthracite coal	Pu'an and Puding	21	90.2–149	Liang et al. (2011)
Outcrop coal	Xishui, Dafang, and Jinsha	38	44–382 (125)	Hong et al. (2016)
Seam channel coal	Zhijin	19	16.6–500 (89.3)	Dai et al. (2004)
Seam channel coal	Western Guizhou Province	50	16.6–500 (83.1)	Dai et al. (2004)
Anthracite coal	Guizhou Province	140	143	Zhong and Luo (2017)
Bituminous coal	Guizhou Province	215	89.0	Zhong and Luo (2017)
Domestic coal	Zhijin	15	89.8 ± 37.6	Li et al. (2004)
Coal and clay	Zhijin, Nayong, and Pingba	9	286–1,190 (507)	Xie et al. (2010)
Coal and clay	Zhijin	9	172–1,498 (828)	Dai et al. (2007)
Coal and clay	Zhijin	45	621–2,816 (1284)	Hong et al. (2015)
Clay	Zhijin	11	452–4,955 (2262)	Dai et al. (2007)
Clay	Zhijin	14	710–2,500 (1415)	Liu and Wang (2005)

[a]Data in the parentheses are the mean value

83.1 mg/kg (Dai et al. 2004). These data were close to the background level of F^- in Chinese coals (130 mg/kg, Dai et al. 2012) and US coals (98 mg/kg, Finkelman 1993), as well as that of the world average (88 mg/kg, Ketris and Yudovich 2009). Recently, a survey on F^- in coals from 90% of the coal-producing areas of Guizhou Province was conducted (more than 1,000 samples were analyzed) (Zhong and Luo 2017). The results indicated that contents of F^- in coal were mostly in the range of 50–180 mg/kg, with mean value of 143 mg/kg in anthracite coal and 89.0 mg/kg in bituminous coal, respectively (Zhong and Luo 2017).

Although content of F^- in local coal is modest, we cannot deny that elevated F^- in roasted foodstuffs comes from domestic coal burning. Because content of F^- in roasted foodstuffs depends not only on the content of F^- in coal, but also on the roasting time and the volatility of F^- from coal. If the roasting time is long enough and F^- in the coal is easy to be volatilized, extremely high level of F^- in roasted foodstuffs can still be expected. Unfortunately, both factors are common in Guizhou Province. Due to the damp weather in cold season, the local residents have to roast their foodstuffs (such as pepper and corn) for a long time, in order to keep their foodstuffs being preserved better. In the meantime, Coal in Guizhou Province is usually abundant in sulfur (Dai et al. 2015; Hong et al. 2015). This facilitates the release of F from coal during the combusting processes (Hong et al. 2018).

2.4.3 Fluorine in Clay

Endemic fluorosis in Guizhou Province was once blamed on high-F coal in early research works (Watanabe et al. 2000; Zhang and Cao 1996). One of the important facts is that extremely high contents of F^- were observed in briquettes collected from the heavily polluted areas (Zhang and Cao 1996). In fact, briquettes used for domestic purpose is actually a mixture of coal and clay (Dai et al. 2004, 2007; Wu et al. 2004). Considering the modest content of F^- in most coal samples from Guizhou Province (Table 5), extremely high level of F^- in briquettes is more likely stemmed from the clay, rather than the coal itself. As displayed in Table 5, clay is true abundant in F^- (Dai et al. 2007; Liu and Wang 2005) and if coal mixed with clay to produce the briquettes, content of F^- would increase sharply in briquette (Xie et al. 2010; Dai et al. 2007; Hong et al. 2015). As early as 1985, investigation on 137 briquette samples has proven that extremely high content of F^- in briquettes do come from the clay (Zhou et al. 1991, and the reference therein), instead of the coal. This is supported by the works conducted in heavily polluted regions (Dai et al. 2004; Wu et al. 2004) and confirmed by the works conducted in the whole province (Hong et al. 2015; Liu and Wang 2005; Luo et al. 2010; Xie et al. 2010).

Based on these information, Sun (2005) has concluded that endemic fluorosis in Guizhou Province is mainly caused by the clay rather than the coal, a viewpoint still being debatable. As mentioned above, if the roasting time is long enough, serious food contamination by F^- can still be possible, even though the content of F^- in coal is modest. In addition, F in the local coal is more easily to be volatilized during the roasting processes (Hong et al. 2018; Liang et al. 2011), because the local coal is usually rich in sulfur (Dai et al. 2015; Duan et al. 2018; Hong et al. 2018). Thus, it can be deduced that even there is no clay; long roasting time may also lead to heavy contamination of F^- in roasted foodstuffs. The mixing of clay in coal is more likely to make things worse, but is not necessary the primary cause for endemic fluorosis in Guizhou Province. Nowadays, it is widely accepted that both coal and clay are the major sources of F in roasted foodstuffs and consumption of roasted foodstuffs is the major cause for endemic fluorosis in Guizhou Province and the adjacent area (Li et al. 2012; Luo et al. 2011b).

2.5 *Real Culprit of the Endemic Fluorosis in Guizhou Province*

Based on the discussion above, it can be concluded that: (1) endemic fluorosis in Guizhou Province is nothing to do with F^- in the local drinking water; (2) geological background of F is indeed high in Guizhou Province, but is not necessary resulting in elevated content of F^- in the fresh food items; (3) consumption of roasted foodstuffs is the major route for the local residents' exposure to F^-; (4) coal is an important source of F in roasted foodstuffs, but content of F in local coal is not so high as expected; and (5) clay mixed in the coal is another important source of F in roasted foodstuffs. Mixing F-rich clay with the coal to produce briquettes can lead to higher exposure risk for the local residents.

Although geological background of F is indeed high in Guizhou Province, it is seldom to be a problem via the natural processes, such as leaching into water and bioaccumulating into crops or vegetables. It is certain that the major exposure route of F^- for local residents is the consumption of roasted foodstuffs and sources of F^- in roasted foodstuffs are coal or briquettes used for domestic purposes. However, there is no direct causal relationship between the endemic fluorosis and the domestic coal burning. Because domestic coal burning activities are common in Chinese rural area, especially in North China (Zhi et al. 2017), while coal-burning type of endemic fluorosis is mainly found in Southwest China and is seldom found in North China. Therefore, it seems unfair to blame the local coal for endemic fluorosis in Guizhou Province.

In fact, the important link between contaminated foodstuffs and F^- in coal/clay briquettes is the roasting processes, an unhealthy way to deal with the foodstuffs. This provides F in coal or briquettes a chance to get into the food chain, resulting in endemic fluorosis via consumption of contaminated foodstuffs. This behavior is usually happened in Southwest China, but is rarely found in other rural areas of China. This is why coal-burning type of endemic fluorosis is mainly happened in Southwest China, but is rarely found in other rural areas of China. Therefore, the real culprit of endemic fluorosis in Guizhou Province should be blamed on the unhealthy lifestyle of the local residents and more precisely, endemic fluorosis in Guizhou Province should be defined as the lifestyle-related endemic disease (Sun 2017). This becomes the theoretical foundation for the control of the endemic fluorosis in Guizhou Province and the adjacent area (Luo et al. 2010; Xu and Xu 2008).

2.6 *Control of the Endemic Fluorosis in Guizhou Province*

Based on the research works, local governments have taken many effective measures to control the endemic fluorosis in Guizhou Province. The specific measures are to: (1) encourage the use of cleaner fuels; (2) help the local residents to ameliorate their household stoves; (3) change the local residents' diet structure; and (4) enhance the related health awareness of the local residents.

Energy source previously used in the rural houses is usually the high-F briquettes, a mixture of local coal and clay. To control the endemic fluorosis, the first step is to encourage the use of cleaner energy. The early engineering design is to mix lime into the briquettes to inhibit the release of F from the briquettes (Sun 2005; Dai et al. 2006a). Nowadays, the use of outcrop coal and briquettes has been banned and seldom found in the rural houses, while commercial coal was still frequently found in rural area of Guizhou Province. The good news is that the coal used for domestic purposes is usually cleaner than it used to be. This can substantially reduce the emission of F into the surrounding environments and mitigate the exposure risk for the local residents. Previously, stoves used in the rural houses are simple and crude and often used in open environments. The second step is to ameliorate the stoves (e.g., adding exhaust pipes to improve the stove's ventilative capacity). Presently, most of the coal-burning stoves used in rural houses have been ameliorated and used in relatively isolated places (Hao et al. 2002; Liu et al. 2016). This can further reduce the residents' exposure risk to F. Formerly, local residents often took roasted corn (usually with elevated F^-) as their staple food. Nowadays, rice has become the staple food for local residents in most rural areas, though consumption of roasted corn still happened occasionally. This is also helpful for reducing the exposure risk to F for the local residents. The last step is to enhance the health awareness of local residents. The key point is to let the local residents known that how the endemic fluorosis happened and how to preserve their foodstuffs free of F contamination (Li et al. 2005a, b; Zhang et al. 2012). These measures have been summarized in the book of "Endemic Disease in China" (Sun 2017).

Since 2005, great efforts have been paid to the control of endemic fluorosis in Guizhou Province. As time passed, prevalence rate of endemic fluorosis in Guizhou Province has declined significantly compared with the historical data. However, prevalence rate of dental fluorosis (32.3%) in Guizhou Province was still high till very recently (Zhang et al. 2017a, b) and 8.79 million cases of dental fluorosis still left by the end of 2018 (Chinese Health Statistics Yearbook 2019). Although most existing cases of dental fluorosis are historically left, how many new cases developed remain unclear. Some administrative regions of Guizhou Province still deserve more attention, including Dafang, Qianxi, Zhijin, Qixingguan, and Nayong, where the prevalence rates of dental fluorosis were still higher than 50% till 2014 (Table 1).

So far, coal-burning type of endemic fluorosis is mainly found in Southwest China, similar health issue may also occur in other parts of the world. The findings on this issue are of great importance to the countries/regions where coal-burning type of endemic fluorosis could develop.

3 Other Health Problems

In addition to endemic fluorosis, other health problems associated with domestic coal burning may also exist in Guizhou Province, such as arsenosis, selenosis, and mercurialism (Finkelman et al. 1999, 2002). Our concerns about the health problems caused by domestic coal burning in Guizhou Province are based on the following facts: (1) commercial coal for domestic purpose is still common in the rural area of Guizhou Province, especially in cold season; (2) stove used for domestic coal burning in rural area is still simple, usually without any decontamination devices (Cheng et al. 2017); and (3) domestic coal used in rural households is usually without any legal or environmental requirements (Zhi et al. 2017). Based on the historical data, consumption rate of domestic coal in China has increased from approximately 50 million tons in 1992 to more than 90 million tons in 2015, and Guizhou Province is the main region of domestic coal utilization, second only to

Hebei Province and Shanxi Province (Zhao and Luo 2018). Although the use of outcrop coal and the briquettes have been banned, commercial coal for domestic purpose may still be widespread in the rural area of Guizhou Province. Coupled with the backward burning mode in the rural houses, local residents may face other health risks via indoor air pollution, which depends largely on the cleanliness of domestic coal.

Unfortunately, coals in Guizhou Province often contain high levels of toxic/harmful elements. Because Guizhou Province is one of the largest low-temperature metallogenic regions in the world and the low-temperature metallogenic processes usually lead to an enrichment of toxic/harmful elements in the coal (Dai et al. 2006b; Li et al. 2006; Song et al. 2007). For example, enrichment of As, Se, Hg, Cd, and Cr in coal is frequently found in Guizhou Province and the adjacent area (Li et al. 2006; Tang et al. 2009). High contents of As (7.6–230 mg/kg), Hg (0.34–19.1 mg/kg), Cr (15.1–155 mg/kg) and Cd (0.9–7.1 mg/kg) were also reported in coal collected from the major coal-producing areas of Guizhou Province (Song et al. 2007). In addition, extremely high levels of As (up to 35,000 mg/kg), Sb (up to 3,860 mg/kg) and Hg (up to 12.1 mg/kg) were observed in coals from Southwest Guizhou Province (Ding et al. 2001; Dai et al. 2006b). As summarized in Table 6, the toxic/harmful elements enriched in coal usually include As, Se, Hg, Cd, and Cr. Therefore, domestic coal burning in Guizhou Province is likely accompanied by the release of toxic/harmful elements and the local residents would face the exposure risk of toxic/harmful elements (especially to As, Se, and Hg), in addition to the common pollutants (such as $PM_{2.5}$ and PAHs).

Table 6 Toxic and harmful elements enriched in coals of Guizhou Province

Coal type	Location	Enriched elements	Reference
Late Permian coal	Xingren and Anlong	As, Sb, and Hg	Ding et al. (2001)
Late Permian/ Triassic Coal	Qianxi	F, As, Se, Hg, V, Mo, and U	Zhang et al. (2004)
Late Permian/ Triassic Coal	Guizhou Province and adjacent area	As, Se, Hg, Cd, and U	Li et al. (2006)
Late Permian coal	Xingren	As, Sb, Hg, and Tl	Dai et al. (2006b)
Late Permian coal	Pu'an	As, Se, Hg, Cd, Cr, Pb, Mo, and U	Yang (2006)
Anthracite coal	Southwest Guizhou Province	S, F, As, Hg, Cd, Cr, and Mo	Song et al. (2007)
Late Permian coal	Guiding	S, Se, Cd, Cr, Ni, V, Mo, and U	Dai et al. (2015)
Late Permian coal	Panxian	As, Hg, Se, Cd, Tl, and Mo.	Xie et al. (2017)
Late Permian coal	Xingren	S, Se, Cr, Cu, Co, V, U, and Mo	Duan et al. (2018)

Over the past few decades, with regard to the health problems caused by domestic coal burning, much attention has been paid to endemic fluorosis. Other health problems were seldom concerned, except that of arseniasis (An and Li 2015; Liu et al. 2007). Nevertheless, clues on other health problems can still be found in early research works (Finkelman et al. 1999, 2002; Liu et al. 2007). Associated health problems include endemic arseniasis, endemic selenosis, and mercury poisoning, as well as the pollution of organic compounds such as PAHs (Liu et al. 2008a, b). Because the health problems mentioned above usually have a long incubation period and strong concealment, it is necessary to carry out systematic research works on the suspected population.

4 Conclusion

Unlike endemic fluorosis in other parts of the world, endemic fluorosis in Guizhou Province is mainly caused by the roasted foodstuffs (especially the roasted pepper and corn), rather than the drinking water. Although geological background of F is high in Guizhou Province, it is not necessary the direct cause for endemic fluorosis. Both coal and clay are the probable sources of F in roasted foodstuffs, but the real culprit for endemic fluorosis is the unhealthy lifestyle of the local residents. The important link between F in coal/clay and endemic fluorosis is the roasting processes, which provides F a chance to get into the food chain.

As the primary energy source in China, coal will continue dominating in the near future (Zhao and Luo 2018). Currently, domestic coal burning is still widespread in rural area of China and can still be a serious environmental and health problem. Because domestic coal used in rural houses is usually without any legal or environmental requirements and coal-burning stoves used in rural area are usually without any decontamination devices. As to Guizhou Province and the adjacent areas, things may be worse owing to the enrichment of toxic/harmful elements in the local coal. It is necessary to determine how serious the situation is and find out the possible solution. To say the least, safe exposure thresholds related to the toxic/harmful elements should be established as soon as possible and effective engineering design should be developed to ensure the exposure doses being safe.

Nowadays, great efforts have been devoted to eradicating poverty in China. One of the main causes of poverty is the health problem. Ensuring the people' health is of great importance for the China's current battle against poverty. As people in other developing countries may also suffer from similar health issues related to the domestic coal burning, same environmental and social issues around the world deserve more attention.

Acknowledgments We are grateful for the financial support from the National Natural Science Foundation of China (91647205, 41877406) and the Joint Project of National Natural Science Foundation of China and Guizhou Karst Scientific Research Center (U1612441). We kindly appreciate the anonymous reviewers' technical comments and valuable suggestions.

Conflict of Interest On behalf of my co-authors and myself, we declare no conflict of interest to this work.

References

Ali S, Thakur SK, Sarkar A, Shekhar S (2016) Worldwide contamination of water by fluoride. Environ Chem Lett 14:291–315

An D, Li D (2015) Epidemiological status and countermeasures of endemic arseniasis of Guizhou Province. Chin J Endem 24(2):214–216 (in Chinese with English abstract)

An D, He G, Wang Q, Hu X (1995) Indoor pollution by coal smoke containing sulfur, arsenic and fluorine and their influence on human's health. J Environ Health 12:167–169 (in Chinese with English abstract)

An D, Yao Z, He G (1996) The medium of fluorine carrier-chili in the pollution of fluorine in Guizhou. Endem Dis Bull 11:55–57 (in Chinese with English abstract)

An D, He P, Li D, Liang Y, Jin Z, Hu X (2009) Study on the changes in the state of endemic fluorosis by coal-burning pollution in Guizhou Province. Chin J Control Endem Dis 24:403–407 (in Chinese with English abstract)

Arif M, Hussain I, Hussain J, Sharma S, Kumar S (2012) Fluoride in the drinking water of Nagaur tehsil of Nagaur district, Rajasthan, India. Bull Environ Contam Toxicol 88:870–875

Ayoob S, Gupta AK (2006) Fluoride in drinking water: a review on the status and stress effects. Crit Rev Environ Sci Technol 36(6):433–487

Cheng MM, Zhi GR, Tang W, Liu SJ, Dang HY, Guo Z, Du JH, Du XH, Zhang WQ, Zhang YJ, Meng F (2017) Air pollutant emission from the underestimated households' coal consumption source in China. Sci Total Environ 580:641–650

Chinese Health Statistics Yearbook (2019) National Health Commission of the People's Republic China, Beijing Union Medical University Press, Beijing (in Chinese)

Dai S, Ren D, Ma S (2004) The cause of endemic fluorosis in western Guizhou Province, Southwest China. Fuel 83:2095–2098

Dai S, Li W, Tang Y, Zhang Y, Peng F (2006a) The source, pathway, and preventive measures of endemic fluorosis in Guizhou. Geol Rev 52:650–654 (in Chinese with English abstract)

Dai S, Zeng R, Sun Y (2006b) Enrichment of arsenic, antimony, mercury, and thallium in a Late Permian anthracite from Xingren, Guizhou, Southwest China. Int J Coal Geol 66:217–226

Dai S, Li W, Tang Y, Zhang Y, Feng P (2007) The sources, pathway, and preventive measures for fluorosis in Zhijin County, Guizhou, China. Appl Geochem 22:1017–1024

Dai S, Ren D, Chou C, Finkelman RB, Seredin VV, Zhou Y (2012) Geochemistry of trace elements in Chinese coals: a review of abundances, genetic types, impacts on human health, and industrial utilization. Int J Coal Geol 94:3–21

Dai S, Seredin VV, Ward CR, Hower JC, Xing Y, Zhang W (2015) Enrichment of U–Se–Mo–Re–V in coals preserved within marine carbonate successions: geochemical and mineralogical data from the Late Permian guiding coalfield, Guizhou, China. Miner Depos 50:159–186

Ding Z, Zheng B, Long J, Belkin HE, Finkelman RB, Chen C, Zhou D, Zhou Y (2001) Geological and geochemical characteristics of high arsenic coals from endemic arsenosis areas in southwestern Guizhou Province, China. Appl Geochem 16:1353–1360

Duan P, Wang W, Sang S, Qian F, Pei S, Xin Z (2018) Partitioning of hazardous elements during preparation of high-uranium coal from Rongyang, Guizhou, China. J Geochem Explor 185:81–92

Finkelman RB (1993) Trace and minor elements in coal. In: Engel MH, Macko S (eds) Organic geochemistry. Plenum, New York

Finkelman RB, Belkin HE, Zheng BS (1999) Health impacts of domestic coal use in China. Proc Natl Acad Sci USA 96:3427–3431

Finkelman RB, Orem W, Castranova V, Tatu CA, Belkin HE, Zheng BS, Lerch HE, Maharaj SV, Bates AL (2002) Health impacts of coal and coal use: possible solutions. Int J Coal Geol 50:425–443

Fordyce FM, Vrana K, Zhovinsky E, Povoroznuk V, Toth G, Hope BC, Iljinsky U, Baker J (2007) A health risk assessment for fluoride in Central Europe. Environ Geochem Health 29:83–102

Gao J, Li D, An D, Liang Y, Zhang R, Ye H, Yao D, Zhang B, Zhang N (2015) Analysis of control and treatment effect on coal-burning endemic fluorosis in Guizhou Province from 2010 to 2014. J Guizhou Med Coll 40:1179–1183 (in Chinese with English abstract)

Guiyang Epidemic Prevention Station et al (1981) Investigation on endemic fluorosis caused by coal-baked food. Chin J Prev Med 15(5):182–283 (in Chinese)

Guiyang Medical College et al (1981) Investigation of endemic food-borne fluorosis in Zhijin County, Guizhou Province. J New Med 12:342–343 (in Chinese)

Hao Y, Sun D, Wei H, Zhao X, Shen Y, Wang L, Sun Y, Yu G (2002) Analysis on the control trends and present state of endemic fluorosis in China mainland. Chin J Endem 21(1):63–68 (in Chinese with English abstract)

He P, An D, Li D, Liang Y, Jin Z (2007) Analysis illness in severe area of coal-burning pollution fluorosis. Guizhou Med J 31:1066–1069 (in Chinese with English abstract)

Hong X, Liang H, Lv S, Wang D, Zhang Y (2015) Potential release of fluoride from clay used for coal-combustion in Hehua village, Zhijin County, Guizhou Province. Geol Rev 61:852–860 (in Chinese with English abstract)

Hong X, Liang H, Zhang Y, Xu D (2016) Distribution of acidity of Late Permian coal in border area of Yunnan, Guizhou and Sichuan. J China Coal Society 41:964–973 (in Chinese with English abstract)

Hong X, Liang H, Zhang Y (2018) Evaluation of acidity in late Permian outcrop coals and its association with endemic fluorosis in the border area of Yunnan, Guizhou, and Sichuan in China. Environ Geochem Health 40:1093–1109

Ketris MP, Yudovich YE (2009) Estimations of clarkes for carbonaceous biolithes: world average for trace element contents in black shales and coals. Int J Coal Geol 78:135–148

Kilborn LG, Outerbridge TS, Lei HP (1950) Fluorosis with report of an advanced case. Can Med Assoc J 62:135–141

Li M, Chen X (1997) Relationship between fluoride environmental geochemistry and endemic fluoride disease in Shuicheng. J Guizhou Univ Technol 26:92–95 (in Chinese with English abstract)

Li F, Yan S (1994) Investigation on coal-burning fluorosis in Liupanshui, Guizhou. Chin J Endem 22:360–361 (in Chinese with English abstract)

Li D, An D, Wang S, Li P, Zhang R, Li Z (2003) Epidemiological study of endemic fluorosis in Guizhou Province. J Endem 22:240–242 (in Chinese with English abstract)

Li D, An D, Zhang R, Chen S, Wang X, Li S, Ren M, Wang D, Wang K, Chen J (2004) Epidemiology of endemic fluorosis in the town central and remote villages of Guizhou Province. Chin J Endem 23:138–141 (in Chinese with English abstract)

Li D, An D, He P (2005a) Epidemiological study of endemic fluorosis of Guizhou Province. J Endem 24:651–654 (in Chinese with English abstract)

Li D, An D, He P, Liang Y, Hu X, Jin Z, Zhang R, Wang X (2005b) Evaluation of the effect of health education on controlling endemic fluorosis in Guizhou Province. Chin J Endem 24 (1):97–100 (in Chinese with English abstract)

Li D, Tang Y, Chen K, Deng T, Cheng F, Liu D (2006) Distribution of twelve toxic trace elements in coals from Southwest China. J China Unit Min Technol 35:15–20 (in Chinese with English abstract)
Li D, An D, He P, Liang Y, Jin Z (2011) Analysis of total intake of fluoride in severe endemic fluorosis regions of Guizhou Province. Chin J Control Endem Dis 26:268–270 (in Chinese with English abstract)
Li L, Luo K, Liu Y, Xu.Y (2012) The pollution control of fluorine and arsenic in roasted corn in "coal-burning" fluorosis area Yunnan, China. J Hazard Mater 229:57–65
Li L, Luo K, Tang Y, Liu Y (2015) The daily fluorine and arsenic intake for residents with different dietaries and fluorosis risk in coal-burning fluorosis area, Yunnan, Southwest China. Environ Sci Pollut Res 22:2031–2040
Li X, Wu P, Han Z, Shi J (2016) Sources, distributions of fluoride in waters and its influencing factors from an endemic fluorosis region in Central Guizhou, China. Environ Earth Sci 75:981
Liang H, Liang N, Gardella JAJ, He P, Yatzor BP (2011) Potential release of F from coal in coal-burning fluorosis areas in Guizhou. Chin Sci Bull 56:2301–2304
Liu J, Wang S (2005) Source of fluorine in coal-burning fluorosis area: a case study in Hehua village, Zhijin County, West Guizhou. Guizhou Geol 22:192–194 (in Chinese with English abstract)
Liu G, Zheng L, Duzgoren-Aydin NS, Gao L, Liu J, Peng Z (2007) Health effects of arsenic, fluorine, and selenium from indoor burning of Chinese coal. Rev Environ Contamin Toxicol 189:89–106
Liu G, Niu Z, Niekerk DV, Jian X, Zheng L (2008a) Polycyclic aromatic hydrocarbons (PAHs) from coal combustion: emissions, analysis, and toxicology. Rev Environ Contamin Toxicol 192:1–28
Liu X, Zheng C, Liu J, Zhang J, Song D (2008b) Analysis on fluorine speciation in coals from Guizhou Province. Proc CSEE 28:46–51 (in Chinese with English abstract)
Liu W, Li J, Guo X, Jiang W (2013) Analysis of fluorine pollution situation and source in poor rural areas of Guizhou Province. Earth Environ 41:138–142 (in Chinese with English abstract)
Liu Y, Ma P, Luo K, Li L (2016) The state and trend of endemic fluorosis from 1991 to 2012 in China. J Chongqing Norm Univ (Nat Sci) 33:142–151 (in Chinese with English abstract)
Luo K, Li H, Niu C (2010) Fluorine and arsenic pollution route of grain in Yunnan-Guizhou "coal-burning" endemic fluorosis area. Geol Rev 56:289–298 (in Chinese with English abstract)
Luo K, Li L, Zhang S, Bi S, Li W, Zhao F, Chen G (2011a) Pathways causing the disease of coal-burning fluorosis in Southwest China. J Hyg Res 40:474–477 (in Chinese with English abstract)
Luo K, Li L, Zhang S (2011b) Coal-burning roasted corn and chili as the cause of dental fluorosis for children in southwestern China. J Hazard Mater 185:1340–1347
Lyth O (1946) Endemic fluorosis in Kweichow, China. Lancet 247(6390):233–237
McKay FS, Black GV (1916) An investigation of mottled teeth: an endemic developmental imperfection of the enamel of the teeth, heretofore unknown in the literature of dentistry. Dent Cosmos 58(1):477–484
Pan Z, Liu X, Meng W, Li C, He S, Yan C, Wang F (2018) Geochemical characteristics of fluorine in soils and its environmental quality in central district of Guiyang. Res Environ Sci 31:87–94 (in Chinese with English abstract)
Pu W, Ding J, Ying W (2013) High fluoride groundwater distribution characteristics and genetic analysis of water exploration engineering in Guizhou. Environ Prot Technol 19:10–15 (in Chinese with English abstract)
Schafer D, Donn M, Atteia O, Sun J, MacRae C, Raven M, Pejcic B, Prommer H (2018) Fluoride and phosphate release from carbonate-rich fluorapatite during managed aquifer recharge. J Hydrol 562:809–820
Schafer D, Sun J, Jamieson J, Siade A, Atteia O, Prommer H (2020) Model-based analysis of reactive transport processes governing fluoride and phosphate release and attenuation during managed aquifer recharge. Environ Sci Technol 54(5):2800–2811

Song D, Zhang J, Zheng C (2007) Geochemistry of hazardous trace elements in coals from Guizhou Province. Coal Convers 30:13–17 (in Chinese with English abstract)
Sun Y (2005) Geochemical study advance about fluorine in coal and coal-burning endemic fluorosis from western Guizhou. Bull Miner Pet Geochem 24:350–356 (in Chinese with English abstract)
Sun D (2017) Endemic disease in China. In: Li L (ed) Public health in China (series II). Springer
Sun D, Shen Y, Zhao X, Wei H, Zheng X, Sun Y, Yu G (2001) Analysis on the disease trend and present state of endemic fluorosis in China mainland. Chinese J Endemiol 20(6):429–433 (in Chinese with English abstract)
Tang J, Xiao T, Wang S, Lei J, Zhang M, Gong Y, Li H, Ning Z, He L (2009) High cadmium concentrations in areas with endemic fluorosis: a serious hidden toxin? Chemosphere 76:300–305
Tian HZ, Wang Y, Xue ZG, Cheng K, Qu YP, Chai FH, Hao JM (2010) Trend and characteristics of atmospheric emissions of hg, as, and se from coal combustion in China, 1980-2007. Atmos Chem Phys 10:11905–11919
Wang X, Li D, Liang Y, Zhang R (2013) Epidemiological investigation of coal-burning endemic skeletal fluorosis in Guizhou. Stud Trace Elem Health 30:42–44 (in Chinese with English abstract)
Watanabe T, Kondo T, Asanuma S, Ando M, Tamura K, Sakuragi S, Rongdi J, Chaoke L (2000) Skeletal fluorosis from indoor burning of coal in southwestern China. Fluoride 33:135–139
WHO (2002) Environmental health criteria fluorides. World Health Organization, Geneva
Wu D, Wang A, Zheng B, Wang D (2004) Investigating the prevalent status on coal-burning pollution endemic fluorosis in some areas in North-Western Guizhou. Chin J Endem 23:454–456 (in Chinese with English abstract)
Wu D, Wu T, Dong R, Li P (2008) Advance in absorption and enrichment of soil fluorine by plants. J Nanchang Univ (Eng Technol) 30:103–111 (in Chinese with English abstract)
Xiao Q, Li F, Wang J, Xu L (2016) Fluorosis effect of coal baked corn and its long-term effect on bone in Shuicheng County, Guizhou. Contemp Med 22:161–163 (in Chinese with English abstract)
Xie X, Yang X, Yang S, Zhang J, Bin Z (2010) A tentative discussion on the source of endemic fluorosis: geo-environmental evidence from three counties in Guizhou Province. Geol China 37:696–703 (in Chinese with English abstract)
Xie P, Zhang S, Wang Z, Wang L, Xu Y (2017) Geochemical characteristics of the late Permian coals from the Yueliangtian coalfield, western Guizhou, southwestern China. Arab J Geosci 10:97–101
Xu M, Xu Y (2008) Investigation on endemic fluorosis: a case study in Guizhou Province. J Anhui Agric Sci 36:4660–4661, 4665 (in Chinese with English abstract)
Yang J (2006) Concentrations and modes of occurrence of trace elements in the late Permian coals from the Pu'an coalfield, southwestern Guizhou, China. Environ. Geochem Health 28:567–576
You CF, Xu XC (2010) Coal combustion and its pollution control in China. Energy 35:4467–4472
Yu Z, Yang S, Chen L, Zeng Q, Wu Y (1994) An elementary survey of fluorine concentration in Huaxi reservoir, Huaxi river and Dabang river. Guizhou Sci 12:60–63 (in Chinese with English abstract)
Zhang Y, Cao SR (1996) Coal burning induced endemic fluorosis in China. Fluoride 29:207–211
Zhang J, Ren D, Zhu Y, Chou CL, Zeng R, Zheng B (2004) Mineral matter and potentially hazardous trace elements in coals from Qianxi fault depression area in southwestern Guizhou, China. Int J Coal Geol 57:49–61
Zhang B, Chang C, Li S, Liang Y, Hu X, Zhang R, Gao J (2012) Investigation and analysis on health education needs of coal burning endemic fluorosis in different economic areas of Guizhou Province. Stud Trace Elem Health 29(1):38–40 (in Chinese)
Zhang X, Zhao B, Tan M, Zhou B, Sun J (2016) Water quality evaluation and environmental significance of ground water in Dafang area Guizhou Province. Ground Water 38:3–6 (in Chinese with English abstract)

Zhang B, An D, Li D, Yao D, Zhang N, Ye H, Gao J, Hu X, Wang X (2017a) An epidemiological investigation of dental fluorosis of children aged 8-12 in coal-burning type endemic fluorosis areas in Guizhou Province. Chin J Endem 36:269–273 (in Chinese with English abstract)
Zhang N, An D, Li D, Yao D, She D, Yang Y, Ye H (2017b) Contents of fluorine and their variation in corn and pepper in heavy polluted area of coal-burning fluorosis in Guizhou Province. Chin J Endem 36:580–582 (in Chinese with English abstract)
Zhao C, Luo K (2018) Household consumption of coal and related sulfur, arsenic, fluorine and mercury emissions in China. Energy Policy 112:221–232
Zheng B, Huang R, Zhi W (1985) Environmental geochemistry of indoor coal-burning fluorosis in Southwest China. Endem Comm 3:49–51 (in Chinese with English abstract)
Zheng Y, Liao L, Liu L (2009) Background fluorine in soils of Guizhou and its relationship with endemic fluorosis. Stud Trace Elem Health 26:39–41 (in Chinese with English abstract)
Zhi G, Zhang Y, Sun J, Cheng M, Dang H, Liu S, Yang J, Zhang Y, Xue Z, Li S, Meng F (2017) Village energy survey reveals missing rural raw coal in northern China: significance in science and policy. Environ Pollut 223:705–712
Zhong F, Luo X (2017) Research on the distribution of fluorine in coal and the occurrence regularity. Coal Qual Technol 5:13–16 (in Chinese with English abstract)
Zhou D, Fu Q, Wang Z, Zhou C (1991) Investigation on the relationship between fluorine content in coal mixed with clay and coal-burning fluorosis. Chin J Control Endem Dis 6:243–244 (in Chinese with English abstract)
Zhu F, Zhang J, Yao S, Zhang M (2000) Fluorine biogeochemical circulation in epidemic fluorosis area. Acta Sci Citsumstantiae 20:111–116 (in Chinese with English abstract)

Dissipation, Fate, and Toxicity of Crop Protection Chemical Safeners in Aquatic Environments

Femi F. Oloye, Oluwabunmi P. Femi-Oloye, Jonathan K. Challis, Paul D. Jones, and John P. Giesy

Contents

F. F. Oloye (✉)
Department of Chemical Sciences, Adekunle Ajasin University, Akungba-Akoko, Nigeria
e-mail: pen2crown@gmail.com; femi.oloye@aaua.edu.ng

O. P. Femi-Oloye
Toxicology Centre, University of Saskatchewan, Saskatoon, SK, Canada

Department of Animal and Environmental Biology, Adekunle Ajasin University, Akungba-Akoko, Nigeria
e-mail: femi-oloye.oluwabunmi@usask.ca

J. K. Challis
Toxicology Centre, University of Saskatchewan, Saskatoon, SK, Canada
e-mail: j.challis@usask.ca

P. D. Jones
Toxicology Centre, University of Saskatchewan, Saskatoon, SK, Canada

School of Environment and Sustainability, University of Saskatchewan, Saskatoon, SK, Canada
e-mail: paul.jones@usask.ca

J. P. Giesy
Toxicology Centre, University of Saskatchewan, Saskatoon, SK, Canada

Department of Biomedical Veterinary Sciences, University of Saskatchewan, Saskatoon, SK, Canada
e-mail: john.giesy@usask.ca

P. de Voogt (ed.), *Reviews of Environmental Contamination and Toxicology Volume 258*,
Reviews of Environmental Contamination and Toxicology 258,
https://doi.org/10.1007/398_2021_70

Abstract Safeners are a group of chemicals applied with herbicides to protect crop plants from potential adverse effects of agricultural products used to kill weeds in monocotyledonous crops. Various routes of dissipation of safeners from their point of applications were evaluated. Despite the large numbers of safeners (over 18) commercially available and the relatively large quantities (~2×10^6 kg/year) used, there is little information on their mobility and fate in the environment and occurrence in various environmental matrices. The only class of safeners for which a significant amount of information is available is dichloroacetamide safeners, which have been observed in some rivers in the USA at concentrations ranging from 42 to 190 ng/L. Given this gap in the literature, there is a clear need to determine the occurrence, fate, and bioavailability of other classes of safeners. Furthermore, since safeners are typically used in commercial formulations, it is useful to study them in relation to their corresponding herbicides. Common routes of dissipation for herbicides and applied safeners are surface run off (erosion), hydrolysis, photolysis, sorption, leaching, volatilization, and microbial degradation. Toxic potencies of safeners vary among organisms and safener compounds, ranging from as low as the LC_{50} for fish (*Oncorhynchus mykiss*) for isoxadifen-ethyl, which was 0.34 mg/L, to as high as the LC_{50} for *Daphnia magna* from dichlormid, which was 161 mg/L. Solubilities and octanol-water partition coefficients seem to be the principal driving force in understanding safener mobilities. This paper provides an up-to-date literature review regarding the occurrence, behaviour, and toxic potency of herbicide safeners and identifies important knowledge gaps in our understanding of these compounds and the potential risks posed to potentially impacted ecosystems.

Keywords Adsorption · Bioavailability · Dichloroacetamide safener · Dissipation · Herbicides · Safeners

1 Introduction

The increased use of reduced tillage systems to minimize erosion and alterations of soil structures, weed resistance, monocropping, and other factors have led to continued, large-scale use of herbicides in agriculture globally (Farenhorst et al. 2011). Knowledge of potential hazards, fate, and possible dissipation of this class of chemicals and their formulations (including safeners) to ground and surface waters is important. Rates of usages of herbicides to control weeds in crops continue to increase. Used along with herbicides is a group of chemicals called safeners, which are used either in pre- or post-application of herbicides, to protect monocotyledonous plants, which are plants that have seeds with one cotyledon, such as sugarcane, corn, and rice from toxic effects of herbicides used to control weeds in these crops.

Safeners act when applied onto crop seeds through either soil treatment, seed treatment, foliar spray or as a mixture with herbicide to prevent, reduce, or suppress the adverse effects (e.g. phyto-toxicity) of herbicides to crops, by physiological or molecular mechanisms, without reducing efficiency of suppressing targeted weeds (Davies and Caseley 1999; Hatzios and Burgos 2004; Sivey et al. 2015; Bolyard et al. 2017; Acharya and Weidhaas 2018). Chemical safeners can thus improve tolerance of herbicides by crops (Behringer et al. 2011).

As a result of chemical safeners being classified as "inert" components of herbicide formulations, there is less information and research regarding their fates in the environment, effects on non-target species, and possible mechanisms of action (Sivey et al. 2015; Acharya and Weidhaas 2018). Safeners are designed to alter biochemistries in crop plants in order to impart protection from herbicide-induced damage, which suggests that they might have biological activity in non-target organisms at ecologically relevant concentrations (Sivey et al. 2015).

Dichloroacetamide safeners have the propensity to be transformed by reduction to herbicide-like products (Sivey and Roberts 2012). Dichlormid has been reported to degrade through dechlorination, dealkylation, oxidation, and hydrolysis (Abu-Qare 1912). Efforts have been made to understand modes and mechanisms of action of safeners on target plants (Sivey et al. 2015; Behringer et al. 2011; Abu-Qare and Duncan 2002); however, many of these mechanisms remain unclear. Same efforts have not been directed towards their fate, mobility, and effects on organisms in various environments, including the aquatic environment.

Safeners are part of the ingredients in formulations of commercial, weed control products (Acharya and Weidhaas 2018). For example, dichloroacetamide safeners contribute up to ~5% by weight of the herbicide formulation (Xu et al. 2020, and the references therein). This class of chemical are in continual usage. For example, the application of dichloroacetamide safeners in 2017 in the USA was estimated to be more than two million kg/year, which exceeded the application of the active ingredients of many common herbicides (Bolyard et al. 2017; Woodward et al. 2018; Xu et al. 2020). Safeners are applied in the same manner as herbicides, either sprayed as a mixture with active ingredient herbicides or as separate seed treatments (Davies and Caseley 1999). Their wide usage and mode of applications suggest that these chemicals will enter the environment like those of active herbicide ingredients and in proportions at which they were applied in the field. In 2004, Syngenta Crop Protection Canada submitted a field data report on application of cloquintocet-mexyl safener for protection of wheat (USEPA 2005) in Alberta, Manitoba, and Saskatchewan, further suggesting the potential for presence of safeners in aquatic environments outside the USA.

Herbicide safeners have diverse applications when it comes to protection of agriculturally important crops (see Table 1). Winter wheat and barley were protected from adverse effect of the herbicide pinoxaden (Axial) by converting its active ingredient into inactive metabolites by the safener cloquintocet-mexyl. However, this safener was ineffective for protecting wild oats or perennial ryegrass from adverse effects of pinoxaden herbicide (Brosnan et al. 2016). Effects of the herbicide carfentrazone-ethyl on spring wheat could be suppressed by simultaneous exposure to sulfonylurea or flucarbazone-sodium (Howatt 2006). In this case, these two herbicides were used as safeners, not herbicides, because they were not effective against weeds, but were used to protect against effects of another herbicide. To

Table 1 Name and structure of herbicide safeners

Safeners	Structure	Common Herbicide	Crops
Benoxacor	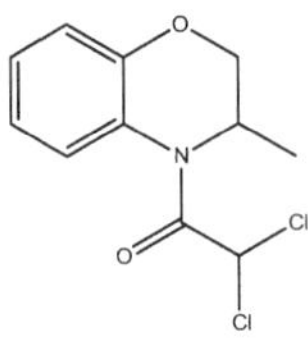	Metolachlor	Corn, Soyabeans and Sorghum
Dichlormid	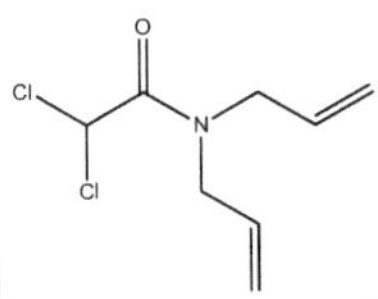	Chloroacetanilide, Thiocarbamate	Corn
Furilazole	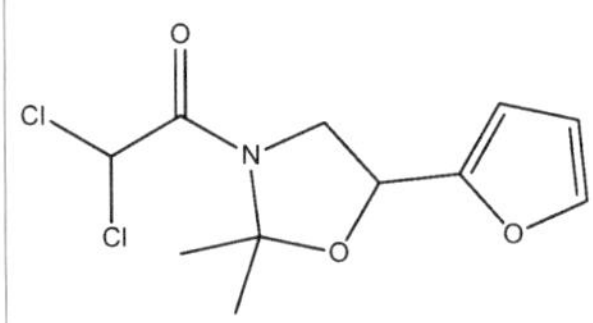	Isoxaflutole	Corn, Rice
AD-67	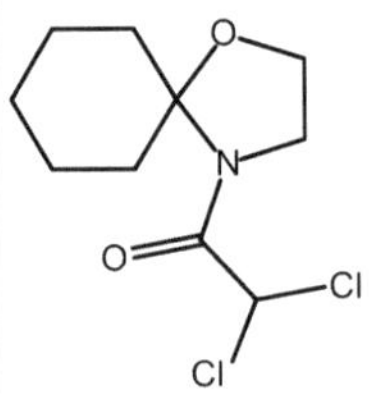	Acetochlor, Butachlor, EPTC	Corn
Mefenpyr-diethyl	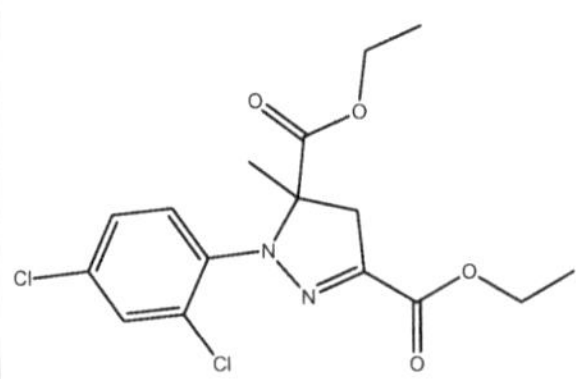	Fenoxaprop-P-ethyl, Iodosulfuron, ACCase inhibitors Sulfonylureas	Corn, Wheat, Rye, Barley, Triticale
Cloquintocet-mexyl	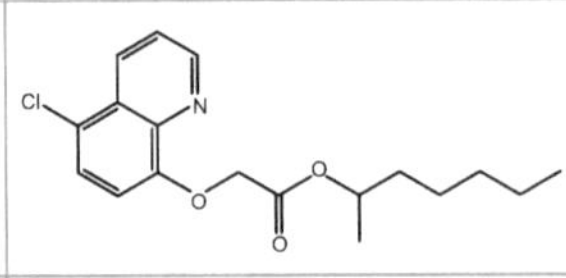	Clodinafop-propargyl, Pinoxaden	Winter wheat, Barley wild oats perennial ryegrass
Cyometrinil	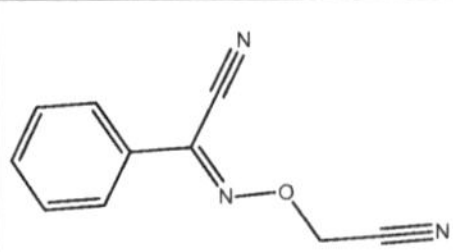	Metolachlor	Sorghum
Cyprosulfamide	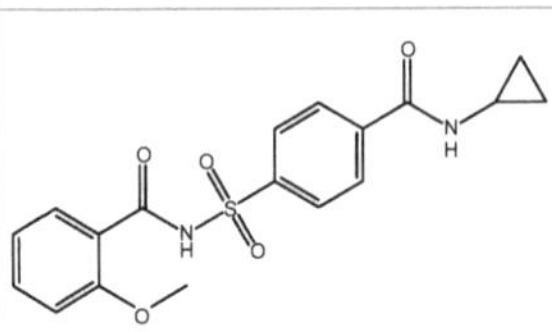	Thiencarbazone-methyl, Isoxaflutole, Tembotrione, Iodo sulfran, Nicosulfuron	Corn

(continued)

Table 1 (continued)

Safeners	Structure	Common Herbicide	Crops
Dicyclonon	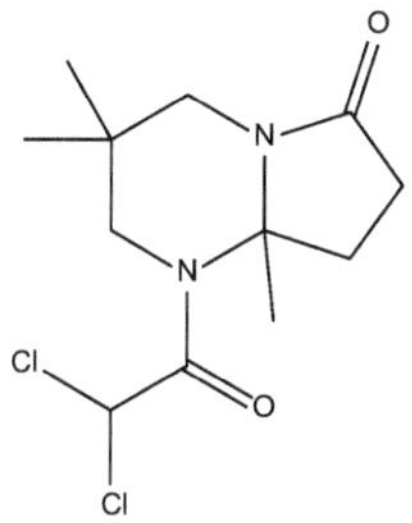	Metazachlor Metalochlor	Corn
Dietholate	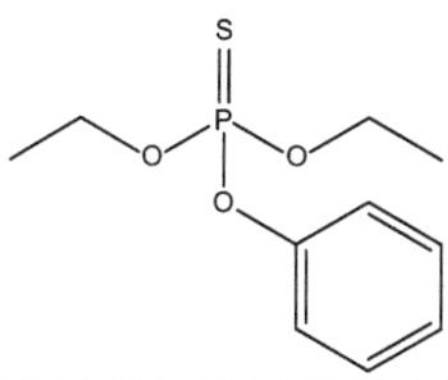	Clomazone	Cotton
Fenchlorazole-ethyl	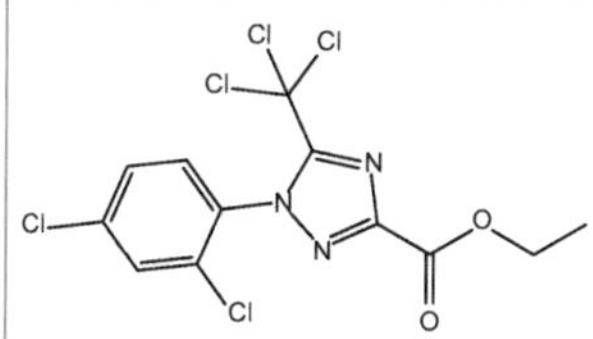	Fenoxaprop-ethyl, Fenoxaprop-P-ethyl	Wheat
Fenclorim	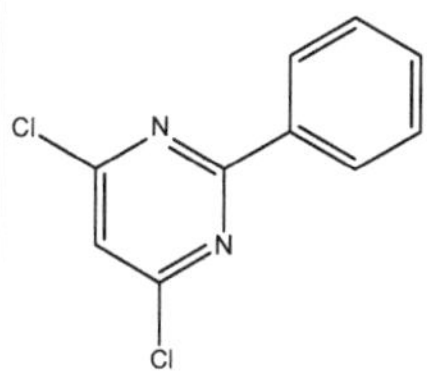	Pretilachlor	Rice
Fluxofenim	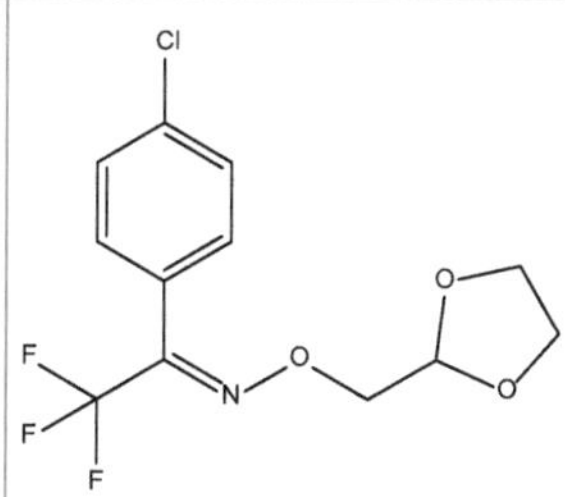	Metolachlor, S-metolachlor	Sorghum
Isoxadifen-ethyl	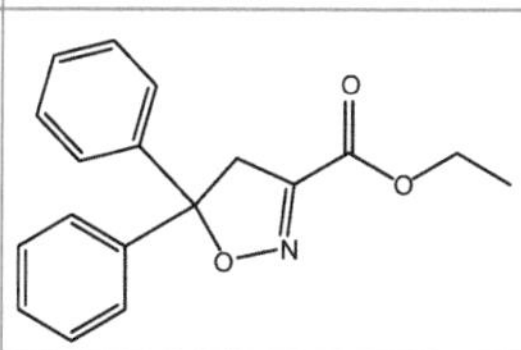	Foramsulfuron, Tembotrione, ACCase inhibitors Sulfonylureas	Corn, Rice

(continued)

Table 1 (continued)

Safeners	Structure	Common Herbicide	Crops
Jiecaowan	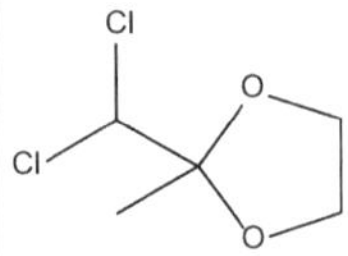		Corn
Mephenate	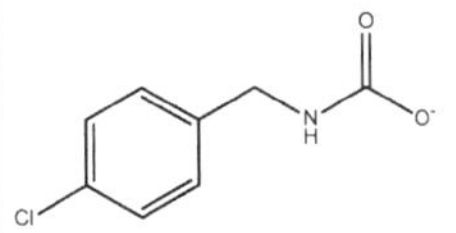		Corn
Metcamifen	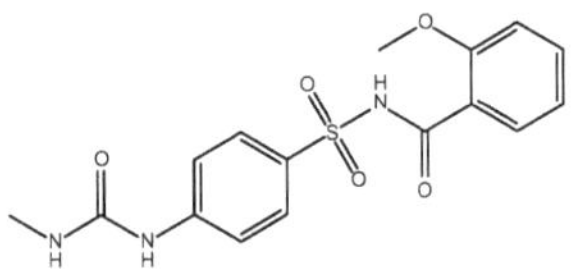		Corn
Naphthalic anhydride	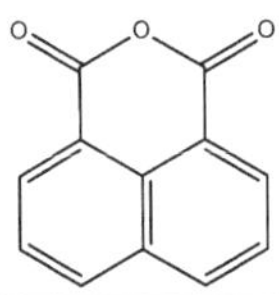	Thiocarbamates	Corn
Oxabetrinil	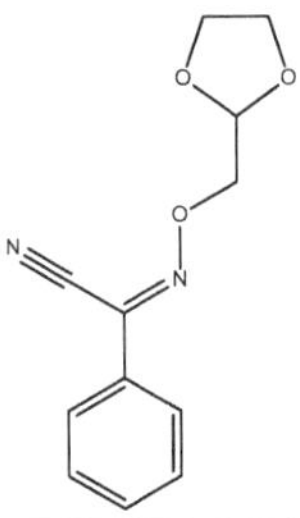	Chloroacetanilides	Sorghum

protect from effects of herbicides, flax seeds were coated with either BASF's Insure Pulse or Vitaflo (Staff 2019). Also, safeners have been used to protect hard, red spring wheat from injury from the herbicide, fenoxaprop (Staff 2019). Dichlormid was also an effective safener when added to S-Ethyl dipropylthiocarbamate (EPTC) and other thiocarbamates in preventing the onset of herbicide harm to maize plants (Abu-Qare 1912). Thus, safeners are used to protect different crops against injurious effects of herbicides.

While some safeners are believed to be crop-specific, there is evidence that mefenpyr-diethyl and cyprosulfamide can protect crops in addition to the target crop (Duhoux et al. 2018) and/or inhibit other herbicides through various modes of action (Ahrens et al. 2013). Cyprosulfamide has been reported to protect all varieties of corn from pre-emergent applications of isoxaflutole (Ahrens et al.

2013) and tembotrione herbicides. Corn and rice are the most common cereal crops protected with safeners against chloroacetanilide, sulfonylurea, imidazolinone, cyclohexanedione, isoxazole, and triketone herbicides (Table 1) (Davies and Caseley 1999; Davies 2001; Guo et al. 2020). Winter cereal crops, such as wheat, also use safeners to protect against effects of post-emergence applications of aryloxyphenoxypropionate and sulfonylurea herbicides on target plants (Hatzios and Burgos 2004). Examples of safeners that protect cereal crops from post-emergence applications of sulfonylurea herbicides are isoxadifen-ethyl and mefenpyr-diethyl (Behringer et al. 2011), which can also work to protect *Arabidopsis thaliana* leaves (Behringer et al. 2011).

As with herbicides, safeners can dissipate in several ways, including leaching, adsorption, volatilization, biotic and abiotic degradation (Fig. 1) (Abu-Qare and Duncan 2002), hydrolysis and for some photolysis (Hertkorn et al. 2010). Dissipations of both herbicides and safeners are functions of their physico-chemical properties that govern behaviours in soil and water (Díez and Barrado 2010). Potential environmental mobilities of safeners are functions of their aqueous solubility and sorption to solid particles (Acharya and Weidhaas 2018). Environmental partitioning such as sorption and leaching from soils, uptake by plants, and accumulation in various environmental compartments can be estimated by use of fugacity based on the octanol-water partitioning coefficient (Log K_{OW}) (Acharya and Weidhaas 2018; Zhang et al. 2016). Therefore, determining Log K_{OW} will give fundamental information on sorption in soil and the potential availability of safeners in the environment.

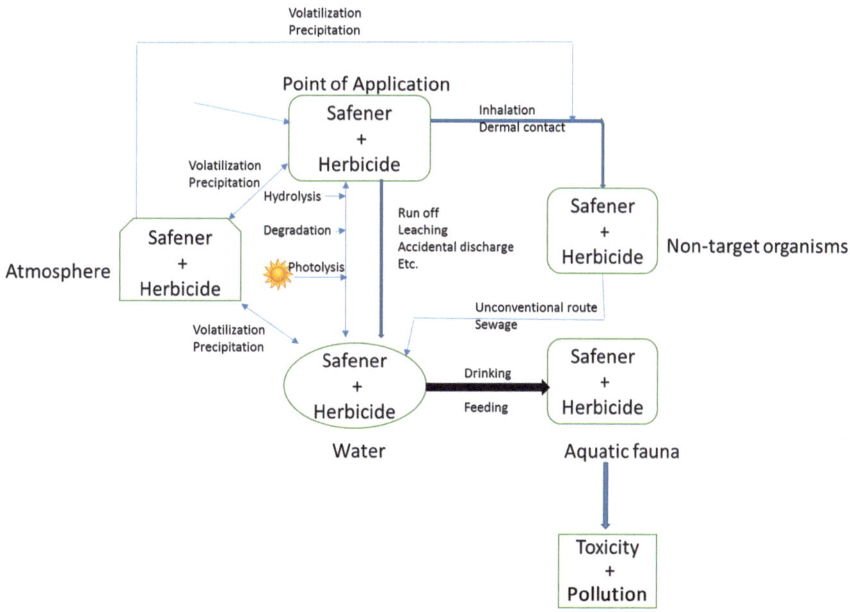

Fig. 1 Possible transportation routes and way of safeners exposure to the aquatic environment

Despite the various sorption mechanisms and degradation pathways which chemical safeners undergo, occurrence in surface waters has been observed, and remains a gap in our understanding of the exposure risks posed by these chemicals. Furilazole, benoxacor, dichlormid, and Ad-67 were detected in Midwestern US rivers between Spring and Summer 2016, with the maximum concentrations ranging from 42 to 190 ng/L (Woodward et al. 2018). In surface waters, chemical safeners are capable of producing a broad range of responses in non-target organisms (USEPA 2006), similar to their often structurally-related herbicide counterparts (Acharya and Weidhaas 2018). Most of the safeners, including benoxacor and dichlormid have *de minimis* mammalian toxicity and a high potential for bioaccumulation (Lewis et al. 2016; PPBD 2020).

There are several factors to consider when assessing our understanding of the fate and behaviour of safeners in the environment. The modes of application of safeners and herbicides and how they enter aquatic environments can affect pathways and rates of transport and dissipation. Additionally, chemistries of safeners might be different when applied together with herbicides than when used alone (Bolyard et al. 2017; Acharya and Weidhaas 2018; Su et al. 2019).

Finally, safeners can be transformed, and their transformation products should also be considered in the assessment of fate, transport, and ultimate hazard to non-target organisms. Also, routes and rates of dissipation of herbicides and safeners in the same formulation may differ. Hence, potential hazardous effects of all types of safeners (Table 1) need to be reviewed to understand the safety of this class of chemical. In this review all classes of known safeners will be discussed with respect to their fate, mobilities, and eco-toxicity. Where experimental data were not available, software and data from manufacturer safety data sheets were used. Some experimental data from our research groups were also used to corroborate some of the existing data.

2 Methods

Properties of safeners, such as Henry law's constant, bioconcentration factors (BCF), aqueous solubility, and octanol-water partition coefficient were estimated with programs within EPI Suite WEB 4.10 (USEPA 2014). The input of chemical compounds was based on the CAS number and Simplified Molecular Information and Line Entry System (SMILES) notation. These data were obtained from either ChemSpider® or from the Pesticide Properties Database (PPDB): University of Hertfordshire. The structures were drawn using Chemdraw® Ultra 12.0 (Table 1). Some data were also sourced from PubChem. For the experimental method presented in this report, see Sect. 11.

3 Leaching

Leaching is one of the primary mechanisms through which safeners and herbicides dissipate from points of application. Aqueous solubility is one important factor that determines the temporal and spatial mobility of safeners and herbicides (Acharya and Weidhaas 2018). Rates of transportation of safeners and herbicides determine their ability to protect target crops as well as the potential for exposure of non-target organisms, such as aquatic organisms. This means that if safeners and herbicides are moving at different rates, safening will not be optimized on crops (Nelson and Penner 2007). Interestingly, a recent study showed that safeners such as benoxacor and the herbicide (S-metolachlor) moved through the soil at a similar rate (Acharya et al. 2020). Unfortunately, not all chemicals move at the same rate in soil and plants. For example, the herbicide, isoxaflutole, leached faster than furilazole and thus required the use of a polymeric carrier to ensure that isoxaflutole and herbicides safeners moved at the same manner (Nelson and Penner 2007). Rates at which chemicals move depend on multiple factors, including their physical and chemical states. Hence, herbicides cannot necessarily be used to predict leaching potential of safeners. Leaching is different from other similar transportation processes, such as surface runoff and erosion, because it involves the downward movement of substances.

When new herbicides are registered, efforts are made by researchers and manufacturers to understand their potential to leach. However, little or no effort is made to understand the leaching of accompanying safeners in the academic community. Indaziflam, an emerging herbicide, was tested under simulated rain and was observed not to leach beyond 30 cm under any conditions tested (Jhala et al. 2012). This observation may not hold true for the corresponding safener and could differ under realistic field conditions, where there is a possibility that interactions between safener and herbicide can affect mobility (Bolyard et al. 2017; Acharya and Weidhaas 2018). Octanol-water partition coefficients can be used to estimate mobilities of chemicals. Smaller Log K_{OW} values indicate that the chemicals will have less potential for sediment and soil sorption, thus greater mobility. For example, the Log K_{OW} values for benoxacor and furilazole were 2.23 ± 0.16 and 1.96 ± 0.22, respectively (Acharya and Weidhaas 2018), suggesting that furilazole will leach more readily than benoxacor, assuming hydrophobic-type interactions are dominating sorption.

Prediction software can be used to estimate chemical mobilities based on predicted Log K_{OW}. These include EPI suite (Sivey and Roberts 2012; Sivey et al. 2015; Bolyard et al. 2017), Marvin Sketch, TOXNET (TOXicology Data NETwork) (ChemIDplus) (Acharya and Weidhaas 2018). The KOWWIN program in EPI Suite software overestimated Log K_{OW} (Table 2) compared to our experimentally measured values and the select data present in the literature of which 25 and 8% were overestimation for benoxacor and furilazole, respectively (Acharya and Weidhaas 2018). Similar overestimations by EPI Suite of 24% for benoxacor and 19% for furilazole were observed in the present study. The reason for overestimation of the EPI suite and the experiment could be attributed to a method error.

Table 2 Log K_{OW}, Log Koa, and solubility of safeners determined using EPI computer simulation and PubChem

Safeners	Log K_{OW} Experimental data based match	Log K_{OW} KOWWIN v1.68 estimate	Solubility Experimental data based match 293 K mg/L	Solubility From PubChem M	Solubility WSKOW v1.42 298 K mg/L	Solubility WSFragment v1.01 estimate 298 K mg/L	Logkoa KOAWIN v1.10 estimate
Benoxacor	2.70 (2.25)[a]	2.38	20.00	0.000077	102.70	750.18	8.21
Dichlormid	1.84 (0.80)[a]	2.28	5000.00	0.02	1067.00	2662.20	6.71
Furilazole	2.12 (0.72)[a]	2.84	197	70,800	254	7,121	10.55
Mefenpyr-diethyl	3.83 (3.78)[a]	4.82	20	0.000054	2.40	4.45	11.15
Cloquintocet-mexyl	5.03	5.28	0.59		0.3816	1.9566	12.51
Cyometrinil	–	2.52	95		196.8	467.66	3.974
Cyprosulfamide	– (0.55)[a]	2.30	–		47.7	13.94	14.44
Dicyclonon	–	1.98	–		273.1	13,826	9.074
Dietholate	3.46	3.91			27.53	17.652	5.973
Fenchlorazole-ethyl		4.54	13	0.000002	0.5799	0.3055	8.916
Fenclorim	4.17	2.99	2.5	0.000011	8.883	19.424	7.529
Fluxofenim	2.90	2.73	30	0.000097	35.96	1.6024	6.701
Isoxadifen-ethyl	2,01[b]	5.66	–		0.1917	0.082283	10.516
Jiecaowan	–	1.22	–		5,383	46,569	–
Mephenate	2.01	1.82	–		992.7	2188.8	8.022
Metcamifen	–	1.47	–		283.7	29.346	17.738
Naphthalic anhydride	3.24[b]	3.24	–		5.878	592.16	7.837
Oxabetrinil	2.76	2.78	20		130.1	376.13	8.365

[a]determined by the authors in the laboratory
[b]from PubChem

The frequency and amount of rainfall affect leaching, as well as the time between rain events and application of safeners and herbicides. Greater rainfall can result in leaching of herbicides, which might result in damage of, for example, citrus tree roots, poor weed control, and groundwater contamination (Jhala et al. 2012). The presence of both herbicides and safeners in surface waters (Woodward et al. 2018) suggests we need a better understanding of leaching behaviours of these chemicals and potentially how the use and management of these herbicide/safener formulations could be optimized to minimize leaching to the environment (Chnirheb et al. 2012). Long- and short-time leaching experiments in laboratory and field trials are essential to determine leaching of both herbicides and safeners, when applied together. A recent study showed that both benoxacor and S-metolachlor leached at the same rate, which was suggested to be influenced by soil texture (Acharya et al. 2020). Leaching of chemicals to aquatic environments is problematic because some chemicals can bio-accumulate and bio-concentrate in either sediments or living organisms within the aquatic system. For example, pre-emergence application of the herbicide bromacil has been banned from use in Florida, USA, because it has been observed to leach and contaminate groundwater (Jhala and Singh 2012; Jhala et al. 2012).

The most commonly used methods for determining potential for leaching in field studies involve chemical concentrations measured in soil, water from tile drains and suction cups and dye tracer experiments (Sarmah et al. 2004). Similarly, laboratory experiments such as soil columns are common methods for qualitative and quantitative measurements of leaching of chemicals (OECD 2004).

4 Sorption

There is limited experimental data describing safener sorption to soil and sediment systems. Therefore, predicted physico-chemical properties, the few data available and data on herbicides will be used as a guide for understanding sorption of safeners. Estimates of chemical mobilities in various environments and rates of dissipation must include an understanding of sorption and desorption properties (OECD 2000). Fate and behaviour (e.g., transportation, or retention) of herbicides in soils are controlled by various factors, including rates of sorption and desorption (Mamy and Barriuso 2007), soil properties, and various processes of transformation, all of which can determine the bioavailable fraction in the environment (Davies and Jabeen 2003). Sorption of chemicals onto soil particles and their relative affinities to those soils can ultimately determine the potential for other types of dissipation.

At a molecular level, sorption is driven and controlled by diffusion, which is a relatively slow and first-order process. Thus, sorption and desorption are often rate-limiting steps in overall dissipation in soils. Desorption rates of chemicals can be used as an estimate of overall dissipation rates and of potential for non-point sources of contamination, for example in groundwaters (El Boukili et al. 2018). A chemical that remains strongly sorbed to the soil will not readily reach ground or surface waters through leaching or runoff. For example, the Freundlich constant for benoxacor and furilazole are 6.4 and 3.4 (mg/g) × (mg/L)(1/n) in granulated activated charcoal (GAC) (Acharya et al. 2020). This suggested that benoxacor sorbed more readily to GAC than furilazole. Similarly, the Freundlich constants for furilazole and dichlormid had been reported to be 0.79–3.5 and 0.25–0.65 (mg/g) × (mg/L)(1/n), respectively (Carr 1990; Subba-Rao 1990). Hence, dichlormid, furilazole, and benoxacor did not sorb to any great extent on agricultural soil (Carr 1990; Subba-Rao 1990; Acharya et al. 2020). Consequently, the sorption of benoxacor and furilazole was found to be reversible (40–90%) (Acharya et al. 2020). Therefore, this class of safeners can be readily leached through the soil.

Characteristics such as pH, texture, cation exchange capacity, crystalline lattice structure, organic matter content, amorphous iron oxides, inorganic matter complexes, temperature, and moisture (Harris and Warren 1964; Gennari et al. 2008; El Boukili et al. 2018; Wiersma et al. 2019) could potentially have large effects on sorption capacities of soils. Most herbicides and all safeners reported here have Log K_{OW} values >1, suggesting a potential to sorb to organic components of soil matrices (Cumming and Rücker 2017). Both soil organic matter and clay are significant constituents affecting the sorption of mefenpyr-diethyl in soils (El Boukili et al. 2018). Similarly, only organic carbon content and adsorption parameter (n) were found to influence adsorption of benoxacor and furilazole (Acharya et al. 2020). Thus, many soil properties might not influence adsorption of safeners.

However, some herbicides such as glyphosate have greater affinity for inorganic minerals in soils (Gennari et al. 2008). Acidic soils will generally enhance herbicide sorption (Gennari et al. 2008). Retention of chemicals by soil depends on the properties of both the soil and chemical. In some cases, retention of chemicals by soils increases/decreases, while in some other cases, it remained constant with changes in soil properties (Mamy and Barriuso 2007). Therefore, each herbicide or safeners needs to be tested to know their sorption capacity under prevailing environmental conditions, ideally through experimentation but also by software prediction, as necessary. A detailed method for quantifying sorption has been developed (OECD 2000) and can be used directly or with little modification for most chemicals.

Studies of sorption of active ingredients of herbicide formulations are common, but reports on sorption of herbicide safeners to soil are limited. Hence, predictions of these properties, including Log K_{OW}, aqueous solubilities, and Log Koa, as estimated from the programs (KOAWIN v1.10) within EPI suite, are given (Table 2). These properties can be used to predict potential mobilities of safeners as their movements in soils, in part, are a function of their aqueous solubility and Log K_{OW} (Acharya and Weidhaas 2018) (Table 2). The aqueous solubilities vary between 0.19 mg/L for isoxadifen-ethyl and 1,070 mg/L for dichlormid. Solubility data varies depending on the source of the data, for example PubChem lists solubility of dichlormid as 5,000 mg/L, WSKOW v1.42 lists it as 1,070 mg/L and WSFragment v1.01 estimate it as 2,662 mg/L. Safeners such as isoxadifen-ethyl, cloquintocet-mexyl, mefenpyr-diethyl, fenchlorazole-ethyl and naphthalic anhydride are less soluble or not readily soluble in aqueous medium, while dichlormid, furilazole, cytometrinail, cyprosulfamide are readily soluble in water and are thus more likely to be mobile in soil environments. For example, isoxadifen-ethyl and

Table 3 Sorption capacity of safeners to organic component of the soil

Safeners	K_{oc} PPDB	K_{oc} MCI method	K_{oc} K_{ow} method	Remarks
Benoxacor	109.0	133.2	267.5	Moderately mobile
Dichlormid	40	120.6	78.6	Mobile
Furilazole	199	252	91.2	Moderately mobile
Mefenpyr-diethyl	634	1,501	739.6	Non-mobile
Cloquintocet-mexyl	9,856	26,130	4,985	Non-mobile
Cyometrinil	–	709.9	1,269	Slightly mobile
Cyprosulfamide	–	10.41	39.68	Mobile
Dicyclonon	–	89.86	84.25	Mobile
Dietholate	–	–	1,543	Non-mobile
Fenchlorazole-ethyl	–	18,780	4,031	Non-mobile
Fenclorim	3,655	3,344	4,267	Slightly mobile
Fluxofenim	–	6,179	222.8	Moderately mobile
Isoxadifen-ethyl	–	81,660	9,777	Non-mobile
Jiecaowan	–	14.69	26.23	Non-mobile
Mephenate	–	–	34.54	Non-mobile
Metcamifen	–	–	233.1	Slightly mobile
Naphthalic anhydride	–	15.16	181.4	Mobile
Oxabetrinil	–	368.9	459.9	Mobile

PPDB Pesticide Properties Database: University of Hertfordshire

fenchlorazole-ethyl (Table 3) have relatively low water solubilities, therefore predicted to be less mobile in soils. Safeners with larger Log K_{OW} values are less mobile. Isoxadifen-ethyl, fenchloazole-ethyl, cloquintocet-mexyl, and mefenpyr-diethyl have Log K_{OW} values of 5.66, 4.54, 5.28, and 4.82, respectively, and are less mobile than cyprosulfamide, benoxacor, dicylonon, cyometrinil, and furilazole with Log K_{OW} values of 2.30, 2.38, 1.98, 2.52, and 2.84, respectively. Predicted Koc values (Table 3), which correlate with Log K_{OW} (Table 2) in terms of mobility and sorption capacity of the chemicals to organic carbon content in soils are provided here. Assuming hydrophobic interactions are governing sorption, mobility will be inversely proportional to Koc. Thus, determining solubility, Log K_{OW}, and K_{OC} properties provide fundamental information on the potential for sorption of chemical safeners to soils. Relationships exist between solubilities and adsorption capacity in one form, and adsorption capacity and molecular size in another (Acharya et al. 2020). Furilazole was determined to have lower adsorption capacity compared to benoxacor because the latter is not very soluble, while furilazole was moderately soluble. Considering their molecular structures, the smaller size of benoxacor might be the driving force for its higher adsorption capacity compared to furilazole. Therefore, both solubility and molecular size might be factors to be considered in sorption of safeners studies.

5 Volatilization

Information on volatilization of chemicals used as herbicides is required by regulatory bodies in many countries (Guth et al. 2004; Boivin and Poulsen 2017). This is because losses of these chemicals from soils, plants, and water bodies can accumulate in the atmosphere and result in long-range atmospheric transport (Dekeyser et al. 2015). Physico-chemical parameters such as Henry's law constant, soil/air and water/air distribution coefficients, and vapour pressure can be used to predict volatilization from water, soil, and plant matrices (Guth et al. 2004) (Tables 2 and 4). Vapour pressure is the best parameter to estimate volatilization of chemicals from plants, while water/air and soil/air distribution coefficients are the best parameters to predict volatilization of chemicals from water and soil, respectively (Guth et al. 2004). Compounds with vapour pressures $<10^{-3}$ to 10^{-4} Pa or Henry's law constant $<5 \times 10^{-5}$ atm-m^3/mol will demonstrate negligible volatility (Guth et al. 2004), therefore safeners such as cyometrinil, and fenclorim, with Henry's law constants of 8.59×10^{-4} and 1.07×10^{-5} atm-m^3/mol, respectively, may be relatively volatile. In contrast, metcamifen with a Henry's law constant of 1.32×10^{-18} atm-m^3/mol would be considered non-volatile (Table 4).

Greater water content of soils normally results in greater vapour pressures of chemicals, which can therefore result in volatilization (Sarmah et al. 2004). Water can facilitate movement of safeners from soil surfaces after displacement from sorption sites. Other activities occurring within the plant, soil, or water can hasten

Table 4 Volatilization from water as estimated using EPI suite

Safeners	Henry Law's constant atm-m^3/mol	Half-life (Lake) days	Half-life (River) Days
Benoxacor	7.58E-008	5,669	519
Dichlormid	3.29E-007	1,173	107
Furilazole	9.2E-011	4,842,000	442,200
Mefenpyr-diethyl	1.16E-009	443,200	40,630
Cloquintocet-mexyl	8.1E-010	602,100	55,190
Cyometrinil	0.000859	6	<1
Cyprosulfamide	1.78E-014	2.893E+010	2.652E+009
Dicyclonon	1.97E-009	231,300	21,200
Dietholate	7.51E-005	11.77	<1
Fenchlorazole-ethyl	1.03E-006	508	46
Fenclorim	1.07E-005	43.25	3.484
Fluxofenim	3.87E-006	128	11.17
Isoxadifen-ethyl	3.41E-007	1,348	123
Jiecaowan	4.45E-008	7,872	721.2
Mephenate	2.38E-008	15,240	1,397
Metcamifen	1.32E-018	3.843E+014	3.523E+013
Naphthalic anhydride	6.19E-007	610.8	55.54
Oxabetrinil	6.08E-008	6,676	611.5

Table 5 Biodegradation of safeners: probability of rapid biodegradation (BIOWIN v4.10)

Safeners	Biowin1 (Linear Model)	Biowin2 (Non-Linear Model)	DT_{50} (typical)[a]	DT_{50} (lab @293 K)[a]
Benoxacor	0.7430	0.6339	50	–
Dichlormid	0.6359	0.2768	–	–
Furilazole	0.0712	0.0008	29	65
Mefenpyr-diethyl	0.3694	0.5283	17.5	–
Cloquintocet-mexyl	0.8198	0.9878	5	1.75
Cyometrinil	1.4015	1.0000	–	–
Cyprosulfamide	0.9114	0.9329	–	–
Dicyclonon	0.4376	0.0510		
Dietholate	1.0723	1.0000		
Fenchlorazole-ethyl	−0.1882	0.0000	2.4	2.5
Fenclorim	0.4036	0.0817	26	
Fluxofenim	−0.7974	0.0000		
Isoxadifen-ethyl	0.8534	0.9916		
Jiecaowan	−0.1409	0.0003		
Mephenate	0.5563	0.3468		
Metcamifen	0.7065	0.5242		
Naphthalic anhydride	0.6532	0.5485		
Oxabetrinil	0.3774	0.3307		

[a]*PPDB* Pesticide Properties Database: University of Hertfordshire

volatilization. For example, cohesion, transpiration, capillary action and translocation in plants may cause the chemicals to move to parts of the plant, where volatilization may be aided (Ramanjaneyulu and Giri 2006). The data presented in Table 5 show that the half-life for volatilization from a river can be different from a lake. The half-life, as a result of volatilization in river and lake of cyometril was estimated to be <1 and 6 days, respectively, while cyprosulfamide's half-life was estimated as 2.7×10^9 and 2.9×10^{10} days, for river and lake, respectively. The proposed differences in volatilization of cyometril and cyprosulfamide could be attributed to the differences in their Henry law's constant. However, mineral components and other factors such as turbulent mixing in rivers might enhance the removal of the safeners.

Other parameters that affect volatility, which are frequently used in volatility experiments, include agricultural practice (rate of application of the chemical), competing processes (uptake by plants and degradation), and environmental conditions (wind speed, air humidity, soil moisture, and temperature) (Dekeyser et al. 2015). The higher the temperature and relative humidity, the higher the volatilization of chemicals. Therefore, the rate of volatilization of the same chemical in different climates will differ. Volatilization rates can be quantified by direct or indirect measurement. The indirect measurement involves conducting a complete mass balance for all relevant environmental compartments (e.g. water, soil, sediments, and plants) and determining rates of volatilization by difference. This indirect approach will often include modelling components such as inverse modelling by the use of the dispersion model (Atmospheric modelling system, ADS) and parameterized PEARL (Pesticide Emission Assessment at Regional and Local scales) model simulation (Dekeyser et al. 2015). The direct measurement can be done by measuring the concentration of air in the treated area over a period while accounting for potential losses due to degradation. The common methods for these measurements are aerodynamic profile (ADP), energy balance (EB), relaxed eddy accumulation (REA), and plume dispersion (PD) (Dekeyser et al. 2015). Herbicides and the corresponding safeners will be present in the atmosphere of the treated sites to varying degrees based on rates of volatilization.

6 Photolysis

Photolysis of safeners is important because they are exposed directly to sunlight either in soil, on plants or crops, during transportation, and in water, if they are reaching surface waters (Balmer et al. 2000). Photolysis of chemicals is possible inside leaves and thus the mechanisms of such processes need to be studied (Schroer et al. 2017). Photodegradation could occur directly as a result of absorption of photons by a chemical or indirectly as a result of photon absorption by other constituents (photosensitizing species) in the environmental compartment (e.g. water, soil), resulting in energy transfer to the target chemical and subsequent degradation (Remucal 2014; Karpuzcu et al. 2016; Lin et al. 2019). The photochemical behaviour of chemicals can be affected by their chemical speciation and light absorption capacity (Challis et al. 2013; Remucal 2014; Karpuzcu et al. 2016). For example, in photolysis, p*K*a values of a chemical and the pH of the environment are important parameters because they control the chemical speciation and dictate the chemical light absorption (Calvayrac et al. 2013; Challis et al. 2013, 2014).

The p*K*a for benoxacor and furilazole are 12.99 and 16.44, respectively, and thus, they will be fully protonated at environmentally relevant pH values (Acharya and Weidhaas 2018). For a chemical to have both protonated and deprotonated species, it must have a p*K*a in the pH range relevant in the environment (i.e., pH 4–9). When this is the case, chemical speciation can have a large impact on rates of photolysis, making the study of pH dependence in photolysis experiments very important (Calvayrac et al. 2013; Challis et al. 2013; Remucal 2014). Most safeners are characterized by at least one aromatic ring, conjugated π system, with 1 or more heteroatoms (such as nitrogen, sulphur) and some functional group. All these aforementioned parameters facilitate direct absorption of solar radiation (Challis et al. 2014). However, Acharya and Weidhaas (2018) observed that both benoxacor and furilazole did not photodegrade (rate coefficient $k = -0.0031\ \mathrm{h}^{-1}$) to any extent under UV light (Halogen S302C lamp (Thorlabs, Newton, NJ) emitting 265 mW with a maximum output $= 200\ \mathrm{W{\cdot}cm^{-2}}$ ($\lambda = 250$–300 nm)). Nevertheless, Kral et al. (2019) noted that benoxacor underwent direct photolysis ($t_{1/2} \sim 10$ min) through photo-initiated ring closure when exposed to simulated natural sunlight (1,000 W

Newport Xe arc lamp, $\lambda = 300\text{–}400$ nm, output $= 8.8 \times 10^{-3}$ W cm^{-2}). The difference in these observations might be because of the experimental setup or the energy and wavelength output of the UV source.

Dichlormid underwent degradation in both water and methanol at 254 nm (313 kcal mol^{-1}), but did not transform under environmentally relevant UV light >290 nm due to the lower energy (Abu-Qare and Duncan 2002). Since the UV wavelength of 254 nm is more energetic than natural sunlight, such results are not directly applicable to environmental photolysis. Photodegradation rate constants determined in the lab at <290 nm usually bear little meaning to environmental photolysis rates; however, they might be of importance in a place where UV light is being employed for water treatment. Compounds that cannot absorb light at >290 nm can still undergo indirect photolysis via photo excited species such as dissolved nitrates, carbonates, iron, and dissolved organic matter, which is often abundant in natural waters, soil, and sediments. Direct photolysis quantum yields can also be measured for chemicals in order to predict photodegradation rates under any light conditions (Challis et al. 2014). Quantum yields are a characteristic property of a chemical and provide information about how efficiently a chemical degrades upon absorption of light (Remucal 2014; Challis et al. 2014; Lu et al. 2015; Lin et al. 2019). For example, benoxacor has a quantum yield of 0.14, which suggests that benoxacor is a highly photo-efficient organic micro-pollutant (Kral 2018). Hence, for photodegradation monitoring to be meaningful, efforts should be made to determine quantum yields and rate constants as a function of p*K*a and consider exposures under both simulated and natural sunlight.

Photodegradation of the herbicide metolachlor was enhanced in the presence of benoxacor on a quartz surface and, to a lesser extent, in water, but not on a soil-simulated (kaolinite) surface (Su et al. 2019). This suggests that the reaction media and chemical mixtures can play a role in photolysis. However, an earlier study showed that dichlormid did not significantly affect s-ethyl dipropylthiocarbamate (EPTC) photodegradation at 254 nm or at >290 nm in water or methanol (Abu-Qare and Duncan 2002). Hence more study is needed to fully understand the role safeners play in the photodegradation of herbicides and potentially the role that herbicides play in the photodegradation of safeners.

In the report of Abu-Qare and Duncan (2002), the roles of singlet oxygen, hydroxyl and peroxide radicals were highlighted. Singlet oxygen is a nucleophile and thus can react with any electron-rich moiety (McNeill and Canonica 2016; Katagi 2018). For example, the furan moiety in the safener furilazole is believed to be reactive with singlet oxygen and was indicated as the major reason for the indirect photolysis of furilazole (Kral 2018; Kral et al. 2019). Other components present in natural waters can facilitate electron transfer, which will aid photodegradation of the chemicals, and might be facilitating degradation of safeners that do not absorb light above 290 nm. These indirect mechanisms are complex and more difficult to predict than direct photolysis due to the multiple ways in which some chemicals can interact with photosensitizers present in natural waters (nitrate, carbonate, dissolved organic matter) (Challis et al. 2014; Remucal 2014; McNeill and Canonica 2016).

Carbonate species are believed to contribute to photodegradation of sulphur containing compounds in natural water (Huang and Mabury 2000) as a result of abstraction of electrons from sulphur. This might play an important role in indirect photolysis of safeners such as cyprosulfamide containing sulphur moieties. Hydroxyl radicals are capable of oxidizing aromatic ring and alkyl chains (Zepp et al. 1987; Katagi 2018); therefore, this is an important oxidizing agent to consider for photolysis in natural waters. For example, indirect photolysis of furilazole and dichlormid in the presence of •OH radicals resulted in 30% and 20% increased decay (Kral et al. 2019). Both furilazole and dichlormid exhibited 80–90% transformation in the presence of nitrate (which is known to mediate production of •OH) during indirect photolysis. Photodegradation products are sometimes assessed when studying photolysis kinetics and can be essential to characterize since in certain scenarios they can be more toxic than their parent chemicals. Mefenpyr-diethyl yielded four transformation products when irradiated under sunlight with a first-order rate constant of 0.580 h^{-1} (El Boukili et al. 2015). Two of these products were a result of hydrolysis (mefenpyrethyl and mefenpyr) (Hertkorn et al. 2010), suggesting that hydrolysis might be an important loss pathway in addition to photolysis (Chovelon et al. 2005).

Variability in kinetic concentration data for safeners has been reported by both Abu-Qare and Duncan (2002) and Acharya and Weidhaas (2018). Kinetic concentration data is important for allowing comparisons across studies and understanding variability in the methods, especially as it relates to rate constants and quantum yields, which can have large uncertainties when compared across studies (Challis et al. 2014). Natural sunlight exposure experiments will provide the most realistic estimation of photolysis rates at the specific latitude and sunlight conditions these experiments are conducted at. However, most researchers conduct these types of photolysis experiments under simulated sunlight using laboratory photo-reactors, which can limit their applicability to environmental systems. Additionally, light attenuation in natural systems can be significant so the result of photolysis based on test tube exposures can only be applied to near surface photolysis (Lu et al. 2015). Despite the reports of dichlormid, benoxacor, furilazole (Kral 2018), and mefenpyr-diethyl's (El Boukili et al. 2015) photolytic fate, more work is still needed on other safeners since photochemical behaviour varies greatly from chemical to chemical (Boreen et al. 2004; Chnirheb et al. 2012).

7 Chemical Reduction and Nucleophilic Substitution

Electron-rich compounds such as hydrogen sulphide may be able to reduce safeners. For example, graphite (black carbon) can aid the transfer of electrons from hydrogen sulphide to dichloroacetamide safeners through nucleophilic substitution pathways (Xu et al. 2020). This process can lead to transformation of dichloroacetamide safeners. Specifically, hydrogen sulphide has been implicated to facilitate the conversion of dichlormid, benoxacor, and AD-67 (Xu et al. 2020). Reductive dechlorination was responsible for dichloroacetamide safener degradation when reduced by Fe (II)-amended goethite in anaerobic abiotic system (Sivey and Roberts 2012). Chemicals such as iron, manganese, and hydrogen sulphide are naturally occurring in the subsurface environment (Ricko et al. 2020).

Different chemicals generally have different abilities to reduce safeners. For example, the half-life for dichlormid was 10 h by graphite-hydrogen sulphide (Xu et al. 2020) and 50 h by Fe (II)-amended goethite (Sivey and Roberts 2012). Also, the increase in molar ratio of Fe (II) to Mn (IV) oxide increased the rate of transformation of benoxacor and furilazole (Ricko et al. 2020). This mixture of Fe (II) and Mn (IV) oxide did not cause transformation of dichlormid after 6 h of reaction. This suggested that individual safeners react differently.

However, the herbicide (S-metolachlor) and three surfactants (Triton X-100, sodium dodecyl sulphate (SDS), and Myristyltrimethylammonium bromide (MyTAB)) did not influence rates of transformation of dichloroacetamide safeners, even in the presence of $Cr(H_2O)_6^{2+}$ (Ricko et al. 2020). This could be an indication that the herbicide (active ingredient) and accompany safeners may not naturally react.

Finally, since safeners can be transformed, their transformation products also need to be considered in the assessment of fate, transport, and ultimate hazard to non-target organisms. For instance, after harvest of rice, no fenoxaprop-P-ethyl or isoxadifen-ethyl was detected in straw or grain from plants, which were previously treated with a formulation containing both chemicals. However, the metabolite of fenoxaprop-P-ethyl, fenoxaprop-P was detected (Lucini and Pietro Molinari 2010, 2011). Isoxadifen-ethyl dissipated completely from the target, while fenoxaprop-P-ethyl was metabolized to a single product, which was still present as a residue.

8 Biological Degradation

Microorganisms, such as bacteria in the soil, can also degrade organic chemicals present in their environment (McGuinness and Dowling (2009); Fenner et al. 2013; Joutey et al. 2013). Soil microorganisms will often treat organic chemicals such as herbicides and safeners as xenobiotics, and thus they can develop adaptive strategies which might result in chemical transformation. Biological degradation of pesticides has not received much attention because most pesticides are not quick to degrade (Parsek et al. 1995; Rama Krishna and Philip 2011; Fenner et al. 2013). Microorganisms initiate chemical degradation by splitting the parent chemicals, then hydrolysing it (Nakamiya et al. 2007). However, soil microorganisms might find dichloroacetamide safeners more difficult to degrade/metabolize because of the additional chlorine atom compared to chloroacetamide herbicides (Xu et al. 2020).

Predictions on biodegradation of safeners using EPI suite are reported here (Table 5). Safeners such as benoxacor are moderately biodegradable based on Biowin model predictions of 0.7430 and 0.6339, for linear and non-linear models, while fenchlorazole-ethyl has biodegradation probabilities of −0.1882 and 0.000, from linear and non-linear models. In general, a half-life (DT_{50}) of a chemical in the soil below 50 days is non-persistent, between 50 and 70 is moderately persistent, $\geq 70 \leq 100$ days is persistent. Benoxacor, furilazole, mefenpyr-diethyl, cloquintocet-mexyl, fenchlorazole-ethyl, and fenclorim have measured DT_{50} values of 50, 29, 17.5, 5, 2.4, and 26 days, indicating that they are non-persistent, consistent with most of the safeners reported in Table 5.

9 Toxicity

Chemical risk assessment is one way from which concerns raised by continuous use and production of chemicals for agricultural purposes could be resolved (Joly et al. 2013). Potential impact of toxic substances including pesticides needs to be assessed by evaluating the responses of living organisms to these substances. Several studies evaluating such effects have been published to better inform the public and help researchers in further toxicity studies. A Microtox study of the effects of the safener benoxacor on bacteria (*V. fischeri*) has shown that the chemical was more toxic considering the concentration capable of inhibiting half (IC_{50}) of the population of V. fischeri (IC_{50} of 93 mg/L) than atrazine (IC_{50} of 197 mg/L) (Joly et al. 2013). Benoxacor has been reported to decrease growth of the freshwater algae, S. *capricornutum* (Day and Hodge 1996). Benoxacor is classified as having high toxic potency with an effective concentration (EC_{50}), exhibiting a maximal response in half of the algae population (EC_{50} of 0.63 mg/L) of freshwater algae (S. *subspicatus*) (Table 6) (European Chemical Agency 2020). Its lethal concentration (LC_{50}) that kills half of the freshwater fish species, *Ictalurus punctatus* is 1.4 mg/L, and the (LC_{50}) for *Lepomis macrochirus* is 6.5 mg/L, this is considered to be a moderately toxic potency to aquatic species (US EPA 2020; Xu et al. 2020; PPBD 2020). Furilazole has lesser toxic potency with an LC_{50} of 4.6 mg/L for freshwater fish and EC_{50} of 85.2 mg/L for algae compared to benoxacor (US EPA 2020; PPBD 2020; SDS 2020), while the LC_{50} of fluxofenim to Rainbow trout was reported to be 0.86 mg/L (SDS 2020). During chronic exposure to benoxacor, no observed effect concentrations (NOEC) of 0.31, and 0.35 mg/L were reported for *Pimephales promelas* (freshwater fish) and *Daphnia magna* (freshwater invertebrates), respectively (European Chemical Agency 2020). Benoxacor also caused a significantly reduced condition index (0.016 mg/L) of rainbow trout (European Chemical Agency 2020). Benoxacor might be moderately toxic to birds, honeybees, earthworms, and most aquatic organisms (Table 6) (European Chemical Agency 2020; PPBD 2020). Mefenpyr-diethyl and cyprosulfamide were stated as harmful to aquatic life with long-lasting effects (SDS 2020). Similarly, Isoxadifen-ethyl, fenchlorazole-ethyl, and cloquintocet-mexyl were described as very toxic to aquatic life with long-lasting effects and fenchlorazole-ethyl may cause cancer (SDS 2020). Alternatively, dichlormid might be relatively non-toxic to birds, fish, and aquatic invertebrates (considering their lethal concentration in Table 6 (PPBD 2020)). As an example of potential hazards, dichloroacetamide (a disinfection by-product) in

Table 6 Ecotoxicity of safeners on selected organisms

	Fish	D-Magna	Algae	Honeybees	Earth worm	Birds	BCF[a]
	96 h LC_{50}	48 h EC_{50}	72 h EC_{50}	72 h LD_{50}	14 days LC_{50}	LD_{50}	(BCFBAF v3.01)
Safeners	Mg/L	Mg/L	Mg/L	μg/bee	Mg/kg	Mg/kg	
Benoxacor	6.5	4.8	0.63	100	1,000	2,000	47.15
Dichlormid	141	161	–	–	–	>5,200	6.34
Furilazole	>6.2	>26	>85	>100	–	>2,000	6.48
Mefenpyr-diethyl	4.2	53	1.65	>700	>1,000	>2,000	30.83
Cloquintocet-mexyl	>14	>100	0.53	>100	1,000	>2,000	88.47
Cyometrinil	>5.6	–	–	–	–	–	25.84
Cyprosulfamide	>106	>102	>99.7	–	–	–	16.15
Dicyclonon	–	–	–	–	–	–	6.85
Dietholate	–	–	–	–	–	–	92.09
Fenchlorazole-ethyl	0.08	1.8	–	>300	–	>2,400	1,028
Fenclorim	0.6	2.2	20.9	20	62.5	500	701.3
Fluxofenim	0.86	0.22	–	–	–	2,000	78.69
Isoxadifen-ethyl	0.34	0.51	1.26	–	–	–	88.43
Jiecaowan	–	–	–	–	–	–	2.389
Mephenate	–	–	–	–	–	–	7.042
Metcamifen	–	–	–	–	–	–	2.718
Naphthalic anhydride	–	–	–	–	–	–	132.6
Oxabetrinil	7.1	8.5	10.7	>20	–	2,500	36.07

[a]*BCF* bioconcentration factor

drinking water has been reported to potentially affect gut microbiota composition (Xue et al. 2019) and cause acute metabolic damage in zebrafish (Yu et al. 2015), a similar observation might be possible for dichloroacetamide safeners. Another safener, naphthalic anhydride, can cause allergic skin and eye irritation in humans (National Center for Biotechnology Information 2020a, b; SDS 2020), depending on the concentration of exposure.

Concentrations of these safeners that are considered toxic are high compared to the amount measured in the environment (Woodward et al. 2018). Therefore there could be possibilities of lethal toxicities of this class of chemical if there were spillage to the environment. Nevetheless more work is needed on the effects of all groups of safeners to understand their various effects.

There are reports that possible synergistic effects exist between safeners and herbicides when used together (Bolyard et al. 2017; Acharya and Weidhaas 2018). However, an earlier study showed that mixtures of benoxacor and s-metolachlor did not alter the toxicity of s-metolachlor to any extent (Joly et al. 2013). The IC_{50} of s-metolachlor before the addition of benoxacor was 178 mg/L, and the IC_{50} of the mixture was 174 mg/L (Joly et al. 2013). More work is needed using other organisms and combinations of different safeners and herbicides to fully understand mixture toxicity of these chemicals.

10 Conclusion

Despite having biological activities, chemical safeners, which are antidotes for the effects of herbicides on target plants, have, for regulatory purposes, been classified as inert. Just as for herbicides with which they are formulated or applied with, safeners can dissipate and be transported to aquatic environments and have been detected in surface waters. However, fates and potential mobilities of safeners in the environment as well as possible adverse effects on non-target organisms, which are affected by estimates of exposure, including magnitude and duration of exposure in addition to toxic potencies have received little attention to date. Some safeners (such as benoxacor) have low aqueous solubility, are quite volatile and, based on their chemical properties, have potential for leaching to groundwater. Here, several possible routes, mechanisms, and rates of dissipation of safeners used with herbicides, including surface runoff, sorption/desorption, photodegradation, leaching and biological transformation, were reviewed. Most of the safeners can be readily leached because of their high solubilities. Laboratory experiments done in the absence of natural materials such as soil and water could not be easily used to predict what might be occurring during and after the application of safeners in the environment. Most of the safeners are not persistent in soil systems. Hazards presented by safeners were also investigated by comparing measured or predicted exposures to aquatic organisms with threshold toxic potencies for effects. It was determined that various safeners exhibit different potentials for exposure in aquatic environments and have a range of toxic potencies among organisms. Most safeners have low mammalian toxicity and moderate potential for bioaccumulation. They are moderately toxic to birds, honeybees, earthworms, and most aquatic organisms.

Considering the presence of safeners in the environment more work is needed to understand their dissipation mechanisms. Baseline information is needed on safeners from various nations of the world so that the exposure of this class of chemical can be used in risk assessment. Furthermore, various degradation products from each of the commonly used safeners should be studied. Also, aggressive efforts should be targeted towards understanding the toxicity of safeners on different organisms. The fate and toxicity of some recently synthesized safeners such as the closely related sulphonamide safeners, metcamifen and cyprosulfamide and their effects on living organisms need to be evaluated as this could help to determine their effects on impacted ecosystems and inform proper use and management of these important agrochemicals.

11 Methods for Laboratory Studies

For the determination of experimental Log K_{OW} a known volume (15 mL) of the test substance (benoxacor, mefenpyr-diethyl, cyprosulfamide, furilazole, and dichlormid) solution was added to an equal amount of octanol which had been saturated with water in a 50 mL polypropylene tube. The mixture was shaken for approximately 24 h at room temperature. Thereafter the mixture was centrifuged at 2400 rpm for 10 min and the two layers separated. Each of the phases was put into LC vials and spiked with atrazine-d5 at 50 ppb as a surrogate standard for LC-MS analysis. Then Log K_{OW} was calculated from the concentration determined by a Vanquish UHPLC and Q-ExactiveTM HF Quadrupole-OrbitrapTM mass spectrometer (Thermo-Fisher). LC separation was achieved with a Kinetex 1.7 μm C18 LC column (100 × 2.1 mm) (Phenomenex, Torrance, CA) using an isocratic elution of 45% H_2O: 55% methanol (each containing 0.1% formic acid) (Fisher Scientific) at a flow rate of 0.2 mL/min and column temperature of 40 °C. Samples were ionized by positive mode heated electrospray ionization (HESI) with the following source parameters: sheath gas flow = 3; aux gas flow = 1; sweep gas flow = 0; aux gas heater = 350 °C; spray voltage = 4.0 kV; S-lens RF = 80; capillary temperature = 320 °C; Aux gas heater temperature = 300 °C. A targeted-SIM and PRM (collision energy was 10, 20, 35, 15, 45, and 35 for cyprosulfamide, furilazole, benoxacor, mefenpyr-diethyl, dichlormid and atrazine, respectively) method at 60,000 resolution, AGC target = 1 × 106, max injection time = 30 ms, and a scan range from 100–1,000 m/z was used to monitor [M+H]+ precursor and product ions of benoxacor (m/z 260.024 → 149.083); mefenpyr-diethyl (m/z 373.071 → 327.029); furilazole (m/z300.016 → 241.975); cyprosulfamide (m/z 375.100 → 254.081, 135.044)); dichlormid (m/z208.029 → 139.966) and atrazine (m/z 216.101 → 174.054) (surrogate standard). Precursor and product ions were used for quantification and confirmation, respectively.

Acknowledgements Prof. Giesy was supported by the Canada Research Chair Program of the Natural Science and Engineering Research Council of Canada. Challis was funded by the Banting Post-Doctoral Fellowship Program, University of Saskatchewan. Oluwabunmi P. Femi-Oloye was sponsored by Nigeria Tertiary Education Trust Fund.

Conflict of Interest On behalf of all authors, the corresponding author states that there is no conflict of interest.

References

Abu-Qare AW (1912) Effect of the safener dichlormid and analogues on the herbicide EPTC. PhD thesis, University of Gasglow, UK

Abu-Qare AW, Duncan HJ (2002) Herbicide safeners: uses, limitations, metabolism, and mechanisms of action. Chemosphere 48(9):965–974

Acharya SP, Weidhaas J (2018) Solubility, partitioning, oxidation and photodegradation of dichloroacetamide herbicide safeners, benoxacor and furilazole. Chemosphere 211:1018–1024. https://doi.org/10.1016/j.chemosphere.2018.08.024

Acharya SP, Johnson J, Weidhaas J (2020) Adsorption kinetics of the herbicide safeners, benoxacor and furilazole, to activated carbon and agricultural soils. J Environ Sci 89:23–34

Ahrens H, Lange G, Müller T, Rosinger C, Willms L, van Almsick A (2013) 4-Hydroxyphenylpyruvatdioxygenase-Inhibitoren plus Safener: Lösungen für eine moderne und nachhaltige Landwirtschaft. Angew Chem 125(36):9558–9569

Balmer ME, Goss KU, Schwarzenbach RP (2000) Photolytic transformation of organic pollutants on soil surfaces an experimental approach. Environ Sci Technol 34(7):1240–1245

Behringer C, Bartsch K, Schaller A (2011) Safeners recruit multiple signalling pathways for the orchestrated induction of the cellular xenobiotic detoxification machinery in Arabidopsis. Plant Cell Environ 34(11):1970–1985

Boivin A, Poulsen V (2017) Environmental risk assessment of pesticides: state of the art and prospective improvement from science. Environ Sci Pollut Res 24(8):6889–6894

Bolyard K, Gresens SE, Ricko AN, Sivey JD, Salice CJ (2017) Assessing the toxicity of the "inert" safener benoxacor toward *Chironomus riparius*: effects of agrochemical mixtures. Environ Toxicol Chem 36:2660–2670. http://doi.wiley.com/10.1002/etc.3814

Boreen AL, Arnold WA, McNeill K (2004) Photochemical fate of sulfa drugs in the aquatic environment: sulfa drugs containing five-membered heterocyclic groups. Environ Sci Technol 38(14):3933–3940

Brosnan JT, Vargas JJ, Armel GR, Elmore MT, Breeden GK (2016) Herbicide safeners increase creeping bentgrass (Agrostis stolonifera) tolerance to pinoxaden and affect weed control. Weed Technol 30(04):919–928

Calvayrac C, Bontemps N, Nouga-Bissoue A, Romdhane S, Coste CM, Cooper JF (2013) Photolysis of tembotrione and its main by-products under extreme artificial conditions: comparison with another β-triketone herbicide. Sci Total Environ 452:227–232

Carr K (1990) Adsorption/desorption studies of MON 13900 and aerobic soil metabolites. Monsanto Agricultural Company. US EPA Report. https://www3.epa.gov/pesticides/chem_search/cleared_reviews/csr_PC-911596_30-Jun-93_i.pdf. Accessed 29 Sep 2020

Challis JK, Carlson JC, Friesen KJ, Hanson ML, Wong CS (2013) Aquatic photochemistry of the sulfonamide antibiotic sulfapyridine. J Photochem Photobiol A 262:14–21. https://doi.org/10.1016/j.jphotochem.2013.04.009

Challis JK, Hanson ML, Friesen KJ, Wong CS (2014) A critical assessment of the photodegradation of pharmaceuticals in aquatic environments: defining our current understanding and identifying knowledge gaps. Environ Sci Process Impacts 16(4):672–696

Chnirheb A, Harir M, Kanawati B, El Azzouzi M, Gebefügi I, Schmitt-Kopplin P (2012) Efficiency and degradation products elucidation of the photodegradation of mefenpyrdiethyl in water interface using TiO_2 P-25 and Hombikat UV100. J Environ Sci (China) 24(9):1686–1693

Chovelon JM, Previtera L, Baudot R, Brigante M, Emmelin C (2005) Abiotic degradation of iodosulfuron-methyl-ester in aqueous solution. J Agric Food Chem 53(13):5347–5352

Cumming H, Rücker C (2017) Octanol–water partition coefficient measurement by a simple 1H NMR method. ACS Omega 2(9):6244–6249

Davies J (2001) Herbicide safeners-commercial products and tools for agrochemical research. Pestic Outlook 12(1):10–15

Davies J, Caseley JC (1999) Herbicide safeners: a review. Pestic Sci 55(11):1043–1058

Davies JED, Jabeen N (2003) The adsorption of herbicides and pesticides on clay minerals and soils. Part 2. Atrazine. J Incl Phenom 46(1–2):57–64

Day KE, Hodge V (1996) The toxicity of the herbicide metolachlor, some transformation products and a commercial safener to an alga (selenastrum capricornurum), a cyanphyte (Anabaena cylindrica) and a macrophyte (Lemna gibba). Water Qual Res J Can 31(1):197–214

Dekeyser D, Spanoghe P, van den Berg F, Houbraken M, De Schampheleire M, Nuyttens D (2015) Volatilisation of pesticides under field conditions: inverse modelling and pesticide fate models. Pest Manag Sci 72(7):1309–1321

Díez C, Barrado E (2010) Soil-dissipation kinetics of twelve herbicides used on a rain-fed barley crop in Spain. Anal Bioanal Chem 397(4):1617–1626

Duhoux A, Pernin F, Desserre D, Délye C (2018) Herbicide safeners decrease sensitivity to herbicides inhibiting acetolactate-synthase and likely activate non-target-site-based resistance pathways in the major grass weed lolium sp. (Rye-grass). Front Plant Sci 8:1–14

El Boukili A, El Bazaoui A, Soulaymani A, Soulaymani-Bencheikh R, Taibi M, EL Hourch A (2015) Photolysis of herbicide safener Mefenpyrdiethyl under sunlight irradiation: an analytical and GC/MS investigation. J Mater Environ Sci 6(9):2386–2391

El Boukili A, Loudiyi N, El Bazaoui A, El Hourch A, Taibi MH, Boukili AE (2018) Adsorption and desorption behavior of herbicide mefenpyr-diethyl in the agricultural soils of Morocco. Mediterr J Chem 7(5):386–395

European Chemicals Agency (2020) European Chemicals Agency. https://echa.europa.eu/web/guest. Accessed 20 Dec 2020

Farenhorst A, Muc D, Monreal C, Florinski I (2011) Sorption of herbicides in relation to soil variability and landscape position. J Environ Sci Heal Part B Pestic Food Contam Agric Wastes 36(4):379–387

Fenner K, Canonica S, Wackett LP, Elsner M (2013) Evaluating pesticide degradation in the environment: blind spots and emerging opportunities. Science 341(6147):752–758

Gennari M, Negre M, Vindrola D (2008) Adsorption of the herbicides imazapyr, emazethapyr and imazaquin on soils and humic acids. J Environ Sci Heal Part B 33(5):547–567

Guo KL, Zhao LX, Wang ZW, Gao YC, Li JJ, Gao S, Fu Y, Ye F (2020) Design, synthesis and bioevaluation of substituted phenyl isoxazole analogs as herbicide safeners. J Agric Food Chem 68(39):10550–10559

Guth JA, Reischmann FJ, Allen R, Arnold D, Hassink J, Leake CR, Skidmore MW, Reeves GL (2004) Volatilisation of crop protection chemicals from crop and soil surfaces under controlled conditions—prediction of volatile losses from physico-chemical properties. Chemosphere 57 (8):871–887

Harris CI, Warren GF (1964) Adsorption and desorption of herbicides by soil. Weeds 12 (2):120–126

Hatzios KK, Burgos N (2004) Metabolism-based herbicide resistance: regulation by safeners. Weed Sci 52(3):454–467

Hertkorn N, Schmitt-Kopplin P, El Azzouzi M, Kanawati B, Harir M, Chnirheb A (2010) Hydrolysis of mefenpyrdiethyl: an analytical and DFT investigation. Anal Bioanal Chem 398 (5):2325–2334

Howatt KA (2006) Carfentrazone-ethyl injury to spring wheat (Triticum aestivum) is minimized by some ALS-inhibiting herbicides 1. Weed Technol 19(4):777–783

Huang J, Mabury SA (2000) The role of carbonate radical in limiting the persistence of sulfur-containing chemicals in sunlit natural waters. Chemosphere 41(11):1775–1782

Jhala AJ, Singh M (2012) Leaching of indaziflam compared with residual herbicides commonly used in Florida citrus. Weed Technol 26(3):602–607. https://doi.org/10.1614/WT-D-11-00161.1

Jhala AJ, Ramirez AHM, Singh M (2012) Leaching of indaziflam applied at two rates under different rainfall situations in Florida Candler Soil. Bull Environ Contam Toxicol 88 (3):326–332

Joly P, Bonnemoy F, Charvy JC, Bohatier J, Mallet C (2013) Toxicity assessment of the maize herbicides S-metolachlor, benoxacor, mesotrione and nicosulfuron, and their corresponding commercial formulations, alone and in mixtures, using the Microtox® test. Chemosphere 93 (10):2444–2450

Joutey NT, Bahafid W, Sayel H, El Ghachtouli N (2013) Biodegradation: involved microorganisms and genetically engineered microorganisms. Biodegrad Life Sci 1:289–320

Karpuzcu ME, McCabe AJ, Arnold WA (2016) Phototransformation of pesticides in prairie potholes: effect of dissolved organic matter in triplet-induced oxidation. Environ Sci Process Impact 18(2):237–245

Katagi T (2018) Direct photolysis mechanism of pesticides in water. J Pestic Sci 43(2):57–72

Kral A (2018) Photochemical transformations of dichloroacetamide safeners. Iowa Research Online, MSc thesis, University of Iowa, pp 1–100

Kral AE, Pflug NC, McFadden ME, LeFevre GH, Sivey JD, Cwiertny DM (2019) Photochemical transformations of dichloroacetamide safeners. Environ Sci Technol 53:6738–6746

Lewis KA, Tzilivakis J, Warner D, Green A (2016) An international database for pesticide risk assessments and management. Hum Ecol Risk Assess Int J 22(4):1050–1064. https://doi.org/10.1080/10807039.2015.1133242

Lin J, Apell JN, McNeill K, Emberger M, Ciraulo V, Gimeno S (2019) A streamlined workflow to study direct photodegradation kinetic and transformation products for persistence assessment of a fragrance ingredient in natural waters. Environ Sci Process Impact 21(10):1713–1721

Lu Z, Challis JK, Wong CS (2015) Quantum yields for direct photolysis of neonicotinoid insecticides in water: implications for exposure to nontarget aquatic organisms. Environ Sci Technol Lett 2(7):188–192

Lucini L, Pietro Molinari G (2010) Residues of the herbicide fenoxaprop-p-ethyl, its agronomic safener isoxadifen-ethyl and their metabolites in rice after field application. Pest Manag Sci 66 (6):621–626

Lucini L, Pietro Molinari G (2011) Detection of the herbicide fenoxaprop-P-ethyl, its agronomic safener isoxadifen ethyl and their metabolites residue in rice. Qual Assur Saf Crop Foods 3 (2):63–68

Mamy L, Barriuso E (2007) Desorption and time-dependent sorption of herbicides in soils. Eur J Soil Sci 58(1):174–187

McGuinness M, Dowling D (2009) Plant-associated bacterial degradation of toxic organic compounds in soil. Int J Environ Res Public Health 6(8):2226–2247

McNeill K, Canonica S (2016) Triplet state dissolved organic matter in aquatic photochemistry: reaction mechanisms, substrate scope, and photophysical properties. Environ Sci Process Impact 18(11):1381–1399

Nakamiya K, Nakayama T, Ito H, Edmonds JS, Shibata Y, Morita M (2007) Degradation of arylarsenic compounds by microorganisms. FEMS Microbiol Lett 274(2):184–188

National Center for Biotechnology Information (2020a) PubChem Database. 1,8-Naphthalic anhydride, CID=6693. https://pubchem.ncbi.nlm.nih.gov/compound/1_8-Naphthalic-anhydride. Accessed 14 Jan 2020

National Center for Biotechnology Information (2020b) PubChem Database. Dichlormid, CID=37829. https://pubchem.ncbi.nlm.nih.gov/compound/Dichlormid. Accessed 14 Jan 2020

Nelson EA, Penner D (2007) Leaching of isoxaflutole and the herbicide safeners R-29148 and furilazole. Weed Technol 21(01):106–109

OCDE (2000) Adsorption - desorption using a batch equilibrium method. OECD Guidel Test Chem, pp 1–44. http://www.oecd-ilibrary.org/environment/test-no-106-adsorption-desorption-using-a-batch-equilibrium-method_9789264069602-en

OECD (2004) Test no. 312: leaching in soil columns, OECD guidelines for the testing of chemicals, section 3. OECD Publishing, Paris. https://doi.org/10.1787/9789264070561-en

Parsek MR, McFall SM, Chakrabarty AM (1995) Microbial degradation of toxic, environmental pollutants: ecological and evolutionary considerations. Int Biodeterior Biodegrad 35 (1–3):175–188

PPDB (2020) Pesticide properties database. University of Hertfordshire. https://sitem.herts.ac.uk/aeru/ppdb/en/Reports/. Accessed 16 Oct 2020

Rama Krishna K, Philip L (2011) Bioremediation of single and mixture of pesticide-contaminated soils by mixed pesticide-enriched cultures. Appl Biochem Biotechnol 164:1257–1277

Ramanjaneyulu AV, Giri G (2006) Phytoremediation -a review phytoremediation - a review. Agric Rev 27(3):216–222

Remucal CK (2014) The role of indirect photochemical degradation in the environmental fate of pesticides: a review. Environ Sci Process Impact 16(4):628–653
Ricko AN, Psoras AW, Sivey JD (2020) Reductive transformations of dichloroacetamide safeners: effects of agrochemical co-formulants and iron oxide+ manganese oxide binary-mineral systems. Environ Sci Process Impact 22:2104–2116
Sarmah AK, Müller K, Ahmad R (2004) Fate and behaviour of pesticides in the agroecosystem - a review with a New Zealand perspective. Aust J Soil Res 42(2):125–154
Schroer HW, Li X, Lehmler HJ, Just CL (2017) Metabolism and photolysis of 2, 4-dinitroanisole in Arabidopsis. Environ Sci Technol 51(23):13714–13722
Safety Data Sheet (SDS) (2020). https://chemicalsafety.com/sds-search/#:~:text=What%20is%20a%20Safety%20Data%20Sheet%3F%20A%20safety,chemical%20handlers%20by%20providing%20a%20Safety%20Data%20Sheet
Sivey JD, Roberts AL (2012) Abiotic reduction of dichloroacetamide safeners: transformations and fate of "inert" agrochemical ingredients. Environ Sci Technol 46:2187–2195
Sivey JD, Lehmler HJ, Salice CJ, Ricko AN, Cwiertny DM (2015) Environmental fate and effects of dichloroacetamide herbicide safeners: "inert" yet biologically active agrochemical ingredients. Environ Sci Technol Lett 2(10):260–269
Staff G (2019) Getting covered: treating flax seed. pp 1–5
Su L, Caywood LM, Sivey JD, Dai N (2019) Sunlight photolysis of safener benoxacor and herbicide metolachlor as mixtures on simulated soil surfaces. Environ Sci Technol 53 (12):6784–6793
Subba-Rao R (1990) R-25788: adsorption and desorption in four soils. ICI Americas, Inc. https://www3.epa.gov/pesticides/chem_search/cleared_reviews/csr_PC-900497_8-Jan-93_e.pdf. US EPA Report. Accessed 29 Sep 2020
US EPA (2005) HED Records Center series 361 science reviews - file R112316. https://www3.epa.gov/pesticides/chem_search/cleared_reviews/csr_PC-700099_1-Aug-05_a.pdf
US EPA (2006) Executive summary. US EPA archive document - EPA archives. https://archive.epa.gov/pesticides/reregistration/web/pdf/copper_red.pdf
US EPA (2014) Estimation programs interface suite for Microsoft® Windows (EPI-Suite), version 4.11, U.S. 452 Environmental Protection Agency, Washington, DC
US EPA (2020) Overview of the ecological risk assessment process in the Office of Pesticide Programs
Wiersma JJ, Durgan BR, Martinson KB (2019) Hard red spring wheat (Triticum aestivum) tolerance to postemergence grass herbicides. Weed Technol 17(2):297–301
Woodward EE, Hladik ML, Kolpin DW (2018) Occurrence of dichloroacetamide herbicide Safeners and co-applied herbicides in Midwestern U.S. streams. Environ Sci Technol Lett 5 (1):3–8
Xu X, Sivey JD, Xu W (2020) Black carbon-enhanced transformation of dichloroacetamide safeners: role of reduced sulfur species. Sci Total Environ 738:139908. https://doi.org/10.1016/j.scitotenv.2020.139908
Xue B, Li C, Wang S, Zhao C, Dai K, Li W, Xi Z, Wang J, Qiu Z, Shen Z (2019) Effects of 2, 2-dichloroacetamide (DCAcAm), an emerging disinfection by-product in drinking water, on the intestinal microbiota of adult zebrafish. J Water Health 17(5):683–690
Yu S, Lin T, Chen W, Tao H (2015) The toxicity of a new disinfection by-product, 2, 2-dichloroacetamide (DCAcAm), on adult zebrafish (Danio rerio) and its occurrence in the chlorinated drinking water. Chemosphere 139:40–46
Zepp RG, Hoigne J, Bader H (1987) Nitrate-induced photooxidation of trace organic chemicals in water. Environ Sci Technol 21(5):443–450
Zhang K, Wiseman S, Giesy JP, Martin JW (2016) Bioconcentration of dissolved organic compounds from oil sands process-affected water by Medaka (Oryzias latipes): importance of partitioning to phospholipids. Environ Sci Technol 50(12):6574–6582

Do Endemic Soil Fauna Species Deserve Extra Protection for Adverse Heavy Metal Conditions?

Herman Eijsackers and Mark Maboeta

Contents

Abstract The concept of Ecotoxicological Species Sensitivity Distributions, as used in EU and US, to derive environmental standards for contaminants, starts from the assumption that by protecting the majority of species (95% confidence interval) all species will be protected. Nevertheless, 5% of the species outside the confidence interval might become harmed; half of it being the most sensitive for the particular compound tested. With respect to protection of rare endemic species it is not clear, however, if contamination is a driving factor for endemicity. The aim of this paper is to explore whether endemic and rare species deserve extra protection from adverse environmental conditions. To this end, a brief overview of the various forms of endemism, their relation to environmental stress factors and the distribution of endemic species is discussed. Further, the sensitivities of these species towards environmental stress factors are analysed, in order to conclude if and how endemic species could be better protected against environmental stress factors. This was achieved by specifically focusing on the potential impacts of metalliferous soils, mining, the treatment of mined soil and the storage of treated mine waste. It is concluded that at present there are some signals about specific sensitivities, but the database is much too small for a definite conclusion about adverse environmental

H. Eijsackers (✉) · M. Maboeta (✉)
Unit for Environmental Sciences and Management, North-West University, Potchefstroom, South Africa
e-mail: herman.eijsackers@wur.nl; mark.maboeta@nwu.ac.za

P. de Voogt (ed.), *Reviews of Environmental Contamination and Toxicology Volume 258*, Reviews of Environmental Contamination and Toxicology 258,
https://doi.org/10.1007/398_2021_72

factors as a threat to endemic species. The data gap has to be filled in with experimental tests with endemic species. This is hampered by the protection status of these endemic, rare species. Recommendations and derived activities are proposed to address this.

Keywords Anthropogenic activities · Contamination · Endemic species · Environmental stress · Risk assessment · Soil fauna · Soil metal

1 Introduction

In order to assess the impacts of contaminants on biodiversity, a method has been developed to calculate the impacts on species diversity by using the Species Sensitivity Distribution (SSD) (Posthuma et al., 2001). Based on the statistical distribution of the sensitivity of a considerable number of test species for a specified compound, the distribution of the sensitivity of "all" species for this compound is calculated. By using a lower percentile of such distribution, often the fifth percentile of an SSD derived from the No Observed Effect Concentration data, it can be reasoned that 95% of the tested species will show no effect at all and 5% only an exceedance of the no-effect level. Thereby it is pragmatically assumed that in nature this offers sufficient protection of structural and functional elements of ecosystems, especially when it is considered that the lower percentile-concentration is often lowered by a safety factor, to account for laboratory-to-field extrapolation, multiple stress and the potential exposure to mixtures. Nevertheless, 5% of the distribution curve, especially the tail ends, is not included. This will inflict, especially, the most sensitive species at one tail end, as Hopkin (1993) already suggested.

Looking at another part of the discussions on rare endemic species, it is not clarified whether contamination stress should be included as driving factor for adaptation leading to endemicity and consequently as part of the considerations on how to protect and manage rare, endemic species.

This paper intends to explore whether rare, endemic species deserve special protection, not only because they are rare but also because they could be more sensitive than the generally occurring species, part of which are used as test species to calculate SSDs and to derive policy limits from these sensitivity distributions. As such it combines three perspectives: ecotoxicological risk assessment, protection of rare species and adaptation to heavy metals leading to more sensitive or resistant species.

The review describes the various forms of endemism and how this applies to environmental stress factors, investigates the gross distribution of endemic species and analyses whether these species could be less or more sensitive to environmental stress factors because of their endemic status. Further, it concludes if and how endemic species could be better protected against environmental stress factors like the potential impacts of heavy metals from metalliferous soils, the treatment of mined soil and the storage of treated mine waste. Heavy metals as such are persistent contaminants that can also occur naturally so that adaptation and selection processes also come into play.

The underlying processes and mechanisms of endemism like described by Kier et al. (2009) are only briefly treated. The review concentrates on soil fauna species as these species show a high species diversity, have a low mobility and dispersal rate and live in very stable and inert ecosystems, all advantageous for endemism. In general, the interest for endemism in soil life is limited. The recently published State of Knowledge of Soil Biodiversity (FAO et al., 2020) and the review towards an integrative understanding of soil biodiversity (Thakur et al., 2020) do not mention nor discuss endemicity. However, given the low dispersal rate of a great number of soil fauna species (Eijsackers, 2010) and the great diversity of soil types, the isolation conditions as well as selecting environmental conditions are high. So endemism is to be expected in a fair amount of soil fauna species, as Janion-Scheepers et al. (2016) illustrate for South Africa.

2 The Various Forms and Descriptions of Endemism

It is important to note that the focus of endemism is not just species but clades at all levels can be endemic (Mishler et al., 2014) and in general, two forms of endemism are distinguished, *viz.* paleo-endemism and neo-endemism (Stebbins and Major, 1965). Paleo-endemism refers to those species that were widespread in the past and presently restricted to a smaller area while neo-endemism refers to species that have recently arisen. Paleo-endemic species can easily become endangered or extinct if their restricted habitat changes.

Principal causes of habitat degradation and loss in highly endemic ecosystems are mainly anthropogenic activities. In addition, the introduction of new species or their expansion by changes in environmental conditions can hamper the competitiveness of these endemic species.

A further point of discussion is the relation between the terms "*endemic*" and "*rare*". Endemic indicates that a species is unique to a limited and defined geographical location. With regards to "*rare*", Rabinowitz (1981) defined seven different forms depending on the geographic range. She describes endemic and rare in one line of reasoning: "*Species with both narrow geographic range and narrow habitat specificity are the classic rarities in the sense of restricted endemics*". But there can be also endemics with a very large range like more than 300 native earthworm species endemic to Australia (Baker et al., 1992; Kingston and Dyne, 1995). Based on this it is clear that the terms endemic and rare cannot be used interchangeably since endemic clades are not necessarily rare.

Moreover, there is a difference between "*rare*" and "*endangered*" or "*threatened*"; the latter two terms are mainly used in relation to conservation. In that context, it means that species are very low in numbers, show significantly declining numbers and/or are suffering under heavy stress as mentioned by the International

Union for Conservation of Nature (IUCN, 2019). However, the status of endangered and threatened species is always area dependent. A species will be rare at the borderlines of its population area, while it is abundant in the centre. Hence, in one country it can get the status of endangered, whereas in a neighbouring country the numbers are so high that they can even be experienced as a pest.

Endemic soil fauna species are mostly reported from isolated areas, islands like Madagascar, New Zealand (Gordon, 2012) and Australia. A considerable number of, e.g., endemic earthworm species has been reported from isolated valleys in Hungary (Zicsi et al., 2011) and the Balkan Peninsula (Trakić et al., 2016). This suggests that mobility and dispersal capacity play an important role. Recently, Janion-Scheepers et al. (2016) reported that in South Africa a great part of soil invertebrates are endemic. Although geographical (spatial) isolation in an evolutionary perspective is understandable for islands including New Zealand, it seems less relevant for areas of southern Africa with no geographical limitations. This begs the question what other factors could be involved next to a combination of spatial isolation and limited species mobility? Partly, this could have resulted from the high biodiversity with many different species and a higher chance for endemic species in South Africa. Further, the extreme environmental conditions in those areas could lead to outcompeting of more sensitive species as with paleo endemicity or better resistance as with neo endemicity.

Various ecology textbooks (Begon et al., 2006; Krebs, 2001; MacDonald and Service, 2007; Smith and Smith, 2001) mention different causal factors for endemism, illustrating that the factors mentioned and defined are related to the perspective by which the issue is approached. In summary the factors mentioned are: limited area of presentation and geographical isolation; low evolutionary dispersal possibilities and patterns; and sensitivity for or collateral reaction to adverse factors like diminution of habitat and adverse environmental factors.

In this paper, we focus on adverse environmental factors, more specifically the impacts of heavy metal contamination on invertebrate soil animals. In relation to heavy metal contamination, paleo-endemism could result from long-term exposition to metalliferous soil while neo-endemism could result from short-term exposition to, e.g., mine waste. The mining of metal rich-soil areas could easily endanger the species that have been able to survive in mining areas, *casu quo* are specialized for this type of selective environmental conditions. Species can become endangered firstly by habitat destruction, e.g. open cast mining. Secondly, they can become endangered by the dispersal of metal-rich dust from mining areas and the disposal of metal-rich waste sediment from which it further disperses by wind and water.

An explicit case of paleo-endemics in heavy metal-containing soils are serpentinite and peridotite soils, collectively called ultramafic or serpentine. Weathered, they contain high amounts of magnesium and sometimes metals like nickel, chromium and cobalt. Moreover, they are very poor and contain hardly any calcium. Due to the low nutrient level in these soils, vegetation is scarce as is the available water. In combination with high temperatures on these partly bare rocky soils the environmental conditions are stressful (Alexander et al., 2007).

Neo-endemics are the result of an adaptation/selection process due to specialised conditions in combination with a limited, confined area. Mining activities, more specifically the redistribution of heavy metal-containing mine waste, can be seen as a driver for the selection of heavy metal adapted specialised species occurring only in heavy metal rich areas. These highly disturbed areas will according to Lehmitz et al. (2012) in theory first be colonized by one or only a few species with high resistance. Given this selection for heavy metal resistance it is not to be expected that neo-endemics have in general a higher sensitivity and vulnerability for toxic conditions. However, this lowered sensitivity or resistance could be limited to one or some few chemically related heavy metals, and not to other contaminants present in these type of surroundings. Moreover, the extreme pH-conditions in some mine sludge deposits partly due to the chemical treatment of mineral ores to extract the heavy metals can be selective. Furthermore, resistance against one or more heavy metals will be achieved to a certain cost for selection which may result in increased sensitivity for other factors. Due to heavy metal contamination, the upper soil layers (Litter, Fermentation and Humus zone) change considerably, not only with respect to chemical factors like pH and CEC, but also physical factors like bulk density, moisture content and WHC (Bengtsson et al., 1994; Tranvik and Eijsackers, 1989). As a consequence, resistance to desiccation may come into play.

In conclusion: based on the ecological descriptions of and arguments for endemic species their limited distribution endemic species will make them rare and as such, from a nature protection point of view, deserve special protection. However, from the viewpoint of environmental toxicology it is not clear whether endemics need extra protection. A discussion on this topic as suggested by Eijsackers et al. (2019) could comprise of the following questions:

1. Where are endemic species to be found and do they occur in metalliferous areas?
2. Do endemic species show a higher or lower sensitivity for adverse environmental factors?
3. Do endemic species deserve extra protection and should ecotoxicological risk assessment procedures be adapted to this different sensitivity of endemic species?

3 Where Are Endemic Species Found and Do They Occur in Metalliferous Areas?

With respect to the first question – where are endemic species to be found – endemicity has received ample attention in bio-geography and evolutionary ecology research (e.g. Howard et al. (2019); Kier et al. (2009)). About soil animals, recently Phillips et al. (2019) reported on the global distribution of earthworms: "*High local species richness, with a higher chance for endemics, was found at mid-latitudes, such as the southern tip of South America, the southern regions of Australia and New Zealand, Europe (particularly North of the Black Sea), and the northeastern United States*". Gordon (2012) reported that for New Zealand 87% of the arthropod species is native and hence endemic on a global scale. These species could be widespread over New Zealand but also limited to small remnants of the original vegetation like observed by Winterbourn et al. (2008) in Canterbury. Native earthworm species had found refuge there on the native vegetation borders of agricultural land, covering less than 0.5% of the total area. At a regional detailed scale, surveys are published by Janion-Scheepers et al. (2016) for South Africa. They estimated that 50 to 90% of the species were native; although it is not fully clear whether these species are still undescribed or described and recognized as native. Even for the well-studied group of earthworms, which have been greatly influenced locally due to import of European and Asian earthworm species by settlers, the percentage of endemic species is 84%. Trakić et al. (2016) conclude that in the Balkan Peninsula 40% of the earthworm species is endemic due to the relatively isolated valleys in the mountainous part of the peninsula. Of these 40%, two-third of the species has a narrow distribution range, so only to be found on one location.

In relation to endemicity and heavy metal contamination, there is quite some literature about endemic plant species: the flora of metalliferous Ultramafic/serpentine soils (Ernst, 1974). Although it seems logical that there is also an endemic soil fauna in these areas, supporting data are extremely scarce. Maleri (2006) searched for earthworms in metalliferous ultramafic soils in South Africa. He found traces of earthworm activity but he could not sample live earthworms. There is one example of very long-term exposition of soil animals to natural Pb contamination from plumbic minerals in galena quartzite bedrock material (Hågvar and Abrahamsen, 1990). The direct vicinity of a water source (with highly lead-contaminated well water) was exclusively inhabited by the springtail *Isotoma olivacea*.

Most literature deals with the impacts of heavy metals in mining areas dating from medieval times in Wales (Hogg, 1895; Ireland, 1979; James et al., 1932; North, 1962) and the Plombière-area on the border of Belgium and The Netherlands (Posthuma and van Straalen, 1993). In these publications, no specific attention is paid to the presence of rare, endemic species or the heavy metal impacts on these species.

When we compare the distribution of the various soil fauna groups in South Africa as provided by Janion-Scheepers et al. (2016) in Fig. 1a with the distribution of the various mining activities in Fig. 1b, the highest numbers of species occur in the area with most mining activities. Although the observations may also be explained by differences in sampling intensity and other technical differences, it is tempting to conclude that the soils in these areas at least are not unfavourable for soil fauna. However, we have to be careful in interpreting this correlation as many environmental factors are involved in these distribution patterns and when we look at different species groups, a number of these show highest species numbers in the coastal areas and not in the mining areas.

Further, there is much research done about the distribution of soil animal species along heavy metal gradients of emitting smelter industries. The primary question there is: are there species present more or less exclusively at the highest metal levels or increasing in numbers with increasing metal concentrations? Studies by Bengtsson and Rundgren (1988) around a copper factory in Gusum (Sweden),

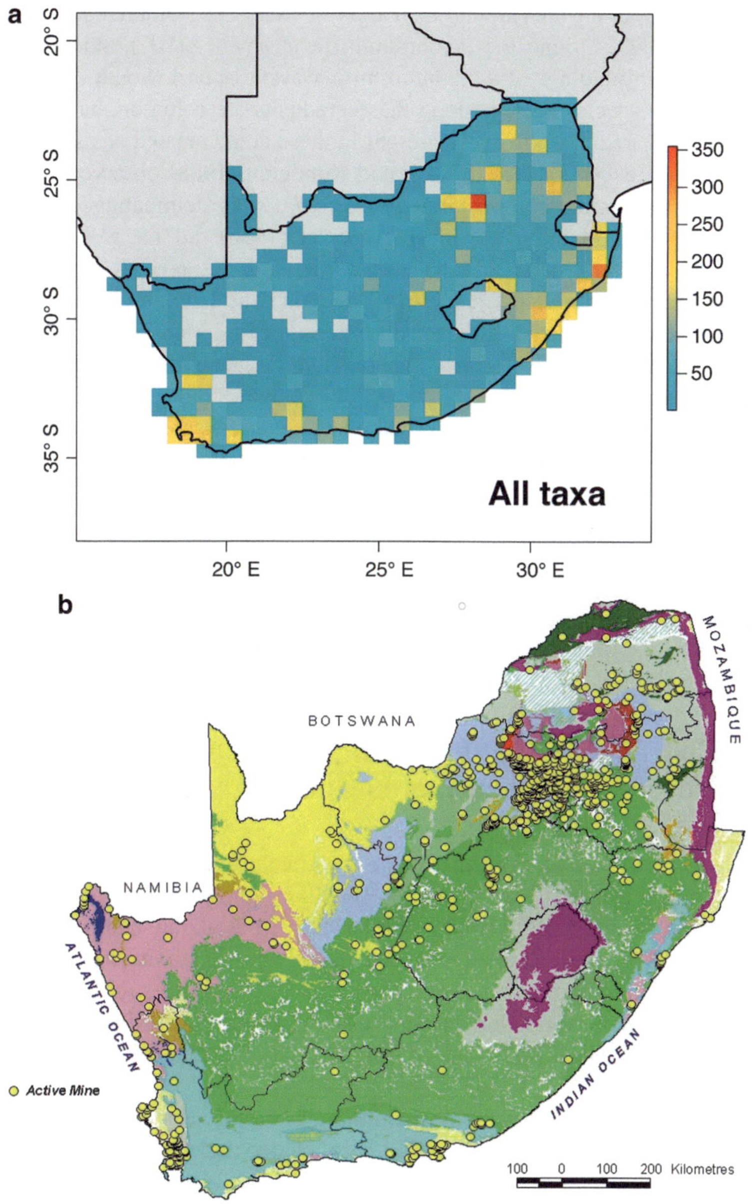

Fig. 1 (**a**) Map indicating the taxonomic richness for all South African soil fauna as discussed in Janion-Scheepers et al. (2016). (**b**) Map (adapted from source) showing active mining areas in South Africa (CGSA, 2018)

Posthuma (1992) in the area around the Budel smelter (The Netherlands), Spurgeon and Hopkin (1999) around the Avonmouth metal smelter (UK) and Wahl et al. (2012) in the surroundings of a platinum mine waste disposal (South Africa) show that most of the species present along these gradients are common, have a general distribution and are not exclusively present in these contaminated areas.

But there are exceptions. Bengtsson and Rundgren (1988) observed that in the Gusum area the springtail *Folsomia fimetarioides* was dominating close to the factory, as well as Haimi and Siira-Pietikainen (1996) did for a Finnish metal contaminated area. In general, this species is a minor species, present in low numbers in Northern European spruce forests. Also, Fountain and Hopkin (2004) found in a rough grassland site used for disposal of heavy metal rich smelter waste that *Folsomia fimetaria* and *Isotomurus palustris* dominated at high Zinc contents (500–10,000 μg g^{-1}).

In conclusion, there are indications that endemic species are limited to, *casu quo* exclusively occurring in, metalliferous areas or that metal-contaminated areas could lead to neo-endemic species. There are some species that dominate in areas with a high heavy metal burden, but in general the species occurring in highly contaminated metal areas have a normal distribution. Nevertheless, it is of interest whether these species will be more or less sensitive to adverse environmental conditions in metal contaminated areas and so could become endemic.

4 Do Endemic Species Show a Higher or Lower Sensitivity for Adverse Environmental Factors?

In theory more as well as less sensitivity is possible. Endemic species should have become reduced to specific areas as they should be unable to compete with other species and so became pushed back in these areas. Especially in nature conservancy this assumption is to be heard, although it is not always clear whether this is due to specialization or out competing. On the other hand, extreme environmental conditions could lead to outcompeting of more sensitive species like with paleo endemicity or better resistance as with neo endemicity, so that endemic species are the harsher ones.

In a review paper on endemic plant communities on special soils, including heavy metal polluted soils, Damschen et al. (2012) concluded that the database is very thin without a clear conclusion on adaptation; only some few references were to be found and of higher resistance as well as higher sensitivity.

In addition to the heavy metal impacts, two further questions are relevant. Firstly, can heavy metal impacts and adverse environmental conditions influence and reinforce each other. Secondly, are these resistant species adapted to these harsh conditions and could this result in processes leading to endemism.

Hirth et al. (2009) observed that native earthworms in Australia, in general, are less common on "disturbed" agricultural soils, suggesting a higher sensitivity in general. But on the other hand, they also conclude that the severe soil acidity is the most likely factor for the better survival of two native earthworm species (Megascolecidae *Spenceriella macleayi* and *S. montana*) compared to three introduced species (Lumbricidae *A. trapezoides*, and Acanthodrilidae *Microcolex dubius* and *M. phophoreus*) suggesting a lower sensitivity of native species.

Zeppelini et al. (2008) in a paper on the restoration of open cast mining in Brazilian sand dune areas state that endemic Collembolan species are more sensitive to disturbance than non-endemic species. In areas sampled 2, 4, 8 and 16 years after restoration the percentages of endemic collembolans increased stepwise 7.1%, 28.5, 28.5, 42.8; although not still as high as the control area with 85.7% endemic species. They refer to Deharveng (1996) who stated that the richness of endemic species is particularly sensitive to environmental disturbance and forest replacement more than non-endemic species. However, both papers do not further substantiate these claims for which factors the species are sensitive.

With respect to the sensitivity of endemic species for heavy metals, Maboeta et al. (2002) studied the effects of Cu-oxide spraying on indigenous *Microchaetus* earthworm species and the introduced European *Allolobophora caliginosa* and observed that indigenous *Microchaetus* species showed higher sensitivity. Field observations by Eijsackers et al. (2005) indicate a similar higher sensitivity. Further confirmation of these field observations by laboratory experiments was not possible so far because of the fragility of these species which makes them difficult to handle.

Wahl et al. (2012) stated that in mine waste deposits with the highest heavy metal contents likely the very hardy species are able to colonize like the prostigmatic mites *Speleorchestes meyeri, Eupodes parafusifer, Pronematus ubiquitus, Coccotydaeolus sp.* and *Bakerdania sp.* Further they observed that species either occurred in the most contaminated sites, or in the less or no contaminated sites, and not in the moderately contaminated sites in between, suggesting already some kind of divergence.

With respect to the interaction of heavy metals and adverse environmental conditions, there has been done quite some work. Holmstrup et al. (2010) published an extensive review and showed that heat stress, desiccation and cold all have a synergistic impact on intoxication by heavy metals for different soil fauna species (springtails, woodlouse and earthworms). Tranvik and Eijsackers (1989) observed that the springtail *Folsomia fimetarioides* dominating at high heavy metal contents had a low desiccation tolerance, possibly a consequence of adaptation to heavy metals. Also, van Capelleveen (1986) found that isopods adapted to heavy metals had a lowered efficiency of water regulation than isopods from a non-adapted population. Donker et al. (1996) concluded in their study on the physiology of metal adaptation in the isopod *Porcellio scaber* that isopods from the studied mine and smelter area in Plombière and Budel allocate more energy in reproductive processes. Donker et al. (1996) concluded consequently that there is less energy for other processes, a disadvantage in harsh conditions like cold and drought.

In conclusion, it seems reasonable to assume that species in contaminated areas, even when they are adapted to heavy metals, suffer indirectly from the adverse environmental conditions for which they are less resistant.

The second question is if adaptation to heavy metals could result in (endemic) speciation.

Examples of adaptation for heavy metals by changes in life history characteristics have been found for springtails in Sweden (Gusum) by Bengtsson and Rundgren (1988) and Tranvik et al. (1993), for isopods in The Netherlands (Budel) by Donker et al. (1993a); Donker et al. (1993b), for springtails in The Netherlands (Plombière and Budel) by Posthuma (1992) and earthworms in Wales by Andre et al. (2010). But the majority of the soil fauna species present at Gusum were not adapted, as was the case for the earthworms occurring in the direct vicinity of the copper factory (Tranvik et al., 1993).

From these studies, it becomes clear that adaptation does not occur for all species present. Moreover, it does not occur for all heavy metals present in the investigated areas. Adaptation can occur locally like the observed earthworms in Wales where two ecotypes occur mixed around a lead-contaminated mine, one of them being metal tolerant. The non-tolerant type avoided the spots with (too) high lead concentrations (Andre et al., 2010). The heterogeneity of the soil with respect to contamination is crucial here, as animals are mobile and can avoid/escape contaminated spots. Sjogren (1997) observed in an experimental gradient of increasingly Cu and Zn contaminated Gusum soil that springtails go to the least contaminated soil. Tranvik and Eijsackers (1989) observed that *Folsomia fimetarioides* was able to distinguish between clean and experimentally with Cu and Zn contaminated substrate and clean and contaminated fungal food. Eijsackers (1981) and Eijsackers et al. (2005) observed that earthworms avoided soil contaminated with, respectively, $CuSO_4$ and copper oxychloride.

With respect to the genetic backgrounds of these possible selecting scenarios, van Straalen and his research group (*pers. comm.*) have investigated the question about endemic, metal adapted soil fauna in mining areas frequently, but always concluded that this is not the case. Species like the springtail *Orchesella cincta* that developed metal tolerance can morphologically not be distinguished from other populations (van Straalen, *pers. comm.*). Moreover, Posthuma (*pers. comm.*) showed that adapted *Orchesella cincta* springtails from both the Budel (smelter) and Plombière (mine) area cross over successfully with animals from uncontaminated reference areas. Van Straalen et al. (1986) could not find truly adapted springtail population exposed to some decades (smelter) or centuries (mining areas) in The Netherlands and Belgium.

In order to assess the evolutionary aspects of adaptation, Bengtsson et al. (1992) carried out a long-term (255 days) culturing experiment. They could not observe any difference in the response of earthworms *Dendrobaena rubida* that had been collected either from soil exposed to heavy metals for 300 years, for 20 years or from reference soil when exposed to each of these soils. There were no indications that these earthworms had evolved resistance by adaptation or acclimatization. In a long-term culture experiments Reinecke and Reinecke (2003) observed that *Eisenia fetida* earthworms cultured for 3 years in a substrate with 1,000 mg kg^{-1} $Pb(NO_3)_2$ showed at fresh exposure a lower sensitivity for the lysosomal neutral red retention test (NRRT) as biomarker, but no acclimatization with respect to survival or weight change compared to exposed non pre-cultured earthworms. On the other hand, the cultured earthworms did show a largely increased storage capacity of lead leading to very high body burdens. An extensive review on the evolutionary aspects of ecotoxicological sensitivity, resistance and adaptation was recently published by Spurgeon et al. (2020) highlighting that increasingly available mechanistic information can be used to predict the sensitivity of species to chemicals.

In conclusion, there are indications from field research that species in contaminated areas are more resistant, but this is not related to genetic changes leading to the endemicity of these species. Adaptation to a certain metal does not mean, however, that a species is also resistant to other metals present. Moreover, environmental conditions in these mining, mine waste and smelter areas are harsh which causes extra stress. Hence, also resistant species could be in jeopardy there.

5 Extra Protection for Endemic Species by Adapted Risk Assessment Procedures

Irrespective of the claim that endemic species are more or less sensitive to adverse environmental factors like heavy metals, or more sensitive to other adverse environmental factors as a consequence of the "ecological costs" of adaptation to the heavy metals, endemic species deserve special protection. This is especially relevant for soil fauna species which represent a cryptic species group which do not get much attention in conservation although their role in and impact on soil organic matter processes is essential (FAO et al., 2020; Filser et al., 2016; Thakur et al., 2020).

Protection of endemic species is relevant primarily based on their rareness and related conservation status and their local distribution. With any change in general environmental factors like climate change, these species have no or reduced possibilities to evade or avoid heavy metal impacted environments. The disappearance of these species will consequently endanger general ecological functions as replacing species will be scarce. According to Zeppelini et al. (2008) endemic species are – despite their greater vulnerability than common species – able to recolonize disturbed areas, but it takes much longer than for common species. This will similarly hold for evasion of disturbed areas.

In the context of environmental risk assessment of endangered species, the EFSA Scientific Committee (EFSA, 2016) came to the following conclusions:

1. Endangered species can be more vulnerable than other species due to particular characteristics related to exposure, recovery and/or sensitivity. In general, they are considered more vulnerable in view of general characteristics such as life-history traits and geographical distribution.
2. There is no convincing scientific evidence that endangered species have in general a higher exposure than general species, with the exception of top predators due to biomagnification.

3. Some few comparisons of the Species Sensitivity of endangered species and taxonomically related species do not provide conclusive evidence that endangered species per se are more sensitive. Some observed differences can be explained on differences between Toxico-dynamics and -kinetics mechanism and traits, because many endangered species are highly specialised.

In addition to these EFSA-conclusions it has to be realised that these advice are aimed at mostly short-termed, toxic impacts by pesticides and biocides that will gradually break down. Impacts of heavy metals due to their persistence in the environment will be long-termed.

With respect to the conclusion on Species Sensitivity Distribution, it seems logical that more sensitive species are in the tail ends of the SSD-curve as used to derive protection limits. Especially in these tail ends the variability will be much higher than in the middle part around LD50. In order to derive these SDD curves databases have been used with many different test species from all over the world, but mainly from Europe and the US where this methodology has been developed (Posthuma et al., 2001). Moreover, these test species have been selected because of their general distribution and, as such, of their representativeness for a great number of other species.

Based on these curves a safety factor of 100 to cover the differences in sensitivity between all species is assumed to be sufficient. Unfortunately, there are no data on heavy metal contamination to check whether this is sufficient, but there are some indications from recent research on the sensitivity of water species for neonicotinoids. Experiments with stone flies already showed that the old limit values for NOEC calculation had to be adapted and it had to be further adapted when the winter generation for this species proved to be even more sensitive (Roessink et al., 2013). Recent experiments with a tropical species revealed even further and far-reaching increased sensitivity levels (Sumon et al., 2018) and this species does not fall in the endemic category.

When we realise that in South Africa, Australia, New Zealand and probably a number of South American countries a great number of soil fauna species groups can be assumed to be native/endemic (Janion-Scheepers et al., 2016) there is still a lot to do. Experimental confirmation of a possibly differing sensitivity of endemic species is urgently needed. In order to do so, we have to surpass the general premises that experimenting with rare species is to be discouraged and dissuaded. By testing a selected number of endemic species we can find out if these species have a similar sensitivity for various groups of contaminating compounds and fall within the established SDD-curves or if they are more sensitive and to what extent they show a higher sensitivity. If the latter is the case, then limits have to be adapted accordingly.

In order to realise this, the different policies and management strategies for natural fauna and flora have to be better attuned to each other. In principle, nature protection and management measures are based on the status of species population: the total numbers of a species, the numbers of local/regional populations and the changes in these numbers. Ecological risk assessment is based on the sensitivities of species for specific compounds in combination with the exposure to these compounds under more or less specific conditions. Nature management wants to exclude any adverse influence as far as possible, while risk assessment calculates to what extent a specific adverse factor does not have an unacceptable impact on a species.

Studies on the adverse impacts of environmental contaminants on flora and fauna are mostly based on correlative relations by comparing numbers of species in a certain area with the amounts of pesticides. See, for instance, Hallmann et al. (2014) about impacts of neonicotinoids on bird species by influencing the numbers of their prey insects. Or they compare the temporal dynamics of numbers with the fluctuations of specific compounds in time like the classical studies on organochlorine impacts on birds of prey. Direct cause-effect studies are almost always missing; understandable because experimenting with protected species is not acceptable on moral grounds. Nevertheless, some breakthrough has to be created here.

In conclusion, contrary to threats to endemic species due to direct habitat destruction, we know still too little about adverse environmental factors as a threat for endemic species, a disappointing conclusion. For plant species on metalliferous soils Damschen et al. (2012) also concluded that the data are sparse. So also in general terms more research is needed.

6 Further Actions

Given the conclusion above, further research could comprise the following activities:

(a) *Define endemic species more precisely*

It has to be more precisely established whether an endemic species is flourishing or endangered. This also means that terms like threatened and endangered have to be redefined in a quantitative manner.

(b) *Define and describe endemic areas more precisely*

So far endemic areas are mainly described on the basis of their geographical features. A more precise description of (deviating?) environmental factors could help us to understand which factors are involved in the preservation of these species. Metalliferous areas as ultramafic soils are not only characterized by the presence of a mixture of heavy metal but also by harsh environmental conditions which make increased sensitivity due to adaptation costs more likely.

(c) *Investigate sensitivity of endemic species for heavy metals more in detail*

This not only means research on the sensitivity of the species present in these areas to single heavy metals but also to the mixture of metals present there and to the combination of heavy metal impacts and adverse environmental conditions like high temperatures, low moisture and high or low acidity levels.

(d) *Adapt ecological risk assessment procedures*

Based on the outcomes of this series of tests with endemic species, the risk assessment procedures have to be adapted. This means that the database has to be expanded with a series of other contaminant groups and other sets of endemic species as endemicity is biogeographically related and differences between different global areas seem to be reasonable.

(e) *Prevent extra stress in endemic areas*

Looking at the level of the local conservation management, the amount of stress in these areas can already be very high, so management should be aimed at as low as a disturbance as possible. So extra activities in or in the vicinity of these areas like building or groundwater withdrawal should be prevented. However, we have to realise that we are facing global warming and derived problems like local water shortages that can hardly be handled or managed at a local scale. As soil animals have in general low mobility, the capacity of these species to evade environmental changes due to global change is low and therefore the prevention of local stress factors becomes more important.

Acknowledgements Funding Statement The authors do not receive any funding from an external source.

Competing Interest Statement On behalf of all authors, the corresponding author states that there is no conflict of interest. All authors have read and agreed to the publication of the manuscript.

References

Alexander EB, Coleman RG, Harrison SP, Keeler-Wolfe T (2007) Serpentine geoecology of western North America: geology, soils, and vegetation. OUP USA

Andre J, King RA, Stürzenbaum SR, Kille P, Hodson ME, Morgan AJ (2010) Molecular genetic differentiation in earthworms inhabiting a heterogeneous Pb-polluted landscape. Environ Pollut 158:883–890. https://doi.org/10.1016/j.envpol.2009.09.021

Baker G, Buckerfield J, Grey-Gardner R, Merry R, Doube B (1992) The abundance and diversity of earthworms in pasture soils in the Fleurieu Peninsula, South Australia. Soil Biol Biochem (24):1389–1395. https://doi.org/10.1016/0038-0717(92)90123-f

Begon M, Townsend C, Harper J (2006) Ecology from individual to ecosystems. Blackwell, Victoria, pp 470–471

Bengtsson G, Rundgren S (1988) The Gusum case: a brass mill and the distribution of soil Collembola. Can J Zool 66:1518–1526. https://doi.org/10.1139/z88-223

Bengtsson G, Ek H, Rundgren S, Bengtsson G (1992) Evolutionary response of earthworms to long-term metal exposure. Oikos 63:289. https://doi.org/10.2307/3545390

Bengtsson G, Rundgren S, Sjögren M, Bengtsson G, Sjogren M (1994) Modelling dispersal distances in a soil gradient: the influence of metal resistance, competition, and experience. Oikos 71:13. https://doi.org/10.2307/3546167

CGSA CfGSA (2018) Active mines South Africa, Lesotho and Swaziland. http://www.geoscience.org.za/images/Maps/activemines.gif. Accessed 9 Oct 2020

Damschen EI, Harrison S, Ackerly DD, Fernandez-Going BM, Anacker BL (2012) Endemic plant communities on special soils: early victims or hardy survivors of climate change? J Ecol 100:1122–1130. https://doi.org/10.1111/j.1365-2745.2012.01986.x

Deharveng L (1996) Soil Collembola diversity, endemism, and reforestation: a case study in the Pyrenees (France). Conserv Biol (10):74–84. https://doi.org/10.1046/j.1523-1739.1996.10010074.x

Donker M, van Capelleveen H, van Straalen N (1993a) Metal contamination affects size-structure and life-history dynamics in isopod field populations. Ecotoxicol Metals Invertebr:383–399

Donker MH, Zonneveld C, van Straalen NM (1993b) Early reproduction and increased reproductive allocation in metal-adapted populations of the terrestrial isopod *Porcellio scaber*. Oecologia 96:316–323. https://doi.org/10.1007/bf00317500

Donker MH, Raedecker MH, van Straalen NM (1996) The role of zinc regulation in the zinc tolerance mechanism of the terrestrial isopod *Porcellio scaber*. J Appl Ecol:955–964

EFSA (2016) Coverage of endangered species in environmental risk assessments at EFSA. EFSA J:14. https://doi.org/10.2903/j.efsa.2016.4312

Eijsackers H (1981) Effecten van koperhoudende varkensmest op regenwormen en op de kwaliteit van grasland. Landbouwkundig Tijdschrift 93:307–314

Eijsackers H (2010) Earthworms as colonisers: primary colonisation of contaminated land, and sediment and soil waste deposits. Sci Total Environ 408:1759–1769

Eijsackers H, Beneke P, Maboeta M, Louw JPE, Reinecke AJ (2005) The implications of copper fungicide usage in vineyards for earthworm activity and resulting sustainable soil quality. Ecotoxicol Environ Saf 62:99–111. https://doi.org/10.1016/j.ecoenv.2005.02.017

Eijsackers H, Reinecke A, Reinecke S, Maboeta M (2019) Heavy metal threats to plants and soil life in southern Africa: present knowledge and consequences for ecological risk assessment. Springer. https://doi.org/10.1007/398_2019_23

Ernst W (1974) Schwermetallvegetation der erde. Gustav Fischer Verlag, Stuttgart

FAO, ITPS, GSBI, SCBD, EC (2020). State of knowledge of soil biodiversity – status, challenges and potentialities, Report 2020. FAO, Rome. https://doi.org/10.4060/cb1928en

Filser J, Faber JH, Tiunov AV, Brussaard L, Frouz J, Deyn GD, Uvarov AV, Berg MP, Lavelle P, Loreau M (2016) Soil fauna: key to new carbon models. Soil 2:565–582

Fountain MT, Hopkin SP (2004) A comparative study of the effects of metal contamination on collembola in the field and in the laboratory. Ecotoxicology 13:573–587. https://doi.org/10.1023/b:ectx.0000037194.66321.2c

Gordon DP (2012) New Zealand inventory of biodiversity. Volume three: kingdoms: Bacteria, Protozoa, Chromista, Plantae, Fungi. Edited by Dennis P. Gordon. Christchurch (New Zealand): Canterbury University Press. NZ$89.95. 616 p. + 16 pl.; ill.; species index. ISBN: 978-1-927145-05-0. 2012 vol 87. The Quarterly Review of Biology, vol 4. University of Chicago Press. doi:https://doi.org/10.1086/668151

Hågvar S, Abrahamsen G (1990) Microarthropoda and Enchytraeidae (Oligochaeta) in naturally lead-contaminated soil: a gradient study. Environ Entomol 19:1263–1277. https://doi.org/10.1093/ee/19.5.1263

Haimi J, Siira-Pietikainen A (1996) Decomposer animal communities in forest soil along heavy metal pollution gradient. Anal Bioanal Chem 354:672–675. https://doi.org/10.1007/s0021663540672

Hallmann CA, Foppen RPB, van Turnhout CAM, de Kroon H, Jongejans E (2014) Declines in insectivorous birds are associated with high neonicotinoid concentrations. Nature 511:341–343. https://doi.org/10.1038/nature13531

Hirth JR, Li GD, Chan KY, Cullis BR (2009) Long-term effects of lime on earthworm abundance and biomass in an acidic soil on the south-western slopes of New South Wales, Australia. Appl Soil Ecol 43:106–114. https://doi.org/10.1016/j.apsoil.2009.06.007

Hogg T (1895) Immunity of some low forms of life from lead poisoning. Chem News 71:1.895

Holmstrup M, Bindesbøl A-M, Oostingh GJ, Duschl A, Scheil V, Köhler H-R, Loureiro S, Soares AMVM, Ferreira ALG, Kienle C, Gerhardt A, Laskowski R, Kramarz PE, Bayley M, Svendsen C, Spurgeon DJ (2010) Interactions between effects of environmental chemicals and natural stressors: a review. Sci Total Environ 408:3746–3762. https://doi.org/10.1016/j.scitotenv.2009.10.067

Hopkin SP (1993) Ecological implications of '95% protection levels' for metals in soil. Oikos 66:137. https://doi.org/10.2307/3545206

Howard C, Flather CH, Stephens PA (2019) What drives at-risk species richness? Environmental factors are more influential than anthropogenic factors or biological traits. Conserv Lett 12: e12624

Ireland MP (1979) Metal accumulation by the earthworms *Lumbricus rubellus, Dendrobaena Veneta* and *Eiseniella tetraedra* living in heavy metal polluted sites. Environ Pollut (1970) 19:201–206. https://doi.org/10.1016/0013-9327(79)90041-7

IUCN (2019) Guidelines for using a global standard for the identification of key biodiversity areas: version 1.0. IUCN, International Union for Conservation of Nature. doi:https://doi.org/10.2305/iucn.ch.2019.kba.1.0.en

James TC, Laurie RD, Newton L (1932) A study of the present condition of the River Rheidol. University of college of Wales

Janion-Scheepers C, Measey J, Braschler B, Chown SL, Coetzee L, Colville JF, Dames J, Davies AB, Davies SJ, ALV D, Dippenaar-Schoeman AS, Duffy GA, Fourie D, Griffiths C, Haddad CR, Hamer M, Herbert DG, Hugo-Coetzee EA, Jacobs A, Jacobs K, Rensburg CJV, Lamani S, Lotz LN, Louw SV, Lyle R, Malan AP, Marais M, Neethling J-A, Nxele TC, Plisko DJ, Prendini L, Rink AN, Swart A, Theron P, Truter M, Ueckermann E, Uys VM, Villet MH, Willows-Munro S, Wilson JRU (2016) Soil biota in a megadiverse country: current knowledge and future research directions in South Africa. Pedobiologia 59:129–174. https://doi.org/10.1016/j.pedobi.2016.03.004

Kier G, Kreft H, Lee TM, Jetz W, Ibisch PL, Nowicki C, Mutke J, Barthlott W (2009) A global assessment of endemism and species richness across island and mainland regions. Proc Natl Acad Sci 106:9322–9327

Kingston T, Dyne G (1995) Potential for agronomic exploitation of Australian native earthworms. In: Temple-Smith MGaP, T (ed) The role of earthworms in agriculture and land management. Department of Primary Industries, Tasmania Technical Report 1/96, Kings Meadow, Tasmania, pp 29–38

Krebs CJ (2001) The experimental analysis of distribution and abundance. 5th edn. Benjamin Cummings, San Francisco

Lehmitz R, Russell D, Hohberg K, Christian A, Xylander WER (2012) Active dispersal of oribatid mites into young soils. Appl Soil Ecol 55:10–19. https://doi.org/10.1016/j.apsoil.2011.12.003

Maboeta MS, Reinecke SA, Reinecke AJ (2002) The relation between lysosomal biomarker and population responses in a field population of Microchaetus sp. (Oligochaeta) Exposedto the fungicide copper oxychloride. Ecotoxicol Environ Saf 52:280–287. https://doi.org/10.1006/eesa.2002.2180

MacDonald D, Service K (2007) Key topics in conservation biology. Blackwell

Maleri RA (2006) The ability of terrestrial Oligochaeta to survive in ultramafic soils and the assessment of toxicity at different levels of organisation. University of Stellenbosch, Stellenbosch

Mishler BD, Knerr N, González-Orozco CE, Thornhill AH, Laffan SW, Miller JT (2014) Phylogenetic measures of biodiversity and neo- and paleo-endemism in Australian Acacia. Nat Commun:5. https://doi.org/10.1038/ncomms5473

North FJ (1962) Mining for metals in Wales. National Museum of Wales Cardiff

Phillips HR, Guerra CA, Bartz ML, Briones MJ, Brown G, Crowther TW, Ferlian O, Gongalsky KB, van Den Hoogen J, Krebs J (2019) Global distribution of earthworm diversity. Science 366:480–485

Posthuma L (1992) Genetic ecology of metal tolerance in Collembola. Free University Amsterdam

Posthuma L, van Straalen NM (1993) Heavy-metal adaptation in terrestrial invertebrates: a review of occurrence, genetics, physiology and ecological consequences comparative biochemistry and physiology part C: pharmacology. Toxicol Endocrinol 106:11–38. https://doi.org/10.1016/0742-8413(93)90251-f

Posthuma L, Suter GW II, Traas TP (2001) Species sensitivity distributions in ecotoxicology. CRC Press

Rabinowitz D (1981) Seven forms of rarity biological aspects of rare plant conservation

Reinecke AJ, Reinecke SA (2003) The influence of exposure history to lead on the lysosomal response in *Eisenia fetida* (Oligochaeta). Ecotoxicol Environ Saf 55:30–37. https://doi.org/10.1016/s0147-6513(02)00086-6

Roessink I, Merga LB, Zweers HJ, van den Brink PJ (2013) The neonicotinoid imidacloprid shows high chronic toxicity to mayfly nymphs. Environ Toxicol Chem 32:1096–1100. https://doi.org/10.1002/etc.2201

Sjogren M (1997) Dispersal rates of Collembola in metal polluted soil. Pedobiologia 41:506–513

Smith R, Smith T (2001) Ecology and field biology. 6th edn. Benjamin Cummings. An imprint of Addison Wesley Longman, USA, San Francisco

Spurgeon DJ, Hopkin SP (1999) Life-history patterns in reference and metal-exposed earthworm populations. Ecotoxicology 8:133–141

Spurgeon D, Lahive E, Robinson A, Short S, Kille P (2020) Species sensitivity to toxic substances: evolution, ecology and applications. Front Environ Sci 8

Stebbins GL, Major J (1965) Endemism and speciation in the California Flora. Ecol Monogr 35:1–35. https://doi.org/10.2307/1942216

Sumon KA, Ritika AK, Peeters ETHM, Rashid H, Bosma RH, Rahman MS, Fatema MK, van den Brink PJ (2018) Effects of imidacloprid on the ecology of sub-tropical freshwater microcosms. Environ Pollut 236:432–441

Thakur MP, Phillips HRP, Brose U, De Vries FT, Lavelle P, Loreau M, Mathieu J, Mulder C, van der Putten WH, Rillig MC (2020) Towards an integrative understanding of soil biodiversity. Biol Rev 95:350–364

Trakić T, Valchovski H, Stojanović M (2016) Endemic earthworms (Oligochaeta: Lumbricidae) of the Balkan Peninsula: a review. Zootaxa 4189:251. https://doi.org/10.11646/zootaxa.4189.2.3

Tranvik L, Eijsackers H (1989) On the advantage of Folsomia fimetarioides over *Isotomiella minor* (Collembola) in a metal polluted soil. Oecologia 80:195–200. https://doi.org/10.1007/bf00380150

Tranvik L, Bengtsson G, Rundgren S (1993) Relative abundance and resistance traits of two Collembola species under metal stress. J Appl Ecol 30:43. https://doi.org/10.2307/2404269

van Capelleveen H (1986) The ecological significance of zinc and cadmium for the woodlouse *Porcellio scaber* Latr. 1: consumption, growth and accumulation. Free University of Amsterdam, Amsterdam

Van Straalen NM, Groot GM, Zoomer HR (1986) Adaptation of Collembola to heavy metal soil contamination. In Proceedings, International Conference on Environmental Contamination, Amsterdam. CEP Consultants, Edinburgh, pp 16–20

Wahl JJ, Theron PD, Maboeta MS (2012) Soil mesofauna as bioindicators to assess environmental disturbance at a platinum mine in South Africa. Ecotoxicol Environ Saf 86:250–260. https://doi.org/10.1016/j.ecoenv.2012.09.027

Winterbourn M, Knox G, Burrows C, Marsden I (2008) The natural history of Canterbury. Canterbury University Press, Christchurch

Zeppelini D, Bellini BC, Creão-Duarte AJ, Hernández MIM (2008) Collembola as bioindicators of restoration in mined sand dunes of Northeastern Brazil. Biodivers Conserv 18:1161–1170. https://doi.org/10.1007/s10531-008-9505-2

Zicsi A, Szlavecz K, Csuzdi C (2011) Leaf litter acceptance and cast deposition by peregrine and endemic European lumbricids (Oligochaeta: Lumbricidae). Pedobiologia 54:S145–S152. https://doi.org/10.1016/j.pedobi.2011.09.004

Aflatoxin M1 in Africa: Exposure Assessment, Regulations, and Prevention Strategies – A Review

Abdellah Zinedine, Jalila Ben Salah-Abbes, Samir Abbès, and Abdelrhafour Tantaoui-Elaraki

Contents

A. Zinedine (✉)
Faculty of Sciences, BIOMARE Laboratory, Applied Microbiology and Biotechnologies, Chouaib Doukkali University, El Jadida, Morocco
e-mail: zinedine.a@ucd.ac.ma; zinedineab@yahoo.fr

J. Ben Salah-Abbes
Laboratory of Genetic, Biodiversity and Bio-Resources Valorization, University of Monastir, Monastir, Tunisia
e-mail: jalila.bensalah@yahoo.fr

S. Abbès
Laboratory of Genetic, Biodiversity and Bio-Resources Valorization, University of Monastir, Monastir, Tunisia

Higher Institute of Biotechnology of Béja, University of Jendouba, Jendouba, Tunisia
e-mail: abb_samir@yahoo.fr

A. Tantaoui-Elaraki
Retired, Department of Food Sciences, Hassan II Institute of Agronomy and Veterinary Medicine - Rabat, Rabat-Instituts, Témara, Morocco
e-mail: abdelrhafour@elaraki.ac.ma

P. de Voogt (ed.), *Reviews of Environmental Contamination and Toxicology Volume 258*,
Reviews of Environmental Contamination and Toxicology 258,
https://doi.org/10.1007/398_2021_73

Abstract Aflatoxins are the most harmful mycotoxins causing health problems to human and animal. Many acute aflatoxin outbreaks have been reported in Africa, especially in Kenya and Tanzania. When ingested, aflatoxin B1 is converted by hydroxylation in the liver into aflatoxin M1, which is excreted in milk of dairy females and in urine of exposed populations. This review aims to highlight the AFM1 studies carried out in African regions (North Africa, East Africa, West Africa, Central Africa, and Southern Africa), particularly AFM1 occurrence in milk and dairy products, and in human biological fluids (breast milk, serum, and urine) of the populations exposed. Strategies for AFM1 detoxification will be considered, as well as AFM1 regulations as compared to the legislation adopted worldwide and the assessment of AFM1 exposure of some African populations. Egypt, Kenya, and Nigeria have the highest number of investigations on AFM1 in the continent. Indeed, some reports showed that 100% of the samples analyzed exceeded the EU regulations (50 ng/kg), especially in Zimbabwe, Nigeria, Sudan, and Egypt. Furthermore, AFM1 levels up to 8,000, 6,999, 6,900, and 2040 ng/kg have been reported in milk from Egypt, Kenya, Sudan, and Nigeria, respectively. Data on AFM1 occurrence in human biological fluids have also shown that exposure of African populations is mainly due to milk intake and breastfeeding, with 85–100% of children being exposed to high levels. Food fermentation in Africa has been tried for AFM1 detoxification strategies. Few African countries have set regulations for AFM1 in milk and derivatives, generally similar to those of the *Codex alimentarius*, the US or the EU standards.

Keywords Aflatoxin M1 · Africa · Exposure assessment · Occurrence · Prevention · Regulations

Abbreviations

AF	Aflatoxin
AF-alb	Aflatoxin-albumin
AFB1	Aflatoxin B1
AFB2	Aflatoxin B2
AFM1	Aflatoxin M1
AFM2	Aflatoxin M2
ALT	Alanine aminotransferase
AST	Aspartate transaminase
b.w.	Body weight
BM	Breast milk
BTA	Bladder tumor antigen

EDI	Estimated daily intake
ELISA	Enzyme-linked immunosorbent assay
GRAS	Generally recognized as safe
HA	Human aflatoxicosis
HBV	Hepatitis B virus
HCC	Hepatocellular carcinoma
HIV	Human immunodeficiency virus
HPLC	High performance liquid chromatography
LAB	Lactic acid bacteria
LC/MS-MS	Liquid chromatography tandem-mass spectrometry
LD50	Lethal dose, 50%
MRL	Maximum regulatory limit
PBS	Phosphate-buffered saline
TDI	Tolerable daily intake
TLC	Thin layer chromatography
UHT	Ultra-heat-treated

1 Introduction

Milk and derivatives play an essential role in a balanced diet and therefore for human health, in particular for babies, young children, and teenagers (Udomkun et al. 2018). However, their role in human nutrition has been increasingly questioned in recent years (Ellen et al. 2013), and their consumption is currently of a scientific debate that gives rise to more concerns regarding mycotoxin contamination since most mycotoxins are probably or confirmed as carcinogenic (Benkerroum 2016).

The presence of mycotoxins has been reported worldwide in milk and derivatives (Benkerroum 2016; Rahmani et al. 2018) and in BM (Cherkani-Hassani et al. 2016). Milk contamination with mycotoxins is not only of a public health concern, but it also induces considerable economic losses, especially for the dairy sector (Abdallah et al. 2019b). Although recent papers have shown the multi-occurrence of several mycotoxins in milk, most of the published investigations focus on the analysis of Aflatoxin M1 (Flores-Flores et al. 2015).

Aflatoxins (AF) are toxic metabolites produced by toxigenic strains of the fungal species *Aspergillus flavus, A. parasiticus,* and *A. nominus* toxigenic species, which are ubiquitous species occurring in various food commodities and animal feed (Bennett and Klich 2003). Aflatoxins M1 (AFM1) and M2 (AFM2) are the hydroxylated metabolites of aflatoxins B1 (AFB1) and B2 (AFB2), respectively (Fig. 1). They were first isolated from the milk of lactating animals fed on diet contaminated with AFB1 and AFB2. They are formed by enzymes associated with liver cytochrome P450 and excreted directly in milk of farm animals, which results in an indirect contamination of dairy products. AFM1 and AFM2 are more water soluble than the parent molecules and, thus, can be detected in biological fluids of the populations exposed to AF, especially breast milk (BM), urine, and serum.

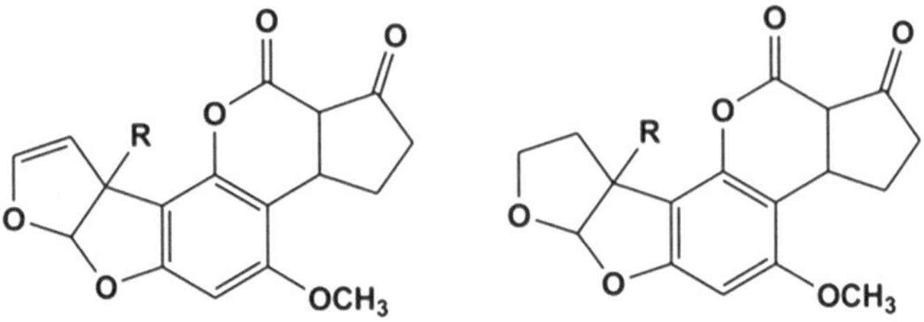

Aflatoxin M1 R=OH
Aflatoxin B1 R=H

Aflatoxin M2 R=OH
Aflatoxin B2 R=H

Fig. 1 Chemical structures of AFB1, AFM1, AFB2, and AFM2

AFM1 is among the most toxic compounds found in milk, and high levels can be consumed by the most vulnerable age groups including babies, children, and the elderly (Ismail et al. 2015). Several factors were reported to affect the conversion of AFB1 into AFM1 such as the feed received, breed, animal health, and lactation stage (Duarte et al. 2013). AFM1 can be detected in milk 12 to 24 h after the first intake of AFB1 and reaches a high level after a few days. When the ingestion of AFB1 is interrupted, AFM1 amounts decrease to become undetectable after 72 h. It has been reported that the carry-over rate of AFB1 into AFM1 in milk of dairy cows varies between 0.3 and 6.2% (Marchese et al. 2018). The presence of AFM1 in milk is considered a public health concern since the toxin is resistant to several heat treatment processes (pasteurization, sterilization, etc.) and therefore, can be detected in numerous milk-based products such as powdered milk, yogurts, cheese, butter, etc. (JECFA 2011). In addition, milk and derivatives were considered the main contributors to the exposure of babies and infants to AFM1, which makes them more sensitive to AF harmful effects (El-Tras et al. 2011).

Africa is the world's second-largest and second-most populous continent, with a total area of 30.3 million km^2, and more than 1.2 billion living people (2016), which represents approximately 16% of the world's human population. This continent is commonly divided into five regions: North Africa, East Africa, Central Africa, West Africa, and Southern Africa. Milk and dairy products are of significant importance in the food tradition of many African ethnic groups.

The presence of mycotoxins in African food and feed has been widely documented in the literature with a special focus on AF (Zinedine and Mañes 2009; Darwish et al. 2014; James and Zikankuba 2018; Misihairabgwi et al. 2019; Alahlah et al. 2020; Mohammedi-Ameur et al. 2020) and high AF exposure through milk has been reported. Chronic health risks of AF are prevalent in Africa because these contaminants occur more frequently under tropical conditions and in the staple diets in some regions of the continent. A strong association between hepatocellular carcinoma (HCC) in African population and the exposure to AF through naturally contaminated foods consumption has been reported. Thus, several African countries have started setting up prevention, control, and surveillance strategies to reduce the AF incidence in foods (Falade 2018).

Despite the lack of research funds for capacity building, insufficient expertise, and limited surveillance systems in Africa, data on the occurrence of AF in food and feed are available. The purposes of this review are, therefore, to highlight the studies carried out on AFM1 in the five African regions, particularly AFM1 levels reported in animal milk and dairy products, and in human biological fluids (BM, serum and urine). Studies on the exposure assessment of some African populations to AFM1 will be considered. A section will be devoted to the prevention strategies to reduce AFM1 contamination in milk and derivatives and another one to AFM1 regulations in Africa as compared to the legislation adopted worldwide (*Codex alimentarius*, EU and US regulations).

2 Toxic Effects of AFM1

Various studies have shown that AFM1 has carcinogenic, immunosuppressive, and genotoxic properties similar to those of AFB1 on both humans and animals, although with a less potent effect (Luongo et al. 2014). An early acute toxicity study on 1-day-old ducklings reported that the LD_{50} of AFM1 and AFM2 was around 16 and 61.4 μg/duckling, respectively, and that ducklings fed with AFM1 showed liver lesions indistinguishable from those induced by AFB1 (Purchase 1967). A longitudinal study conducted on 515 patients with chronic hepatitis showed that the primary liver cancer year-incidence of AF-exposing people increased with the rising urine excretion of AFM1 which was also related with the abnormal liver function (Lu et al. 2010). Thus, AFM1 remains a highly toxic substance, which can cause health problems for milk and dairy consumers with low immunity levels such as babies, young children, and the elderly. According to WHO report, studies carried out on human hepatocytes have shown that AFM1 is cytotoxic, and its acute toxicity in several species is similar to that of AFB1. Because of its higher polarity compared to AFB1 and its retention in the digestive tract, AFM1 showed also higher intestinal carcinogenicity as reported by earlier studies (Cullen et al. 1987). In addition, carcinogenicity studies have shown that in vitro genotoxicity of AFM1 is similar to that of AFB1 (WHO 2002). The metabolism of AFM1 in HBV-associated HCC patients and its ability to form DNA adduct demonstrates a strict correlation between HBV-associated HCC and urinary AFM1 levels (Marchese et al. 2018).

The International Agency for Research on Cancer (IARC) referred AFM1 as tenfold less toxic than AFB1, but still categorized it as “a group 1” human carcinogen on the basis of its toxicity and prevalence levels in milk (IARC 2002), although no study is available on the association between the risk of liver cancer and the dietary AFM1 intake. The Joint FAO/WHO Expert Committee on Food Additives (JECFA) estimated the mean AFM1 levels in milk in African diets to be 1.8 ng/L, as compared to 23, 22, 360, and 5 ng/L estimated for the European, Latin American, Far Eastern, and Middle Eastern diets, respectively (WHO 2001). Given the carcinogenic potential of AF, JECFA has not set a tolerable daily intake (TDI) for these toxins in food, but recommended, in order to guarantee safe levels of AF in foods, to reduce their levels according to the *ALARA* (“As Low As Reasonably Achievable”) principle (JECFA 2011).

3 AFM1 in Milk and Dairy Products

3.1 North Africa

North Africa is a region that includes six Arab countries: Algeria, Egypt, Libya, Morocco, Sudan, and Tunisia. It is surrounded by the Atlantic Ocean, the Mediterranean, and the Red seas. Data regarding the presence of AFM1 in milk and dairy products from the five African regions are summarized in Table 1.

The first study published in Algeria showed that AFM1 was detected in 11% (5/47) of the milk samples at levels ranging from 9 to 103 ng/L. Only one sample exceeded the maximum regulatory limit (MRL) of 50 ng/L set by the European Union (EU) for AFM1 in liquid milk (European Commission 2006). AFM1 incidence was higher in imported powdered milk (29%) as compared to raw milk (5%). AFM1 incidence in powdered milk imported especially from Argentina and consumed in northeast Algeria reached 75% (3/4) (Redouane-Salah et al. 2015). Later, another study showed an AFM1 incidence of 46.43% (39/84) in raw milk (mean value of 71.92 ng/L) collected in three regions of Algeria, and a seasonal variation of AFM1 levels was noted, with high levels in spring than in autumn (Mohammedi-Ameur et al. 2020).

In Tunisia, AFM1 was detected in 60.7% (68/112) of cow raw milk from the province of "Beja" (median 13.6 ± 1.4 μg/L), with AFM1 levels in 4.4% of total samples higher than the MRL of 50 ng/L (Abbès et al. 2012).

In North-Western Libya, AFM1 contamination frequencies were high; i.e., 71.4% (35/49) and 75% (15/20) in milk and soft cheese, respectively (Elgerbi et al. 2004).

In Morocco, a first study reported a high AFM1incidence of 88.8% (48/54) in pasteurized milk produced by five dairy industries, with levels of 1–117 ng/L (mean value of 18.6 ng/L), and 7.4% of milk samples (4/54) exceeding the MRL of 50 ng/L (Zinedine et al. 2007). Another survey revealed an AFM1 incidence of 27% (13/48) in raw milk collected in "Fez" city, with levels of 10–100 ng/L and 8% of samples above the MRL of 50 ng/L (El Marnissi et al. 2012). A seasonal variation in AFM1 occurrence in milk was observed in this study, with a highest incidence (58%) and high levels of AFM1 in autumn than in spring. More recently, AFM1 was detected in 100% (7/7) of powder milk and 35% (14/40) of UHT milk from Northern Morocco, with average levels of 25.50 ± 12.06 ng/kg and 14.76 ± 10.21 ng/kg, respectively. None of the samples exceeded the MRL of 50 ng/L (Alahlah et al. 2020).

A study on milk collected in dairy farms and vendors in "Khartoum" (Sudan) showed an AFM1 incidence as high as 95.45% (42/44), with contamination levels ranging from 220 to 6,900 ng/L and an average concentration of 2070 ng/L (Elzupir and Elhussein 2010). In another Sudanese survey, AFM1 was found in 100% of raw milk (35/35) and imported powdered milk (12/12) sold in "Khartoum" state, with levels of 100–2,520 and 10–850 ng/kg, respectively. Regarding regulations, 50% of the powdered milk and 100% of the raw milk exceeded 50 ng/kg, whereas 33% of the powdered milk and 77% of the raw milk samples exceeded 500 ng/kg (Ali et al. 2014).

Table 1 AFM1 presence in milk and dairy products from the five African regions

African Region	Country	Milk product	Sampling City or area	Total samples	Positive samples and frequency (%)	Mean level ± RSD (ng/L or ng/kg)	Range (ng/L or ng/Kg)	Above 50 ng/Kg (%)	Reference
North Africa	*Algeria*	Reconstituted and powdered milk	"Constantine"	47	5 (10.6%)	–	9–103	2.1	Redouane-Salah et al. (2015)
		Raw milk	North-East, North-Center, North-West	84	39 (46.43%)	71.92	95.6–557.2	–	Mohammedi-Ameur et al. (2020)
	Morocco	Pasteurized Milk	Different areas	54	48 (89%)	18.6	1–117	7.4	Zinedine et al. (2007)
		Raw Milk	"Fez"	48	13(27%)	–	10–100	8	El Marnissi et al. (2012)
		UHT milk	Northern area	40	14 (35%)	14.76 ± 10.21	5.1–44.42	–	Alahlah et al. (2020)
		Powdered Milk		7	7(100%)	25.50 ± 12.06	15.2–39.9	–	
	Libya	Milk	North-West	49	35(71.4%)	–	30–3,130	69.3	Elgerbi et al. (2004)
		Soft cheese		20	15(75%)	–	110–520	15	
	Tunisia	Raw cow milk	"Beja"	112	68 (60.7%)	1,360	–	4.4	Abbès et al. (2012)
	Sudan	Raw Milk	"Khartoum"	35	35(100%)	920	100–2,520	100	Ali et al. (2014)
		Powdered Milk		12	12 (100%)	290	10–850	50	
		Milk		44	42 (95.45%)	2070	220–6,900	27.5	Elzupir and Elhussein (2010)

(continued)

Table 1 (continued)

African Region	Country	Milk product	Sampling City or area	Total samples	Positive samples and frequency (%)	Mean level ± RSD (ng/L or ng/kg)	Range (ng/L or ng/Kg)	Above 50 ng/Kg (%)	Reference
	Egypt	Camel milk	–	25	24 (96%)	–	30–850	–	Balata and Bahout (1996)
		Buffalo's milk	"Ismailia"	50	32(64%)	–	Up to 270	48%	Motawee et al. (2009)
		Cow's milk		50	26(52%)	–	Up to 220	34%	
		Goat milk		50	18(36%)	–	Up to 230	26%	
		Camel milk		25	9(36%)	–	Up to 210	20%	
		Cow's milk	"Cairo-Giza"	15	3(20%)	6,300	5,000–8,000	–	El-Sayed et al. (2000)
		Dried milk		10	1(10%)	5,000	Up to 5,000	–	
		Hard cheese		10	2(20%)	4,600	3,000–6,000	–	
		Soft cheese		10	1(10%)	500	Up to 500	–	
		Infant milk powder	"Minoufiya"	125	54 (43.2%)	9.796 ± 1.036	0.3–21.8	–	El-Tras et al. (2011)
		Raw cow milk	"Alexandria"	50	19(38%)	49.74 ± 17.26	23–73	52.6	Amer and Ibrahim (2010)
		Soft cheese		50	20(40%)	70.63 ± 18.42	52–87.6	100	
		Hard cheese		50	19(38%)	132.2 ± 57.48	51.6–182	100	
		Processed cheese		50	11 (22%)	52.52 ± 13.56	51.8–54	100	
		Raw milk	"Alexandria"	30	12(40%)	32.5 ± 5.98	8.30–85.00	58	Aiad and Abo El-Makarem (2013)
		"Kareish" cheese		30	14(46%)	28.6 ± 4.66	6.20–70.26	71	
		"Damietta" cheese		30	16(53%)	53.05 ± 8.01	16.50–133.2	75	

		“Ras” cheese		30	17(56%)	56.04 ± 6.29	7.40–111.50	88	
		Yoghurt		30	8(26%)	28.41 ± 4.78	11.40–98.80	63	
		Fresh milk[a]	“Assiut”	88	18 (20.45%)	–	–	–	Mwanza et al. (2015)
		Fresh milk[b]		88	72 (81.8)	–		–	
		Fresh milk[c]		88	65 (73.9)	25.79	8.52–78.06	20.45	
		Raw milk	“Aswan”	75	37 (49%)	100.3 ± 0.008	53–207	49	Zakaria et al. (2019)
		UHT milk		15	–	–	–	–	
		Raw milk	“Assiut”	20	20(100%)	–	20–190	70	Abdallah et al. (2019a)
East Africa	*Kenya*	“Lala”[d,e]	“Nairobi”	8	–	111	10–340	62.5	Lindahl et al. (2018)
		Boiled Milk[e]		13	–	46	14–88	38.5	
		Pasteurized Milk[e]		18	–	126	Up to 740	72.2	
		Raw Milk[e]		62	–	131	Up to 1,100	74.2	
		UHT Milk[e]		17	–	46	7.3–84	41.2	
		Yoghurt[e]		17	–	96	26–270	76.5	
		“Lala”[f]		27	–	48	12–160	29.6	
		Pasteurized Milk[f]		53	–	55	7.6–210	49.1	
		UHT Milk[f]		55	–	58	Up to 470	29.1	
		Yoghurt[f]		21	–	134	17–1,100	57.1	
		Cow milk	“Kasarani”	84	83 (98.8%)	83.66	Up to 255.96	64	Kagera et al. (2018)
		Milk	“Dagoretti”	200	–	128.7	Up to 1674.9	55	Kirino et al. (2016)
		Raw milk	“Bomet”	150	150 (100%)	–	Up to 2,930	52	Langat et al. (2016)

(continued)

Table 1 (continued)

African Region	Country	Milk product	Sampling City or area	Total samples	Positive samples and frequency (%)	Mean level ± RSD (ng/L or ng/kg)	Range (ng/L or ng/Kg)	Above 50 ng/Kg (%)	Reference
		Processed milk		35	6 (17.1%)	–	Up to 690	8.6	
		Cow milk	Different villages	512	203 (39.7%)	34.9 ng/L	2–6,999	10.4	Senerwa et al. (2016)
		Milk	"Nairobi"	96	96 (100%)	290.3 ± 663.4	15.4–4,563	66.4	Kuboka et al. (2019)
	Tanzania	Milk	"Singida"	37	31 (83.8%)	–	Up to 2007	83.8	Mohammed et al. (2016)
	Ethiopia	Milk	"Addis Ababa"	110	110 (100%)	410	28–4,980	93	Gizachew et al. (2016)
Central Africa	*Cameroon*	Cow milk	Various areas	63	10 (15.9%)	–	6–525	9.5	Tchana et al. (2010)
	Burundi	Fresh milk	"Gitega" and "Cibitoke"	10	10 (100%)	31.4	8.4–82.8	40	Udomkun et al. (2018)
		Yogurt		6	6 (100%)	32.5	8.2–63.2	33.3	
	Eastern DRC[g]	Fresh milk	"Kabare", "Bukavu" and "Uvira"	3	3 (100%)	37.3	25.6–49.8	–	
		Yogurt		2	3 (66.6%)	18	4.8–26.0	–	
		Cheese		5	5 (100%)	170.0	18.5–261.1	80	
West Africa	*Nigeria*	Cow milk	"Ogun" state	22	3 (13.6%)	–	Up to 2040	–	Atanda et al. (2007)
		Ice cream		6	2 (30.3%)	–	Up to 2,230	–	
		Fresh milk	"Bida"	10	10(100%)	665 ± 0.190	407–952	100	Okeke et al. (2012)
		"Nono"		10	10(100%)	924 ± 0.626	248–2,510	100	
		"Kindirmo"		10	10(100%)	575 ± 0.341	139–1,238	100	
		Raw milk	"Ogun" state	100	75 (75%)	108.15	9 to 456	73	Oluwafemi et al. (2014)
		Fresh milk	"Minna"	20	16(80%)	530.8 ± 0.0938	10.9–1354.3	15	Makun et al. (2016)

		Commercial milk		20	5(25%)	58.4 ± 0.0052	46.4–99.2	8	
		Cheese		20	8(40%)	588 ± 0.1296	104.5–1530.2	7	
		"Nono"		20	7(35%)	592.9 ± 0.0867	234.2–1251.6	4	
		Yoghurt,		20	2(10%)	615.2 ± 0.0101	583.5–647.0	2	
		Powdered milk	South-Western	125	67 (53.6%)	–	20–460	–	Oyeyipo et al. (2017)
Southern Africa	*South Africa*	Milk[h]	Local farms	42	42 (100%)	–	30–1,320	–	Dutton et al. (2012)
		Milk[i]		39	39 (100%)	–	Up to 1,540	–	
		Raw milk[j,k]	"Mpumalanga", "Limpopo" and "Gauteng"	85	85(100%)	150	10–2,850	85.6	Mulunda and Mike (2014)
		Raw milk[k,l]		125	98(78.4%)	140	10–2,850	79	
		Fresh milk[a]	"Ngaka" "Modiri" and "Molema"	50	8 (16%)	–	–	–	Mwanza et al. (2015)
		Fresh milk[b]		50	34 (68%)	17.06	–	–	
		Fresh milk[c]		50	34 (68%)	39	–	12	
	Zimbabwe	Raw milk	"Harare"	48	38 (79.2%)	1800	590 to 4,510	100	Stewart et al. (2016).

[a]Milk analyzed by AFM1 Strip Kit,
[b]Milk analyzed by TLC
[c]Milk analyzed by ELISA
[d]Locally fermented milk
[e]Samples from "Dagoretti"
[f]Samples from "Westlands"
[g]Eastern Democratic Republic of Congo
[h]Farm samples during winter
[i]Farm samples during summer
[j]Commercial milk
[k]Data from HPLC analysis
[l]Rural milk

Much more data are available in the literature on AFM1 occurrence in milk and dairy products in Egypt. A first study showed the presence of AFM1 in 96% (24/25) of camel milk samples with levels ranging from 30 to 850 ng/L (Balata and Bahout 1996). Later, AFM1 was found in 20% (3/15) of cow milk and one sample of dried milk (El-Sayed et al. 2000). In "Assuit" province, AFM1 was found in 58.5% of raw milk samples from six dairy farms (Salem 2002). In "Ismailia," 48.6% (85/175) of milk samples were positive for AFM1, with maximum levels of 210, 220, 230, and 270 ng/L in camel, cow, goat, and buffalo's milk, respectively (Motawee et al. 2009). In "Alexandria," positive raw milk and cheese samples exceeded the MRL set by Egyptian regulations, stating that milk and dairy products should be free of AFM1, while processed cheese were the least contaminated (Amer and Ibrahim 2010). An AFM1 incidence of 43.2% was found in infant formula milk powder samples from "Minoufiya," but the MRL of 25 ng/L was not exceeded (El-Tras et al. 2011). In "Alexandria" city, AFM1 was found in 40% (60/150) of samples at levels of 8.3–85 ng/kg, detected in 46, 53, and 56% of *"Kareish," "Damietta,"* and *"Ras"* cheese samples, respectively. This toxin was also detected in 26% of yoghurt samples (11.40–98.80 ng/kg). A maximum level of 111.5 ng/kg was registered in a sample of "Ras" cheese (Aiad and Abo El-Makarem 2013). More recently, AFM1 was found in 73.9% (65/88) of the milk samples collected from dairy farms in the Governorate of "Assiut" (8.52–78.06 ng/L), and 20.45% (18/88) of samples analyzed by ELISA had AFM1 levels above 50 ng/L (Mwanza et al. 2015). In 2019, 49% (44/90) of milk (15 UHT and 75 raw milk) collected from different dairy shops in "Aswan" city was found positive with AFM1 levels of 0.053–0.207 ppb (Zakaria et al. 2019). In another recent investigation, 100% (20/20) of raw dairy milk samples collected from "Assiut" city were contaminated with AFM1 (0.02–0.19 ng/g), and 70% exceeded the MRL of 50 ng/L (Abdallah et al. 2019a).

3.2 *East Africa*

East Africa is an area known for high risk of exposure to mycotoxins. Various studies have reported the association between AF exposures and high HCC incidence in Kenya, Tanzania, and Uganda (Kimanya 2015). In Kenya the informal food sales sector is dominant. Mycotoxin hazards, particularly those associated with AF, are considered of food safety priority in the country (Sirma et al. 2018). On the other hand, milk consumption rate in this country is around 145 l per capita per year, which is almost higher than in some West African and sub-Saharan countries and AFM1 levels in Kenyan milk are higher than those found in the neighboring countries (Kirino et al. 2016; Ahlberg et al. 2018). An association between exposure to AFM1 in milk and reduced growth in children in "Nairobi" low-income areas has been reported (Kiarie et al. 2016). Moreover, high AFM1 levels have been found in milk from various urban and peri-urban areas of Kenya including pasteurized and UHT milk (Mutegi et al. 2018). In "Nairobi," AFM1 was found in 45.5% (178/391) of the milk samples taken from urban smallholder dairies and 49% exceeded the MRL of 500 ng/kg (Kang'ethe et al. 2007).

The presence of AFM1 was also investigated in 613 milk samples obtained from four urban centers in the country (dairy farmers, medium and large scale farmers, and pasteurized marketed milk). The toxin was found in 77.3% (474/613) of the samples (Kang'ethe and Lang'a 2009). AFM1 was detected also in milk from 200 retailers in "Dagoretti Division-West Nairobi" with a maximum level of 1,675 ng/kg. Overall 55% of the samples exceeded 50 ng/kg and 6% exceeded 500 ng/kg (Kirino et al. 2016). A cross-sectional survey revealed that AFM1 occurred in 100% of the milk randomly collected from two low-income areas of "Nairobi" and 63% of the samples had AFM1 content higher than 50 ng/kg (Kiarie et al. 2016). On the other hand, AFM1 occurred in 84.32% (156/185) of whole milk samples collected from "Bomet County" in Kenya, with 43.8% of samples that exceeding 50 ng/kg (Langat et al. 2016).

Furthermore, AFM1 was detected in 39.7% (203/512) of the milk samples collected from 282 farms in agro-ecological zones selected from eight villages in Kenya, with levels of 2–6,999 ng/L, and 10.4% of samples above 50 ng/L (Senerwa et al. 2016). More recently, 99% (83/84) of milk collected from small holder dairy farms in Kenyan "Kasarani sub-county" was contaminated with AFM1 levels up to 255.96 ng/kg, with 64% of samples exceeding 50 ng/kg (Kagera et al. 2018). Besides that, the analysis of 291 samples of milk and dairy products in "Nairobi" (raw, pasteurized, UHT milk, yoghurt and "*lala*") showed AFM1 levels as high as 1,000 ng/kg, which is higher than every other levels observed before in Kenya, and more than 50% of the samples exceeded 50 ng/kg (Lindahl et al. 2018). Finally, AFM1 was found in 100% (96/96) of informally marketed milk in peri-urban "Nairobi" (mean level of 290.3 ng/kg); with 66.4% of samples above 50 ng/kg and 7.5% above 500 ng/kg (Kuboka et al. 2019).

In Tanzania, AFM1 was found in 83.8% (31/37) of raw milk samples collected from different locations in "Singida" region, with levels up to 2007 ng/L. All positive samples exceeded the limit of 50 ng/L set by Tanzanian authorities, while 16.1% of the samples exceeded 500 ng/kg (Mohammed et al. 2016).

In Ethiopia, AFM1 was found in 100% (110/110) of milk obtained from dairy farms from the Greater "Addis Ababa" at levels of 28–4,980 ng/L, while 93% (102/110) and 26.3% (26/110) of samples exceeded 50 ng/L and 500 ng/kg, respectively (Gizachew et al. 2016).

3.3 West Africa

West African countries are characterized by a tropical climate. They are characterized by high ambient temperature and relative humidity, which provide optimal conditions for the growth of fungi and mycotoxin production. AF in food products has gained prominence in scientific literature from the West African region (Bankole and Adebanjo 2015). However, few reports are published on AFM1 in milk and dairy products from this region except for Nigeria where AF was a general problem as it was highlighted by previous investigations (Oluwafemi 2012).

Nigeria has experienced high recorded AF exposure levels in humans and has also the highest estimated number of cases of HCC attributable to AF worldwide (Atanda et al. 2013). Contamination with AF has been widely reported in a variety of agricultural crops such as maize, peanuts (groundnuts), rice, sorghum, melon seeds (Ezekiel et al. 2018). An earlier investigation conducted in 1979 found no detectable levels of AFM1 in 92 samples of milk from a dairy farm in Kano State (Atanda et al. 2007). However, a nationwide survey reported AFM1 concentrations of 150 and 170 ng/L in yoghurt and ice-cream, respectively (Ogunbanwo 2005). Later, a study carried out on milk and locally produced dairy products in "Ogun" State showed high levels of AFM1 in cow milk up to 2040 ng/L and ice-cream up to 2,230 ng/L (Atanda et al. 2007). Fresh milk and dairy products (*nono* and *kindirmo*) collected in "Bida" revealed the presence of AFM1 in 100% of them, with levels of 139–2,510 ng/L (Okeke et al. 2012). Moreover, AFM1 was found in 75% (75/100) of raw milk samples obtained from "Abeokuta" (Ogun state) at levels of 9.0–456.0 ng/L with a mean level of 108.15 ng/L (Oluwafemi et al. 2014). More recently, AFM1 was found at different frequencies of 80, 40, 35, 25, and 10% in fresh milk, commercial milk, yoghurt, cheese, and "*nono*" samples taken in "Minna" (Nigeria), respectively (Makun et al. 2016). Finally, AFM1 was detected in 53.6% (67/125) of locally repacked powdered milk samples, from five states in the South-Western region of Nigeria, at levels 0.02–0.46 μg/kg (Oyeyipo et al. 2017).

3.4 Central Africa

In Cameroon, foods recommended as high protein sources for children (cow milk and eggs) were collected during 4 years (1991–1995) from various climatic areas and analyzed for AFM1 contamination. This toxin was detected in 15.9% of cow milk samples with levels of 6–525 ng/L and in 16.1% of the egg samples (Tchana et al. 2010). In Burundi and Eastern Democratic Republic of Congo (DRC), AFM1 was detected in 100% of dairy products (fresh milk, yogurt and cheese), with levels of 4.8–261.1 ng/kg (Udomkun et al. 2018).

In samples originating from Burundi, AFM1 was detected in 100% (10/10) of fresh milk at levels of 8.4–82.8 ng/kg and in 100% (6/6) of yogurt samples at levels of 8.2–63.2 ng/kg (Udomkun et al. 2018). In Eastern DRC, AFM1 was detected in 100% (3/3) of fresh milk at levels of 25.6–49.8 ng/kg, in 66.6% (2/3) of yogurt at levels of 4.8–26.0 ng/kg, and in 100% (5/5) of cheese at levels of 18.5–261.1 ng/kg (Udomkun et al. 2018).

3.5 Southern Africa

In South Africa, a survey revealed that 100% of cow milk from farmsteads was positive for AFM1, with levels of 20–1,500 ng/L, while retail milk was also frequently contaminated with AFM1, at levels of 10–3,100 ng/L (Dutton et al. 2012). Another investigation revealed higher incidences of AFM1 in raw milk from rural subsistence and commercial farms in "Mpumalanga," "Limpopo," and "Gauteng" provinces, with mean levels of 150 ng/kg and 140 ng/kg in rural dairy farms and commercial dairy farms, respectively (Mulunda and Mike 2014). In "Ngaka Modiri Molema" district of South Africa, AFM1 occurred in 68% of milk with mean levels of 17.06 and 39 ng/L found by ELISA and HPLC, respectively, and 12% (6/50) of samples tested by ELISA were above the South African regulatory limits (Mwanza et al. 2015).

In Zimbabwe, a South African country with very rare information with regard to mycotoxin contamination of food commodities, the current status of mycotoxin contamination of food commodities has been recently reviewed (Nleya et al. 2018). The fisrt study in this country assessed AFM1 levels in animal milk with an overall contamination of 79.2% of total samples and toxin levels varying between 590 and 4,510 ng/L. All AFM1 levels in positive samples exceeded the EU and *Codex alimentarius* regulations and were on average 30 times higher than the MRL of 50 ng/L (Stewart et al. 2016).

4 AFM1 in Human Biological Fluids: Breast Milk, Serum, and Urine

Data on AFM1 levels in BM, serum, and urine from African countries are summarized in Table 2. AFM1 is frequently observed in blood and urine of individuals exposed to AF, and also in the BM of nursing mothers. According to some reports, the presence of AFM1 in human urine indicates a recent exposure to AF contaminated food because AFM1 is easily excreted from the body (Kang'ethe et al. 2017). Although urinary AFM1 may reflect the ingestion of AFM1 with milk or other contaminated foods, it could be regarded as a subsequent metabolite of AFB1, rather than being from direct transfer (Polychronaki et al. 2008).

4.1 North Africa

Most AFM1 monitoring data in biological fluids in North Africa come from Egypt. AFM1 was found to be prevalent in cirrhotic patients with higher levels in serum and urine in patients from Upper Egypt as compared to patients from Delta (Mokhles et al. 2007). AFM1 was detected in 16.7% (7/42) of BM with a mean value of

Table 2 AFM1 levels in human biological fluids from African countries

African Region	Country	Biological fluid	Total samples	Positive samples (frequency)	Mean value (ng/mL)	Range (ng/mL)	References
North Africa	*Egypt*	Breast milk	42	7 (16.7%)	254	–	El-Sayed et al. (1998)
			10	2 (20%)	2.75	0.5–5	El-Sayed et al. (2000)
			120	66 (55%)	0.3 ± 0.53	0.02–2.09	El-Sayed Abd Alla et al. (2002)
			388	138 (36%)	0.0135	0.0056–5.131	Polychronaki et al. (2006)
			443	248 (56%)	–	Up to 0.889	Polychronaki et al. (2007)
			150	98 (65.33%)	7.1 ± 5.0	0.2–19	Tomerak et al. (2011)
			125	87 (69.6%)	0.074 ± 7.070	0.0073–0.3286	El-Tras et al. (2011)
		Blood	20	5 (25%)	1.18	0.10–2.1	El-Sayed Abd Alla et al. (2002)
		Urine	20	1 (5%)	3.13	–	
		Urine	93	44 (47.3%)	5.48	4.1–408.6	Piekkola et al. (2012)
		Urine	50	4(8%)	0.0055	0.0050–0.0062	Polychronaki et al. (2008)
	Morocco	Breast milk	82	43 (52.4%)	5,750 ± 3.44	Up to 13,330	Cherkani-Hassani et al. (2020)
	Sudan	Breast milk	99	37 (37.4%)	–	0.003 to 0.084	Coulter et al. (1984)
			94	51 (54.2%)	0.401 ± 0.525	0.007–2.561	Elzupir et al. (2012)
Central Africa	*Cameroon*	Urine	220	31 (14%)	0.33	0.06–4.7	Njumbe Ediage et al. (2013)
		Urine	175	16 (9%)	0.05	Up to 1.38	Abia et al. (2013)
		Breast milk	62	3 (4.8%)	–	0.005–0.625	Tchana et al. (2010)
East Africa	*Ethiopia*	Urine	200	14 (7%)	–	0.06–0.07	Ayelign et al. (2017)
	Kenya	Breast milk	98	85 (86.7%)	0.00846	0.000215–0.0475	Kang'ethe et al. (2017)
		Breast milk	67	38 (56.7%)	0.00002	0.000003–0.0037	

		Urine[a]	377	298 (79%)	0.91056	0.72131–1.09978	
		Urine[b]	362	300 (83%)	0.51860	0.40058–0.63662	
	Tanzania	Urine	84	72 (86%)	0.0365	0.015–2.840	Chen et al. (2018)
		Breast milk	143	143 (100%)	–	0.01–0.55	Magoha et al. (2014)
West Africa	*Sierra Leone*	Breast milk	113	35 (31%)	0.8 ± 2.7	0.2–99	Jonsyn et al. (1995)
	Guinea	Urine	50	34 (68%)	0.097	0.008–0.801	Polychronaki et al. (2008)
	Ghana	Urine	91	83 (91.2%)	1.5	Up to 17.2	Jolly et al. (2006)
	Nigeria	Breast milk	28	5 (17.8%)	4	–	Atanda et al. (2007)
			120	17 (14.1%)	–	2–187	Oluwafemi (2012)
			50	41 (82%)	–	Up to 0.09214	Adejumo et al. (2013)
			40	31 (77.5%)	0.0662 ± 0.0178	0.0010–0.6012	Makun et al. (2016)
			75	1 (1,33%)	–	Up to 0.087	Braun et al. (2018)
		Urine	22	22 (100%)	0.235 ± 0.072	–	Oyeyemi et al. (2018)
			120	87 (72.5%)	0.04	0.001–062	Šarkanj et al. (2018)
			84	83 (99%)	0.27 ± 0.08	0.06–0.51	Ezekiel et al. (2018)
Southern Africa	*Zimbabwe*	Breast milk	54	6 (11%)	–	Up to 0.05	Wild et al. (1987)
		Urine	2,553	153 (6%)	–	–	Nyathi et al. (1989)
		Breast milk	54	6 (11%)	–	–	
		Urine	1,228	1,007(82%)	4.2	–	Nyathi et al. (1987)
		Urine[c]	1,580	484 (30%)	162	31–6,046	Smith et al. (2017)

[a]Urine samples from "Makueni"
[b]Urine samples from "Nandi"
[c]Data are expressed in "pg of AFM1 per mg creatinine"

254 ng/mL (El-Sayed et al. 1998). Another study reported 20% (2/10) of BM with AFM1, with levels of 0.5–5 ng/mL and a mean value of 2.75 ng/mL (El-Sayed et al. 2000). In "Cairo," AFM1 was present in 55% of the BM samples with amounts ranging from 0.02 to 2.09 ng/mL, and a mean value of 0.3 ± 0.53 ng/mL (El-Sayed Abd Alla et al. 2002). Furthermore, AFM1 was detected in 35.5% (138/388) of BM with AFM1 levels ranging from 0.0056 to 5.131 ng/mL (Polychronaki et al. 2006). On the other hand, AFM1 was measured for 12 months in positive milk of 50 women and detected in 56% (248/443) of BM with higher rates during summer months, albeit the greatest AFM1 level (0.889 ng/mL milk) was registered in April (Polychronaki et al. 2007). A study from "Cairo" showed that 65.33% (98/150) of BM had AFM1 levels between 0.2 and 19 ng/mL, and that liver enzymes of infants and mothers (ALT and AST) were significantly higher in days having AFM1-positive BM, which might represent a real hazard toward HCC development in Egypt (Tomerak et al. 2011). Finally, AFM1 was found in 69.6% (87/125) of BM from "Minoufiya" in Egypt at levels of 0.0073–0.3286 ng/mL and a mean of 0.0744 ng/mL (El-Tras et al. 2011).

In Morocco, 52.4% (43/82) of BM obtained from nursing mothers living in Rabat were found positive, with AFM1 levels up to 13,330 ng/mL (mean of 5,750 ± 3.44 ng/mL). A statistically significant association was found between maternal AFM1 level in BM and dietary habits (Cherkani-Hassani et al. 2020).

In Sudan, 37.4% of BM samples (37/99) from "Khartoum" were contaminated with levels of 0.003–0.084 ng/mL (AFM1 + AFM2); 13 milk samples with AFM1, 11 with AFM2, and 13 with both AFM1 and AFM2 (Coulter et al. 1984). More recently, 54.2% (51/94) of BM from "Omdurman" were found positive with levels of 0.007–2.561 ng/g and a mean level of 0.401 ± 0.525 ng/g. It was observed from this survey that 70.6% of samples contained AFM1 levels 20-fold higher than 25 ng/kg, and 27.5% of the samples were above 50 ng/kg (Elzupir et al. 2012), which is extremely higher than previous data on AFM1 in human milk from "Khartoum" (Coulter et al. 1984, 1986a).

In Egypt, AFM1 was found in 25% of blood from patients at levels of 0.10–2.1 ng/mL and a mean of 1.18 ng/mL, while one urine sample contained an AFM1 level of 3.13 ng/mL (El-Sayed Abd Alla et al. 2002). Another study showed the presence of AFM1 both in the urine and serum of 60 malnourished Egyptian infants suffering from kwashiorkor and marasmus (Hatem et al. 2005). Later, a study carried out by Polychronaki et al. (2008) revealed the presence of at least one AF in 38% of the urine collected from 50 children (30 fully weaned, 20 partially breastfed), and that 100% of the AFM1 positive samples came from fully weaned children. This study showed also that 8% of urine samples contained detectable levels of AFM1 ranging from 5.0 to 6.2 pg/mL, with mean value of 5.5 pg/mL. Furthermore, AFM1 was found in 47.3% (44/93) of urine obtained from pregnant women (Piekkola et al. 2012).

In Sudan, the prevalence of AF in biological fluids of target populations and its relationship with nutritional diseases like kwashiorkor started since the early 1980s (Hendrickse et al. 1982). Indeed, AFM1 was detected in 6.1% of urine of Sudanese malnourished children suffering from kwashiorkor (Coulter et al. 1986a), and AFM2 detected in liver biopsies of Sudanese children with miscellaneous liver diseases (Coulter et al. 1986b).

4.2 Eastern Africa

The first published papers on the correlation between AF exposure and HCC incidence in the Eastern Africa area appeared since the 1960s, particularly in Kenya, Tanzania, and Uganda. This region has been experiencing outbreaks of acute aflatoxicosis from the year 1967 and several studies found an association between AF exposure and child growth (Kimanya 2015). A survey in Kenya showed that 86.7% and 56.7% of BM samples were positive for AFM1 in "Makueni" and "Nandi" counties, respectively; with 10.2% of BM collected in "Makueni" above 25 ng/L (Kang'ethe et al. 2017). In Northern Tanzania, 100% of BM were found contaminated by AFM1 at levels of 0.01–0.55 ng/mL (Magoha et al. 2014), with 90% and 76% of the BM above 25 and 50 ng/L, respectively. A significant inverse association between AFM1 exposure and infant growth was observed, which suggests that AF affects child nutritional status at the very early life period in the country.

In Kenya, Kang'ethe et al. (2017) showed that AFM1 occurred in 79% (298/377) and 83% (300/362) of urine of Kenyan children from "Makueni" and "Nandi" counties, respectively; and that children in "Makueni" had higher levels of AFM1 in urine than those of the same age in "Nandi."

In Ethiopia, AFM1 was detected in 7% of the urine of 200 children (aged 1–4 years) with a range of 0.06–0.07 ng/mL (Ayelign et al. 2017). Finally, AFM1 was detected in 86% of urine samples obtaiend from 84 Tanzanian children at levels of 15–2,840 pg/mL (Chen et al. 2018).

4.3 West Africa

A study from Gambia investigated child exposure to AFM1 through BM intake and reported that the percentage of dietary AF excreted as AFM1 in milk ranged from 0.09 to 0.43% (Zarba et al. 1992). In Sierra Leone, AFM1 and AFM2 were detected in 31% (35/113) and 62% (70/113) in BM at levels of 0.2–99 ng/mL and 0.07–77.5 ng/mL, respectively (Jonsyn et al. 1995).

In Nigeria, high AFM1 levels (up to 4.0 ng/mL) were reported in BM from local governments of "Ogun" State (Atanda et al. 2007). Later, an AFM1 incidence of 14.1% (17/120) was found in BM collected from two Nigerian cities with levels of 2–187 ng/mL (Oluwafemi 2012). Another study showed that 82% (41/50) of BM of mothers from "Ogun" state were contaminated with AFM1 (up to 0.09214 ng/mL), with a significant positive correlation between AFM1 content of BM of mothers and their dietary exposure to AFB1 (Adejumo et al. 2013). High AFM1 incidence of

77.5% (31/40) was also observed in BM ("Minna," Nigeria), with 37.5% having levels above 50 ng/L (Makun et al. 2016). More recently, AFM1 occurred only in 1% of BM in "Ogun" state with a maximum level of 0.087 ng/mL (Braun et al. 2018).

In Ghana, AFM1 was detected in 91.2% of the urine of 91 participants from "Ejura Sekyedumase" villages, with levels up to 17.2 ng/mL (mean value of 1.5 ng/mL). The urinary AFM1 levels were associated with AF uptake in the participants diet (Jolly et al. 2006).

In Guinea, 64% (64/100) of children (2–4 years) urine contained detectable levels of AFM1 that ranged from 8.0 to 801 pg/mL with a mean value of 97.0 pg/mL (Polychronaki et al. 2008). On the other hand, 100% (22/22) of participants from "Ogun" State, Nigeria were tested positive to AFM1 in urine with a mean level of 0.235 ± 0.072 ng/mL (Oyeyemi et al. 2018). The exposure assessment of Nigerian children, adolescents, and adults showed that AFM1 occurred in 72.5% (87/120) of urine, with levels of 0.001–0.62 ng/mL (Šarkanj et al. 2018). Finally, an exposure bio-monitoring study in "Ogun" state revealed the presence of AFM1 in 98% (82/84) of urine at levels of 0.06–0.51 ng/mL (Ezekiel et al. 2018).

4.4 Central Africa

Few data are available about AFM1 occurrence in biological samples from Central Africa. Most AF monitoring data in body fluids in this region come from Cameroon. Since early studies reported the presence of AFB1 and its metabolites in food and feed in this country at higher levels than the international regulatory limits, some studies investigated body fluids of target populations for AF occurrence. AFM1 was detected in 4.8% of BM of mothers from "Yaounde" with levels of 0.005–0.625 ng/mL (Tchana et al. 2010). Later, AFM1 was also found in 14% (31/220) of urine of children from six villages in Cameroon, with a mean level of 1.43 ng/mL in urine of the partially breastfed children higher than those of the fully weaned children (0.282 ng/mL), which suggested that weaning status rather than age could be a strong determinant of AFM1 exposure in children less than 5 years (Njumbe Ediage et al. 2013). Bio-monitoring of AF biomarkers in Cameroon showed that AFM1 was present in 9% (16/175) of urine, with a strong association between high AFM1 levels in urine and contaminated diets intake (Abia et al. 2013).

4.5 Southern Africa

A study from Zimbabwe reported the presence of AFM1 in 11% (6/54) of BM collected from women in rural villages, with levels up to 50 ng/L (Wild et al. 1987). AFM1 was also found in 82% (1,007/1228) of urine samples collected from different centers in the country (Nyathi et al. 1987). Later, another investigation reported that

AFM1 occurred in 6% (153/2553) of urine collected from donors of different ages and sexes at centers throughout Zimbabwe (Nyathi et al. 1989). More recently, AFM1 was detected in 30% (474/1580) of urine collected from pregnant women in rural Zimbabwe, with levels that ranged from 31 to 6,046 pg/mg creatinine, indicating that women had recently been exposed to AF, with season, spatial, and dietary practices variability (Smith et al. 2017).

5 Exposure Level Estimates in Africa

AFM1 levels in human milk, urine, and serum have been reported as molecular biomarkers of AF exposure. AFM1 exposure level estimates of some African populations through breastfeeding or milk and dairy products intake are summarized in Table 3. Investigations showed that 85–100% of African children are highly exposed to AFM1 through human milk intake. Furthermore, the highest levels of AF exposure observed worldwide have been reported in the West African countries (Okoth 2016). This exposure is generally associated with child stunting, child mortality, immune suppression, and childhood neurological impairment. A correlation between AFM1 in human milk, AFB1 in food and socioeconomic status of mothers was also reported in some West African countries (Adejumo et al. 2013).

Human exposure to AFM1 is due to the consumption of contaminated milk and dairy products and AFM1 daily intake could be highly variable in the world. Infants are considered the most exposed population due to their high consumption either of bovine milk and derivatives in their diet, or from BM intake. The dietary intake of AFM1 was estimated from data on the concentration of AFM1 in milk as reported by many countries and established by JECFA based on AFM1 mean concentrations in milk, on the one hand, and the milk consumption data in the GEMS/Food regional diets on the other hand (WHO 1998). The intake of AFM1 from milk was calculated to be 0.1 ng/person/day for the African diet (JECFA 2011). These values are below the intakes calculated for Europe, Latin America, the Far East, and the Middle East (6.8, 3.5, 12, and 0.7 ng/person/day, respectively).

5.1 *Exposure Through Milk Intake*

Milk and derivatives have the greatest potential for introducing AFM1 into the human diet. There is no evidence that the toxin is degraded or removed by the conventional technological processes used at industrial scale such as heat treatments, drying and fermentation, etc. (Duarte et al. 2013).

In North Africa, few studies are available on the AFM1 exposure levels through milk and derivatives consumption. In Morocco, the AFM1 daily intake through pasteurized milk consumption was estimated to be 3.26 ng/person/day, which is similar to that in the Latin American diet (3.5 ng/person/day), and approximately

Table 3 Exposure levels of African populations to AFM1 through breastfeeding and milk intake

	Country	Studied population	Milk product	Exposure estimates	Reference
Milk and dairy products	Egypt	Newborns	Infant formula milk powder	8.170 ng/day	El-Tras et al. (2011)
	Kenya	Children	Milk	3.7 ng/day (>100 ng/day)[a]	Kiarie et al. (2016)
		Children	Milk	93.8 ng/day	Kirino et al. (2016)
		Adults	Milk	120.0 ng/day	
		Children <5 years[b]	Milk	0.4 ng/kg b.w./day	Kang'ethe et al. (2017)
		Other[c]	Milk	0.1 ng/kg b.w./day	
		Children <5 years[d]	Milk	0.1 ng/kg b.w./day	
		Other[e]	Milk	0.04 ng/kg b.w./day	
		Adults	Milk	46 ng/day	Ahlberg et al. (2018)
		Children	Milk	3.5 ng/kg b.w./day	
		Adults	Milk	0.2 ng/kg of b.w./day	Sirma et al. (2018)
	Morocco	Adults	Pasteurized milk	3.26 ng/day	Zinedine et al. (2007)
Breastfeeding	Egypt	Infants	Breast milk	6.75 ng/day	Polychronaki et al. (2006)
		Newborns	Breast milk	52.684 ng/day	El-Tras et al. (2011)
	Nigeria	Infants	Breast milk	22.0 μg/day[f]	Oluwafemi (2012)
	Tanzania	Infants <6 months	Breast milk	1.13–66.79 ng/kg b.w./day	Magoha et al. (2014)
	Kenya	Nursing children	Breast milk	6 ng/kg b.w./day (Makueni) 0.001 ng/kg b.w./day (Nandi)	Kang'ethe et al. (2017)
	Morocco	Newborns	Breast milk	Up to 1.16 ng/kg b.w./day	Cherkani-Hassani et al. (2020)

[a]A maximum exposure level was found for some children consuming more than 100 ng AFM1 from milk per day
[b]Children from "Makueni" county
[c]Persons aged between 5 and 59 years, and over 60 years from "Makueni" county
[d]Children from "Nandi" county
[e]Persons aged between 5 and 59 years, and over 60 years from "Nandi" county
[f]Exposure levels ranged was from 1.0 to 93.5 μg/day

32-fold higher than the value estimated for the African diet (Zinedine et al. 2007). While in Egypt, the average daily exposure of newborns to AFM1 via consumption of infant formula milk powder was estimated to 8.170 ng/day (El-Tras et al. 2011).

In Eastern Africa, most available data on the AFM1 exposure levels through milk intake are from Kenya. Indeed, it was reported that milk contributed daily a median of 3.7 ng AFM1 to the Kenyan children, with some children consuming more than 100 ng AFM1 from milk per day (Kiarie et al. 2016). Another study estimated the exposure of children and adults to AFM1, as based on milk consumption of retailers' households, to be 93.8 ng/day for children and 120.0 ng/day for adults. These estimates suggest that members of milk retailers' households have an AFM1 intake 1,000 times higher than the estimate of 0.1 ng/capita/day for the African diet (Kirino et al. 2016). The AFM1 exposure levels through milk were determined in two Kenyan counties ("Makueni" and "Nandi"). In "Makueni," AFM1 exposure levels of children (<5 years), persons aged between five and 59 years, and persons over 60 years were estimated to be 0.4, 0.1, and 0.1 ng/kg body weight (b.w.)/day, respectively. While in "Nandi," AFM1 exposure levels were estimated to be 0.1, 0.04 and 0.04 ng/kg b.w./day for the same age groups, respectively (Kang'ethe et al. 2017). Based on analytical data, the average AFM1 exposure in milk was estimated for Kenyan adults at 0.2 ng/kg b.w./day, which corresponds to five HCC cases and deaths in the country, by using the quantitative risk model (Sirma et al. 2018).

Recently, Ahlberg et al. (2018) conducted a predictive risk assessment for AFM1 exposure, cancer risk, and stunting among urban milk consumers during 2013–2016 in Nairobi. This study showed that the total AFM1 exposure estimation was 46 ng/day on average (0.8 ng/kg b.w./day), and the daily exposure of children below 3 years to AFM1 was estimated at 3.5 ng/kg b.w., which is significantly higher than in adults because of the high average milk consumption and low body weight. It was also established that consumers purchasing dairy products from informal markets are more likely to be exposed (1.2 ng/kg b.w./day) to AFM1 than middle-income consumers purchasing processed products (0.7 ng/kg b.w./day).

5.2 *Exposure Through Breast Milk and Urine Bio-Monitoring*

AFM1 is a marker of dietary exposure of mothers to AFB1, and therefore of exposure of breastfeeding infants to this BM contaminant (Zarba et al. 1992). It was assumed that the average daily AFM1 intake of infants based on BM exposure is not easy to estimate with precision, due to the fluctuation of BM levels and to the possible variation of maternal AF intake.

In Egypt, the average infant intake of AFM1 was estimated at 6.75 ng/day, and the maternal intake of AFB1 estimated at 1570–7500 ng/day, supporting that maternal AF exposures are occurring at moderate levels in this country as compared with some of the high risk regions of Africa and South East Asia (Polychronaki et al.

2006). However, an AFM1 exposure level as high as 52.684 ng/day was determined for Egyptian newborns via the consumption of maternal BM (El-Tras et al. 2011). In Nigeria, a study on the prevalence of AFM1 in 120 samples of human BM showed that the average intake of AFM1 by the infants was 22.0 μg/day and the range between 1.0 and 93.5 μg/day (Oluwafemi 2012). In Cameroon, the AFM1 exposure levels in urine of adults among HIV sub-populations were estimated at 42 ng/kg b.w./day with a maximum of 1.150 ng/kg b.w./day (Abia et al. 2013). In Tanzania AFM1 exposure levels for 143 infants under 6 months of age through breastfeeding were reported to range from 1.13 to 66.79 ng/kg b.w./day (Magoha et al. 2014). In Kenya, the exposure of nursing children through BM was estimated at 6 and 0.001 ng/kg b.w./day in "Makueni" and "Nandi," respectively (Kang'ethe et al. 2017).

More recently, it was reported from a Moroccan survey on AFM1 in BM that the detected amounts of this toxin were low as compared to those reported worldwide. Thus, the estimated daily intake (EDI) of AFM1 following infants' consumption of BM varied between immeasurable and a maximum of 1.16 ng/kg b.w. Additionally, the EDI was greater than 0.5 ng/kg b.w. for 22 babies and greater than 1 ng/kg b.w. for 3 babies (Cherkani-Hassani et al. 2020).

Although exposure of infants to AFM1 from human BM has been reported in developing African countries, limited data are available on the subject in the Southern African region. Studies are, therefore, necessary to assess the exposure level of suckling babies to mycotoxins and to determine the consequences of AFM1 ingestion from BM and/or from the milk of livestock in Southern Africa (Misihairabgwi et al. 2019).

6 Detoxification Strategies of AFM1

Physical, chemical, and biological methods are the three major detoxification strategies described for AF (Peng et al. 2018). Given its toxicity, AFM1 is strictly regulated and ongoing investigations at an international level, as in Africa, aim to propose validated protocols for AFM1 reduction which can be accepted by official authorities and applicable on the industrial scale. To improve food safety and prevent economic losses, several decontamination strategies have been investigated for their effectiveness to remove or inactivate AFM1 from contaminated milk and dairy products by using multiple physical, biological, and chemical methods (Benkerroum 2016).

It is well established that heat treatments (pasteurization, sterilization and microwave heating) do not cause an appreciable decline in the amount of AFM1 in dairy products, despite the fact that milk pasteurization is globally applied as a preventive method to reduce pathogens and eliminate putative mycotoxin-producing molds (Ashiq 2015; Ismail et al. 2015). In Egypt, a study on the fate of AFM1 through "*Domiati*" cheese processing revealed a reduction of 64% of AFM1 with the combination of milk pasteurization and a cheese storage period of 3 months

(Motawee 2013), while a reduction of 50% of AFM1 originally present in the raw milk was obtained during "*feta*" cheese manufacturing after milk pasteurization and a cheese storage in brine during at least 50 days, which reduces the exposure of the consumers (Motawee and McMahon 2009).

Decontamination of AFM1 with clay particles without affecting the nutritional properties of milk and the use of clay-based products as enterosorbents for AF to reduce AF exposure in animals has been reported. Especially, the inclusion of a calcium montmorillonite clay (NS) in animal feed reduced the adverse effects associated with AF exposure in various animal species and decreased the level of AFM1 in milk from lactating dairy cows and goats (IARC 2015). A saponite-rich bentonite showed the highest adsorbent capacity to reduce AFM1 in contaminated bovine milk (up to 80 ng/L) to a safe level (50 ng/L for adults and 25 ng/L for lactants) with moderate alteration of the nutritional properties of milk (Carraro et al. 2014).

Several chemical methods have been reported for mycotoxin removal, including acidification. Indeed, the decrease of mycotoxins levels can occur naturally in many fermented dairy products, as a result of the combined activity of acidity and the antagonist effects of LAB starters (Benkerroum 2016).

An earlier study from Egypt investigated the use of hydrogen peroxide (H_2O_2) under various conditions to inactivate AFM1 in artificially contaminated raw milk. The maximum inactivation of AFM1 (27.8%, 28.8% and 45.1%) was obtained using 1% H_2O_2 followed by heat treatment at 36 °C for 30 min, 75 °C for 15 s, and boiling for 5 min, respectively, but hydrogen peroxide has no effect on AFM1 in milk when it is used without being followed by heat treatment (Aman 1992), which suggests that the heat-processing of milk is the principal factor that could decrease AFM1 levels in this reduction.

Natural substances were also tried to reduce AF toxicity in animal models. For example, a study from Côte d'Ivoire showed that the leaf extract of *Alchornea cordifolia*, a widespread plant used in tropical Africa for the treatment of many ailments, significantly reduced the formation of AFB1 biomarkers, including 51.75% of AFM1 reduction, which is a consequence of an AF detoxification process, due to the strong antioxidant activity and the high content in phenolic compounds (Adépo et al. 2019). Nevertheless, more research is highly recommended to investigate the potential interactions of *A. cordifolia* with other compounds, to investigate the long-term effects of prolonged administration of these extracts to the exposed subjects.

The use of microorganisms and their components (cell wall, proteins, polysaccharides, etc.) in biological approaches for the bio-decontamination of AFM1 has been increasingly investigated by scientists in recent decades, particularly in the dairy sector. Many studies reported the ability of the cell wall of certain microorganisms to bind AFM1. Several strains of lactic acid bacteria (LAB) (particularly *Lactobacillus acidophilus* and *Bifidobacterium spp.*) can bind AFM1 and furthermore eliminate it. Although the precise mechanism of this attachment is still unknown, it is mainly physical, due to a weak non-covalent complex binding to peptidoglycan and polysaccharides present in the bacterial cell wall (Mahmood

Fashandi et al. 2018). LAB are nowadays extensively investigated for their detoxification and binding capacity of AFM1 in milk during fermentation because of their wide distribution and occurrence in foodstuffs and in the environment and are *Generally Recognized As Safe* (GRAS) (Ismail et al. 2015). The possible reasons for AFM1 decrease in fermented milk by using LAB could be due to the pH decrease and the organic acids formation. However, the effect of low pH and organic acids on AFM1 removal is not clear yet. The ability of some LAB strains to remove AFM1 from liquid media suggested their use in dairy products as means for decontaminating AFM1 from milk (Pierides et al. 2000). Among LAB, *Lactobacillus acidophilus* leads to the most considerable reduction of AFM1 quantities and showed the best binding capacity of AFM1 among all the probiotic strains (Mahmood Fashandi et al. 2018).

In Africa, some traditional food processing techniques may potentially be useful for detoxifying foods from mycotoxin contamination. They could, therefore, be considered as interesting solutions to reduce mycotoxin hazards for local consumers (Fandohan et al. 2005; Tantaoui-Elaraki et al. 2018). For example, a mixed fermentation using a symbiotic association between LAB and yeasts has been shown to improve organoleptic properties of fermented foods and to reduce mycotoxins. However, the real mechanism of reduction and their role in mitigating risk from contamination need further investigations (Okoth 2016).

During the processing of Moroccan "*Lben,*" a widely consumed traditional fermented milk known in Arab countries, lactic acid bacteria (*Lactococcus lactis* alone or associated with *L. diacetylactis*) eliminate completely AFM1 added to milk at 5 μg/L, while a very limited decrease was observed in the case of yoghurt fermented with *Streptococcus thermophilus* and *Lactobacillus bulgaricus* strains (Khaddor et al. 2003). However, no information is available on the molecular mechanism of this elimination, on the conversion of AFM1 during "*Lben*" processing, and on possible hazardous effects of AFM1 metabolites produced by *Lactococcus lactis* alone or in combination with *L. diacetylactis*.

In Tunisia, a study tested probiotic bacteria originating from dairy products to degrade or bind AFM1 in vitro and data showed that *Lactobacillus plantarum* MON03 and *L. rhamnosus* GAF01 effectively bind AFM1 in PBS (78.6%) and reconstituted milk (95.1%) (Abbès et al. 2013). The ability of *L. plantarum* MON03 to effectively remove AFM1 from solutions and to mitigate immunotoxicity in mice suggested its use for a major goal of mycotoxin detoxification (Jebali et al. 2015), albeit the detoxification mechanism remains unexplained and requires further investigations.

In Kenya, some studies showed a significant change in AFM1 concentration during milk storage. Thus, fermentation experiments using a mesophilic starter culture (*Streptococcus lactis* and *Leuconostoc mesenteroides ssp. mesenteroides*) demonstrated the ability to reduce 71.8% of AFM1 through processing milk to "*lala,*" a traditional fermented drink, after incubation at room temperature for 15 h; while thermophilic yogurt starter culture (*Lactobacillus delbrueckii* ssp. *bulgaricus* and *Streptococcus salivarius* ssp. *thermophilus*) reduced 73.6% of AFM1 after incubation at 45 °C for 4 h (Kuboka et al. 2019). In Egypt, numerous investigations assayed LAB and *Bifidobacteria* strains to remove AFM1 and *L. plantarum* was the best strain at removing AFM1 from yoghurt (Elsanhoty et al. 2014). Moreover, it was reported from this study that yoghurt fermented by 50% yoghurt culture (*S. thermophilus* and *L. bulgaricus*) and 50% *L. plantarum* recorded the highest reduction in the level of AFM1 at the end of storage period. The combination of non-viable probiotic bacterial and yeast species (*L. plantarum, L. acidophilus, Bifidobacterium bifidum, Kluyveromyces lactis,* and *Saccharomyces cerevisiae*) succeeded in reducing AFM1 for more than 90% (Abdelmotilib et al. 2018). In a recent investigation, it was confirmed that a mixture of starter cultures of *L. bulgaricus* and *S. thermophilus* used in yogurt manufacture has beneficial effect on the reduction of AFM1 level during milk fermentation (Zakaria et al. 2019). Although the true milk decontamination mechanism with LAB strains has not been fully understood, some studies have suggested the toxin's ability to bind to the cell wall of bacteria during fermentation. Some components of the bacterial cell wall also play a known role in the detoxification process. Currently, several studies in Africa have shown that traditional starters originating from extreme biotopes could be composed of active LAB strains that could attenuate the toxic effects of certain mycotoxins. These suggestions open up real research perspectives in this area to deepen knowledge and study the possible interactions between LAB strains not with one mycotoxins (such as AFM1), but with several mycotoxins simultaneously.

7 AFM1 Regulations in Africa

AFM1 is widely spread in milk and milk products, particularly in the developing countries. Thus, strict regulations should be adopted in order to control this mycotoxin in the dairy sector. The FAO global survey of 2003 indicated that only 15 African countries had known AF regulations (FAO 2004). The non-existence of specific legislations regarding MRL of mycotoxin in the rest of African countries might be due to lack of data on the prevalence of certain toxins; and to the poor capacities and resources to obtain toxicological and exposure data and enforce regulations (Misihairabgwi et al. 2019).

Regarding the general standards set for contaminants and toxins in food and feed by *Codex alimentarius* amended in 2019, the MRL of AFM1 in milk is 500 ng/kg (FAO/WHO 2019), which is identical to the US regulation (FDA 2005). However, the European legislation adopted an MRL as low as 50 ng/L for AFM1 in liquid milk (raw milk, heat-treated milk, and milk for the manufacture of milk-based products), and 25 ng/kg for infant formulae and dietary foods (European Commission 2006).

Regarding Africa, few countries have set MRL for AFM1 in milk and derivatives. Table 4 gives a comparison of the regulatory limits set in African countries with the MRL adopted worldwide for this toxin (*Codex alimentarius*, EU and US regulations).

Egyptian authorities decided on a zero-tolerance strategy for the presence of AFM1 in milk and derivatives to ensure maximum consumer health protection at

Table 4 AFM1 legislation in some African countries compared to the *Codex alimentarius*, EU, and US regulations

Country/ region	MRL (ng/kg or ng/L)	Milk Product	Reference
Egypt	0	Milk and derivatives	Egyptian-Regulations (1990)
Kenya	500	Dairy products	Kibugu et al. (2019)
Morocco	50	Raw milk, heat-treated milk, and milk intended for manufacture of dairy products	Bulletin officiel (2016)
	25	Infant formulae and dietary foods	
Nigeria	500	Milk	Oyeyipo et al. (2017)
Tanzania	50	Milk	Mohammed et al. (2016)
Tunisia	500	Milk	Abbès et al. (2012)
South Africa	50	Milk and milk products	Mulunda and Mike (2014)
Europe	50	Raw milk, heat-treated milk, and milk for the manufacture of milk-based products	European Commission (2006)
	25	Infant formulae and dietary foods	
USA	500	Whole milk, low fat milk, skim milk	FDA (2005)
Codex alimentarius	500	Milk	WHO (2006)

the expense of milk and cheese producers (Egyptian-Regulations 1990). However, according to some authors, this target (0 ng/L) is difficult to achieve as long as AF contamination in animal feed remains an issue (Mwanza et al. 2015).

Tunisian regulations set MRL of AFM1 in milk at 500 ng/kg, which is similar to that of the *Codex alimentarius* standards and US regulations (Abbès et al. 2012).

In Nigeria, the maximum permitted level of AFM1 in milk considered by the National Agency for Food and Drug Administration and Control (NAFDAC) is set at 500 ng/kg (Oyeyipo et al. 2017), which is similar to the MRL (500 ng/kg) set for AFM1 in processed dairy products by Kenyan authorities (Kibugu et al. 2019).

In South Africa, the MRL is set at 50 ng/kg for AFM1 in milk and milk products (Mulunda and Mike 2014) which is similar to the MRL (50 ng/kg) set by Tanzanian authorities for AFM1 regulations in milk (Mohammed et al. 2016).

Morocco has recently adopted new regulations for mycotoxins, and the MRL of AFM1 are set at 50 ng/kg for raw milk, heat-treated milk, and milk intended for manufacture of dairy products, and at 25 ng/kg for infant formulae and dietary foods (Bulletin Officiel 2016).

It should also be noted that most of African countries suffer from a significant lack of qualified human resources and specific material to control mycotoxin. This situation implies an urgent implementation of capacity building actions, setup AF control programs in Africa, and reduce human exposure.

8 Conclusion and Recommendations

Through this review paper, we have emphasized that AFM1 widely occurs in milk and derivatives available for consumers in all regions of the African continent. The registered contamination levels often exceed the guidelines standards set at the international level (*Codex alimentarius*, US, EU, and regional regulations, etc.). Several articles have been published considering the issue of AFM1 contamination in BM, animal milk, dairy products, and AF risk assessment for African populations. Based on the reported data on AFM1 occurrence, on contamination and exposure levels, we can consider that AFM1 is a major concern in the continent, although is not objective to compare the levels found in different samples of animal and human milk collected in different African regions.

In this context, several studies determined AFM1 levels by using immunochemical and chromatographic techniques (ELISA, HPLC, and LC/MS-MS). The development by scientists of rapid, reliable, and less expensive procedures for the analysis of AFM1 in milk and derivatives and their introduction by decision-makers in peripheral laboratories in countries suffering from AFM1 contamination problems would be a correct approach to detect contaminated batches and take the right decisions to protect the health of consumers.

Currently, few African countries have adopted strict and specific regulations for AF presence in food and feed especially for milk and derivatives. There is an urgent need for AF legislation at national and/or regional level (harmonized regulations) to improve food and feed control, reduce AF exposure, and protect African populations from their acute and chronic harmful effects.

In addition to the legislation aspect, specific awareness programs should be put in place for milk and derivatives producers, processors, and consumers regarding hazards due to the occurrence of AFM1 in milk, as well as practical prevention strategies to reduce the contamination.

In order to reduce human exposure to AF and to lower AFM1 levels in animal milk and dairy products, it is urgent to control AFB1 levels in stored food commodities and animal feed, to privilege planting of crops resistant toxigenic fungi. In addition, good agricultural practices are needed, including crop rotation, reduction of grain damage, and insect attacks.

Finally, another health threat is represented by moldy foods considered as improper for human intake and used as feed for lactating animals. For example, in some African countries, dairy cows are fed moldy bread and grains in order to allegedly improve milk production, while such a practice undoubtedly leads to milk contamination with mycotoxins.

Acknowledgments This paper has been supported by the PHC Maghreb "09MAG20" project. The authors are most grateful to the Moroccan Ministry of Higher Education and Scientific Research, the National Center for Scientific and Technical Research (CNRST) of Morocco, and the Tunisian Ministry of Higher Education and Scientific Research for the support given.

Conflicts of Interest The authors have no conflicts of interest to declare regarding this paper.

References

Abbès S, Ben Salah-Abbès J, Bouraoui Y et al (2012) Natural occurrence of aflatoxins (B_1 and M_1) in feed, plasma and raw milk of lactating dairy cows in Beja, Tunisia, using ELISA. Food Addit Contam B 5:11–15. https://doi.org/10.1080/19393210.2011.640756

Abbès S, Ben Salah-Abbès J, Sharafi H et al (2013) Ability of Lactobacillus rhamnosus GAF01 to remove AFM1 in vitro and to counteract AFM1 immunotoxicity in vivo. J Immunotoxicol 10:279–286. https://doi.org/10.3109/1547691X.2012.718810

Abdallah M, Girgin G, Baydar T (2019a) Mycotoxin detection in maize, commercial feed, and raw dairy milk samples from Assiut City, Egypt. Vet Sci 6:57. https://doi.org/10.3390/vetsci6020057

Abdallah M, Maarten A, De Saeger S et al (2019b) Biological control of mycotoxigenic fungi and their toxins: an update for the pre-harvest approach. In: Mycotoxins – impact and management strategies. IntechOpen

Abdelmotilib N, Hamad G, Elderea H et al (2018) Aflatoxin M1 reduction in milk by a novel combination of probiotic bacterial and yeast strains. Eur J Nutr Food Saf 8:83–99. https://doi.org/10.9734/EJNFS/2018/39486

Abia WA, Warth B, Sulyok M et al (2013) Bio-monitoring of mycotoxin exposure in Cameroon using a urinary multi-biomarker approach. Food Chem Toxicol 62:927–934. https://doi.org/10.1016/j.fct.2013.10.003

Adejumo O, Atanda O, Raiola A et al (2013) Correlation between aflatoxin M1 content of breast milk, dietary exposure to aflatoxin B1 and socioeconomic status of lactating mothers in Ogun State, Nigeria. Food Chem Toxicol 56:171–177. https://doi.org/10.1016/j.fct.2013.02.027

Adépo J-BA, Manda P, Ngbé JV et al (2019) Study of aflatoxicosis reduction: effect of Alchornea cordifolia on biomarkers in an aflatoxin B1 exposed rats. Drug Chem Toxicol 42:243–251. https://doi.org/10.1080/01480545.2017.1416625

Ahlberg S, Grace D, Kiarie G et al (2018) A risk assessment of aflatoxin M1 exposure in low and mid-income dairy consumers in Kenya. Toxins 10:348. https://doi.org/10.3390/toxins10090348

Aiad A, Abo El-Makarem H (2013) Aflatoxin M1 levels in milk and some dairy products in Alexandria City. Assiut Vet Med J 59:93–98

Alahlah N, El Maadoudi M, Bouchriti N et al (2020) Aflatoxin M1 in UHT and powder milk marketed in the northern area of Morocco. Food Control 114:107262. https://doi.org/10.1016/j.foodcont.2020.107262

Ali MAI, El Zubeir IEM, Fadel Elseed AMA (2014) Aflatoxin M_1 in raw and imported powdered milk sold in Khartoum state, Sudan. Food Addit Contam B 7:208–212. https://doi.org/10.1080/19393210.2014.887149

Aman I (1992) Reduction of aflatoxin M1 in milk using hydrogen peroxide and hydrogen peroxide plus heat treatment. Zentralbl Veterinarmed B 39:692–694. https://doi.org/10.1111/j.1439-0450.1992.tb01223.x

Amer AA, Ibrahim MAE (2010) Determination of aflatoxin M1 in raw milk and traditional cheeses retailed in Egyptian markets. J Toxicol Environ Health 2:50–52

Ashiq S (2015) Natural occurrence of mycotoxins in food and feed: Pakistan perspective. Compr Rev Food Sci Food Saf 14:159–175. https://doi.org/10.1111/1541-4337.12122

Atanda O, Oguntubo A, Adejumo O et al (2007) Aflatoxin M1 contamination of milk and ice cream in Abeokuta and Odeda local governments of Ogun state, Nigeria. Chemosphere 68:1455–1458. https://doi.org/10.1016/j.chemosphere.2007.03.038

Atanda O, Makun HA, Ogara IM et al (2013) Fungal and mycotoxin contamination of nigerian foods and feeds. In: Mycotoxin and food safety in developing countries. InTech

Ayelign A, Woldegiorgis AZ, Adish A et al (2017) Assessment of aflatoxin exposure among young children in Ethiopia using urinary biomarkers. Food Addit Contam A 34:1606–1616. https://doi.org/10.1080/19440049.2017.1350290

Balata M, Bahout A (1996) Aflatoxin M1 in camel's milk. Vet Med J Giza-Egypt 44:109–111

Bankole SA, Adebanjo A (2015) Mycotoxins in food in West Africa: current situation and possibilities of controlling it. Afr J Biotechnol 2:254–263. https://doi.org/10.5897/ajb2003.000-1053

Benkerroum N (2016) Mycotoxins in dairy products: a review. Int Dairy J 62:63–75. https://doi.org/10.1016/j.idairyj.2016.07.002

Bennett JW, Klich M (2003) Mycotoxins. Clin Microbiol Rev 16:497–516. https://doi.org/10.1128/CMR.16.3.497-516.2003

Braun D, Ezekiel CN, Abia WA et al (2018) Monitoring early life mycotoxin exposures via LC-MS/MS breast milk analysis. Anal Chem 90:14569–14577. https://doi.org/10.1021/acs.analchem.8b04576

Bulletin Officiel (2016) Arrêté conjoint du Ministre de l'Agriculture et de la Pêche Maritime et du Ministre de la Santé n°1643-16 du 30 mai 2016 fixant les limites maximales de contaminants autorisées dans ou sur les produits primaires et les produits alimentaires

Carraro A, De Giacomo A, Giannossi ML et al (2014) Clay minerals as adsorbents of aflatoxin M1 from contaminated milk and effects on milk quality. Appl Clay Sci 88–89:92–99. https://doi.org/10.1016/j.clay.2013.11.028

Chen G, Gong YY, Kimanya ME et al (2018) Comparison of urinary aflatoxin M1 and aflatoxin albumin adducts as biomarkers for assessing aflatoxin exposure in Tanzanian children. Biomarkers 23:131–136. https://doi.org/10.1080/1354750X.2017.1285960

Cherkani-Hassani A, Mojemmi B, Mouane N (2016) Occurrence and levels of mycotoxins and their metabolites in human breast milk associated to dietary habits and other factors: a systematic literature review, 1984–2015. Trends Food Sci Technol 50:56–69. https://doi.org/10.1016/j.tifs.2016.01.024

Cherkani-Hassani A, Ghanname I, Zinedine A et al (2020) Aflatoxin M1 prevalence in breast milk in Morocco: associated factors and health risk assessment of newborns "CONTAMILK study". Toxicon 187:203–208. https://doi.org/10.1016/j.toxicon.2020.09.008

Coulter JBS, Lamplugh SM, Suliman GI et al (1984) Aflatoxins in human breast milk. Ann Trop Paediatr 4:61–66. https://doi.org/10.1080/02724936.1984.11748311

Coulter JB, Hendrickse R, Lamplugh S et al (1986a) Aflatoxins and kwashiorkor: clinical studies in Sudanese children. Trans R Soc Trop Med Hyg 80:945–951. https://doi.org/10.1016/0035-9203(86)90266-x

Coulter JB, Suliman GI, Lamplugh SM et al (1986b) Aflatoxins in liver biopsies from Sudanese children. Am J Trop Med Hyg 35:360–365. https://doi.org/10.4269/ajtmh.1986.35.360

Cullen JM, Ruebner BH, Hsieh LS et al (1987) Carcinogenicity of dietary aflatoxin M1 in male Fischer rats compared to aflatoxin B1. Cancer Res 47:1913–1917

Darwish WS, Ikenaka Y, Nakayama SMM, Ishizuka M (2014) An overview on mycotoxin contamination of foods in Africa. J Vet Med Sci 76:789–797. https://doi.org/10.1292/jvms.13-0563

Duarte SC, Almeida AM, Teixeira AS et al (2013) Aflatoxin M1 in marketed milk in Portugal: assessment of human and animal exposure. Food Control 30:411–417. https://doi.org/10.1016/j.foodcont.2012.08.002

Dutton MF, Mwanza M, de Kock S, Khilosia LD (2012) Mycotoxins in South African foods: a case study on aflatoxin M1 in milk. Mycotoxin Res 28:17–23. https://doi.org/10.1007/s12550-011-0112-9

Egyptian-Regulations (1990) Maximum limits for mycotoxin in foods. Part L Aflatoxins, Egypt

El Marnissi B, Belkhou R, Morgavi DP et al (2012) Occurrence of aflatoxin M1 in raw milk collected from traditional dairies in Morocco. Food Chem Toxicol 50:2819–2821. https://doi.org/10.1016/j.fct.2012.05.031

Elgerbi AM, Aidoo KE, Candlish AAG, Tester RF (2004) Occurrence of aflatoxin M 1 in randomly selected North African milk and cheese samples. Food Addit Contam 21:592–597. https://doi.org/10.1080/02652030410001687690

Ellen M, Anthony B, Deirdre M (2013) Chapter 1. Introduction. In: Muehlhoff E, Bennett A, McMahon D (eds) Milk and dairy products in human nutrition. FAO, Rome, pp 1–10

Elsanhoty RM, Salam SA, Ramadan MF, Badr FH (2014) Detoxification of aflatoxin M1 in yoghurt using probiotics and lactic acid bacteria. Food Control 43:129–134. https://doi.org/10.1016/j.foodcont.2014.03.002

El-Sayed Abd Alla A, Soher EA, Neamat-Allah AA (2002) Human exposure to mycotoxins in Egypt. Mycotoxin Res 18:23–30. https://doi.org/10.1007/BF02946136

El-Sayed A, Wahba S, Sayed Ahmed K, Bostany E (1998) Aflatoxins in breast milk, cord blood and sera of pregnant wome. Role of food consumption habits. Gaz Gypt Paed 46:165–176

El-Sayed AMAA, Neamat-Allah AA, Soher EA (2000) Situation of mycotoxins in milk, dairy products and human milk in Egypt. Mycotoxin Res 16:91–100. https://doi.org/10.1007/BF02946108

El-Tras WF, El-Kady NN, Tayel AA (2011) Infants exposure to aflatoxin M1 as a novel foodborne zoonosis. Food Chem Toxicol 49:2816–2819. https://doi.org/10.1016/j.fct.2011.08.008

Elzupir AO, Elhussein AM (2010) Determination of aflatoxin M1 in dairy cattle milk in Khartoum State, Sudan. Food Control 21:945–946. https://doi.org/10.1016/j.foodcont.2009.11.013

Elzupir AO, Abas A, Rouf A, Fadul MH et al (2012) Aflatoxin M1 in breast milk of nursing Sudanese mothers. Mycotoxin Res 28:131–134. https://doi.org/10.1007/s12550-012-0127-x

European Commission (2006) Commission Regulation (EC) No 1881/2006 of 19 December 2006 setting maximum levels for certain contaminants in foodstuffs

Ezekiel CN, Oyeyemi OT, Oyedele OA et al (2018) Urinary aflatoxin exposure monitoring in rural and semi-urban populations in Ogun state, Nigeria. Food Addit Contam A 35:1565–1572. https://doi.org/10.1080/19440049.2018.1475752

Falade T (2018) Aflatoxin management strategies in sub-Saharan Africa. In: Mycotoxins – socio-economic and health impact as well as pre- and postharvest management strategies [working title]. IntechOpen

Fandohan P, Zoumenou D, Hounhouigan DJ et al (2005) Fate of aflatoxins and fumonisins during the processing of maize into food products in Benin. Int J Food Microbiol 98:249–259. https://doi.org/10.1016/j.ijfoodmicro.2004.07.007

FAO (2004) Worldwide regulations for mycotoxins in food and feed in 2003. Food and Agriculture Organization of the United Nations. FAO Food and Nutrition Paper, 81

FAO/WHO (2019) General standard for contaminants and toxins in food and feed. Codex Standard 193-1995. Revised in 2019

FDA (2005) Compliance policy guide, CPG Sec. 527.400 whole milk, lowfat milk, skim milk e Aflatoxin M1, U.S. Food and Drug Administration

Flores-Flores ME, Lizarraga E, López de Cerain A, González-Peñas E (2015) Presence of mycotoxins in animal milk: a review. Food Control 53:163–176. https://doi.org/10.1016/j.foodcont.2015.01.020

Gizachew D, Szonyi B, Tegegne A et al (2016) Aflatoxin contamination of milk and dairy feeds in the Greater Addis Ababa milk shed, Ethiopia. Food Control 59:773–779. https://doi.org/10.1016/j.foodcont.2015.06.060

Hatem NL, Hassab HMA, Abd Al-Rahman EM et al (2005) Prevalence of aflatoxins in blood and urine of Egyptian infants with protein-energy malnutrition. Food Nutr Bull 26:49–56. https://doi.org/10.1177/156482650502600106

Hendrickse RG, Coulter JB, Lamplugh SM et al (1982) Aflatoxins and kwashiorkor: a study in Sudanese children. BMJ 285:843–846. https://doi.org/10.1136/bmj.285.6345.843

IARC (2002) Some traditional herbal medicines, some mycotoxins, naphthalene and styrene/IARC working group on the evaluation of carcinogenic risks to humans. WHO/IARC, Lyon

IARC (2015) Mycotoxin control in low and middle income countries. International Agency for Research on Cancer, Lyon

Ismail A, Akhtar S, Levin RE et al (2015) Aflatoxin M1: prevalence and decontamination strategies in milk and milk products. Crit Rev Microbiol 42:1–10. https://doi.org/10.3109/1040841X.2014.958051

James A, Zikankuba VL (2018) Mycotoxins contamination in maize alarms food safety in sub-Sahara Africa. Food Control 90:372–381. https://doi.org/10.1016/j.foodcont.2018.03.018

Jebali R, Abbès S, Ben Salah-Abbès J et al (2015) Ability of Lactobacillus plantarum MON03 to mitigate aflatoxins (B_1 and M_1) immunotoxicities in mice. J Immunotoxicol 12:290–299. https://doi.org/10.3109/1547691X.2014.973622

JECFA (2011) Joint FAO/WHO Expert Committee on Food Additives. Safety evaluation of certain mycotoxins in food. Series N°47. Geneva

Jolly P, Jiang Y, Ellis W et al (2006) Determinants of aflatoxin levels in Ghanaians: sociodemographic factors, knowledge of aflatoxin and food handling and consumption practices. Int J Hyg Environ Health 209:345–358. https://doi.org/10.1016/j.ijheh.2006.02.002

Jonsyn FE, Maxwell SM, Hendrickse RG (1995) Ochratoxin A and aflatoxins in breast milk samples from Sierra Leone. Mycopathologia 131:121–126. https://doi.org/10.1007/BF01102890

Kagera I, Kahenya P, Mutua F et al (2018) Status of aflatoxin contamination in cow milk produced in smallholder dairy farms in urban and peri-urban areas of Nairobi County: a case study of Kasarani sub county, Kenya. Infect Ecol Epidemiol 9:1547095. https://doi.org/10.1080/20008686.2018.1547095

Kang'ethe E, Lang'a KA (2009) Aflatoxin B1 and M1 contamination of animal feeds and milk from urban centers in Kenya. Afr Health Sci 9:218–226

Kang'ethe EK, M'Ibui GM, Randolph TF, Lang'at AK (2007) Prevalence of aflatoxin M1 and B1 in milk and animal feeds from urban smallholder dairy production in Dagoretti Division, Nairobi, Kenya. East Afr Med J 84:S83–S86

Kang'ethe EK, Gatwiri M, Sirma AJ et al (2017) Exposure of Kenyan population to aflatoxins in foods with special reference to Nandi and Makueni counties. Food Qual Saf 1:131–137. Ministry of Agriculture Livestock and Fisheries, Veterinary Laboratories Kabete, Private Bag Kabete. https://doi.org/10.1093/fqs/fyx011

Khaddor M, Tantaoui-Elaraki A, Benajiba A et al (2003) Destruction de l'aflatoxine M1 par les bactéries lactiques du lben marocain et du yaourt. Bull Soc Pharm Bord 142:101–112

Kiarie G, Dominguez-Salas P, Kang'ethe S et al (2016) Aflatoxin exposure among young children in urban low-income areas of Nairobi and association with child growth. Afr J Food Agric Nutr Dev 16:10967–10990. https://doi.org/10.18697/ajfand.75.ILRI02

Kibugu JK, Mburu D, Munga LK et al (2019) Food-borne mycotoxin hazards in the Kenyan market-a retrospective study. bioRxiv. https://doi.org/10.1101/773747

Kimanya ME (2015) The health impacts of mycotoxins in the eastern Africa region. Curr Opin Food Sci 6:7–11. https://doi.org/10.1016/j.cofs.2015.11.005

Kirino Y, Makita K, Grace D, Lindahl J (2016) Survey of informal milk retailers in Nairobi, Kenya and prevalence of aflatoxin M1 in marketed milk. Afr J Food Agric Nutr Dev 16:11022–11038. https://doi.org/10.18697/ajfand.75.ILRI05

Kuboka MM, Imungi JK, Njue L et al (2019) Occurrence of aflatoxin M1 in raw milk traded in peri-urban Nairobi, and the effect of boiling and fermentation. Infect Ecol Epidemiol 9:1625703. https://doi.org/10.1080/20008686.2019.1625703

Langat G, Tetsuhiro M, Gonoi T et al (2016) Aflatoxin M1 contamination of milk and its products in Bomet County, Kenya. Adv Microbiol 06:528–536. https://doi.org/10.4236/aim.2016.67053

Lindahl JF, Kagera IN, Grace D (2018) Aflatoxin M1 levels in different marketed milk products in Nairobi, Kenya. Mycotoxin Res 34:289–295. https://doi.org/10.1007/s12550-018-0323-4

Lu P, Wang J, Zhang Q et al (2010) Longitudinal study of aflatoxin exposure in the development of primary liver cancer in patients with chronic hepatitis. Zhonghua Yi Xue Za Zhi 90:1665–1669

Luongo D, Russo R, Balestrieri A et al (2014) In vitro study of AFB 1 and AFM 1 effects on human lymphoblastoid Jurkat T-cell model. J Immunotoxicol 11:353–358. https://doi.org/10.3109/1547691X.2013.848250

Magoha H, Kimanya M, De Meulenaer B et al (2014) Association between aflatoxin M1 exposure through breast milk and growth impairment in infants from Northern Tanzania. World Mycotoxin J 7:277–284. https://doi.org/10.3920/WMJ2014.1705

Mahmood Fashandi H, Abbasi R, Mousavi Khaneghah A (2018) The detoxification of aflatoxin M1 by lactobacillus acidophilus and Bifidobacterium spp.: a review. J Food Process Preserv 42: e13704. https://doi.org/10.1111/jfpp.13704

Makun H, Apeh D, Mwanza M et al (2016) Aflatoxin M1 in breast milk, cow milk and milk products in Minna, Nigeria and their predisposing factors. Biochem Anal Biochem 5:1–6. https://doi.org/10.4172/2161-1009.1000303

Marchese S, Polo A, Ariano A et al (2018) Aflatoxin B1 and M1: biological properties and their involvement in cancer development. Toxins 10:214. https://doi.org/10.3390/toxins10060214

Misihairabgwi JM, Ezekiel CN, Sulyok M et al (2019) Mycotoxin contamination of foods in Southern Africa: a 10-year review (2007–2016). Crit Rev Food Sci Nutr 59:43–58. https://doi.org/10.1080/10408398.2017.1357003

Mohammed S, Munissi JJE, Nyandoro SS (2016) Aflatoxin M1 in raw milk and aflatoxin B 1 in feed from household cows in Singida, Tanzania. Food Addit Contam B 9:85–90. https://doi.org/10.1080/19393210.2015.1137361

Mohammedi-Ameur S, Dahmane M, Brera C et al (2020) Occurrence and seasonal variation of aflatoxin M1 in raw cow milk collected from different regions of Algeria. Vet World 13:433–439. https://doi.org/10.14202/vetworld.2020.433-439

Mokhles M, Abd El Wahhab MA, Tawfik M et al (2007) Detection of aflatoxin among hepatocellular carcinoma patients in Egypt. Pak J Biol Sci 10:1422–1429. https://doi.org/10.3923/pjbs.2007.1422.1429

Motawee M (2013) Reduction of aflatoxin M1 content during manufacture and storage of Egyptian domaiti cheese. Int J Vet Med Res Rep 2013:1–11. https://doi.org/10.5171/2013.207299

Motawee MM, McMahon DJ (2009) Fate of aflatoxin M1 during manufacture and storage of feta cheese. J Food Sci 74:T42–T45. https://doi.org/10.1111/j.1750-3841.2009.01158.x

Motawee MM, Bauer J, McMahon DJ (2009) Survey of aflatoxin M1 in cow, goat, buffalo and camel milks in Ismailia-Egypt. Bull Environ Contam Toxicol 83:766–769. https://doi.org/10.1007/s00128-009-9840-3

Mulunda M, Mike D (2014) Occurrence of aflatoxin M1 from rural subsistence and commercial farms from selected areas of South Africa. Food Control 39:92–96. https://doi.org/10.1016/j.foodcont.2013.11.011

Mutegi CK, Cotty PJ, Bandyopadhyay R (2018) Prevalence and mitigation of aflatoxins in Kenya (1960-to date). World Mycotoxin J 11:341–357. https://doi.org/10.3920/WMJ2018.2362

Mwanza M, Abdel-Hadi A, Ali AM, Egbuta M (2015) Evaluation of analytical assays efficiency to detect aflatoxin M1 in milk from selected areas in Egypt and South Africa. J Dairy Sci 98:6660–6667. https://doi.org/10.3168/jds.2014-9220

Njumbe Ediage E, Diana Di Mavungu J, Song S et al (2013) Multimycotoxin analysis in urines to assess infant exposure: a case study in Cameroon. Environ Int 57–58:50–59. https://doi.org/10.1016/j.envint.2013.04.002

Nleya N, Adetunji M, Mwanza M (2018) Current status of mycotoxin contamination of food commodities in Zimbabwe. Toxins 10:89. https://doi.org/10.3390/toxins10050089

Nyathi CB, Mutiro CF, Hasler JA, Chetsanga CJ (1987) A survey of urinary aflatoxin in Zimbabwe. Int J Epidemiol 16:516–519. https://doi.org/10.1093/ije/16.4.516

Nyathi CB, Mutiro CF, Hasler JA, Chetsanga CJ (1989) Human exposure to aflatoxins in Zimbabwe. Cent Afr J Med 35:542–545

Ogunbanwo B (2005) Incidence of mycotoxins in local and processed foods marketed in Nigeria. Laboratory training on mycotoxin contamination of agricultural commodities and processed foods. In: National Agency for Food and Drug Administration and Control (NAFDAC)/International Atomic Energy Agency (IAEA) Regional Workshop. Lagos, Nigeria

Okeke H, Abdullahi I, Makun H, Mailafiya S (2012) A preliminary survey of aflatoxin M1 in dairy cattle products in Bida, Niger State, Nigeria. Afr J Food Sci Technol 3:273–276

Okoth S (2016) Improving the evidence base on aflatoxin contamination and exposure in Africa. CTA, Wageningen, pp 1–113

Oluwafemi F (2012) Aflatoxin M1 levels in lactating mothers in two Nigerian cities. Arch Clin Microbiol 3:1–6. https://doi.org/10.3823/257

Oluwafemi F, Badmos AO, Kareem SO et al (2014) Survey of aflatoxin M1 in cows' milk from free-grazing cows in Abeokuta, Nigeria. Mycotoxin Res 30:207–211. https://doi.org/10.1007/s12550-014-0204-4

Oyeyemi O, Ezekiel C, Ayeni K et al (2018) A pilot biomonitoring study of bladder tumor antigen (BTA) in aflatoxin exposed Nigerian villagers. Afr J Urol 24:152–156. https://doi.org/10.1016/j.afju.2018.02.003

Oyeyipo OO, Oyeyipo FM, Ayah IR (2017) Aflatoxin (M1 and B1) contamination of locally repacked Milk powder in South-Western Nigeria. Sokoto J Med Lab Sci 2:175–187

Peng Z, Chen L, Zhu Y et al (2018) Current major degradation methods for aflatoxins: a review. Trends Food Sci Technol 80:155–166. https://doi.org/10.1016/j.tifs.2018.08.009

Piekkola S, Turner PC, Abdel-Hamid M et al (2012) Characterisation of aflatoxin and deoxynivalenol exposure among pregnant Egyptian women. Food Addit Contam A 29:962–971. https://doi.org/10.1080/19440049.2012.658442

Pierides M, El-nezami H, Peltonen K et al (2000) Ability of dairy strains of lactic acid bacteria to bind aflatoxin M1 in a food model. J Food Prot 63:645–650. https://doi.org/10.4315/0362-028X-63.5.645

Polychronaki N, Turner PC, Mykkänen H et al (2006) Determinants of aflatoxin M 1 in breast milk in a selected group of Egyptian mothers. Food Addit Contam 23:700–708. https://doi.org/10.1080/02652030600627222

Polychronaki N, West RM, Turner PC et al (2007) A longitudinal assessment of aflatoxin M1 excretion in breast milk of selected Egyptian mothers. Food Chem Toxicol 45:1210–1215. https://doi.org/10.1016/j.fct.2007.01.001

Polychronaki N, Wild CP, Mykkänen H et al (2008) Urinary biomarkers of aflatoxin exposure in young children from Egypt and Guinea. Food Chem Toxicol 46:519–526. https://doi.org/10.1016/j.fct.2007.08.034

Purchase IFH (1967) Acute toxicity of aflatoxins M1 and M2 in one-day-old Ducklings. Food Cosmet Toxicol 5:339–342. https://doi.org/10.1016/S0015-6264(67)83060-8

Rahmani J, Alipour S, Miri A et al (2018) The prevalence of aflatoxin M1 in milk of Middle East region: a systematic review, meta-analysis and probabilistic health risk assessment. Food Chem Toxicol 118:653–666. https://doi.org/10.1016/j.fct.2018.06.016

Redouane-Salah S, Morgavi DP, Arhab R et al (2015) Presence of aflatoxin M1 in raw, reconstituted, and powdered milk samples collected in Algeria. Environ Monit Assess 187:375. https://doi.org/10.1007/s10661-015-4627-y

Salem D (2002) Natural occurrence of aflatoxin in feedstuffs and milk of dairy farm in Assuit Province, Egypt. Wien Tierarztl Monatsschr 89:86–91

Šarkanj B, Ezekiel CN, Turner PC et al (2018) Ultra-sensitive, stable isotope assisted quantification of multiple urinary mycotoxin exposure biomarkers. Anal Chim Acta 1019:84–92. https://doi.org/10.1016/j.aca.2018.02.036

Senerwa D, Sirma A, Mtimet N et al (2016) Prevalence of aflatoxin in feeds and cow milk from five counties in Kenya. Afr J Food Agric Nutr Dev 16:11004–11021. https://doi.org/10.18697/ajfand.75.ILRI04

Sirma AJ, Lindahl JF, Makita K et al (2018) The impacts of aflatoxin standards on health and nutrition in sub-Saharan Africa: the case of Kenya. Glob Food Sec 18:57–61. https://doi.org/10.1016/j.gfs.2018.08.001

Smith LE, Mbuya MNN, Prendergast AJ et al (2017) Determinants of recent aflatoxin exposure among pregnant women in rural Zimbabwe. Mol Nutr Food Res 61:1601049. https://doi.org/10.1002/mnfr.201601049

Stewart RJ, Choga O, Chiriseri B, Pfukenyi DM (2016) Detection and levels of aflatoxin M_1 in raw milk of dairy cows from selected small scale and commercial farms in Harare, Zimbabwe. Zimb Vet J 34:1–6

Tantaoui-Elaraki A, Riba A, Oueslati S, Zinedine A (2018) Toxigenic fungi and mycotoxin occurrence and prevention in food and feed in northern Africa – a review. World Mycotoxin J 11:385–400. https://doi.org/10.3920/WMJ2017.2290

Tchana A, Moundipa P, Tchouanguep F (2010) Aflatoxin contamination in food and body fluids in relation to malnutrition and cancer status in Cameroon. Int J Environ Res Public Health 7:178–188. https://doi.org/10.3390/ijerph7010178

Tomerak RH, Shaban HH, Khalafallah OA, El Shazly MN (2011) Assessment of exposure of Egyptian infants to aflatoxin M1 through breast milk. J Egypt Public Health Assoc 86:51–55. https://doi.org/10.1097/01.EPX.0000399138.90797.40

Udomkun P, Mutegi C, Wossen T et al (2018) Occurrence of aflatoxin in agricultural produce from local markets in Burundi and Eastern Democratic Republic of Congo. Food Sci Nutr 6:2227–2238. https://doi.org/10.1002/fsn3.787

WHO (1998) GEMS/food regional diets: regional per capita consumption of raw and semi-processed agricultural commodities/prepared by the Global Environment Monitoring System/ Food Contamination Monitoring and Assessment Programme (GEMS/Food). Food Safety Department. WHO, Geneva

WHO (2001) Evaluation of certain mycotoxins in foodifty-sixth report of the Joint FAO/WHO Expert Committee on Food Additives. World Health Organization. WHO Technical Report Series. Geneva

WHO (2002) Evaluation of certain mycotoxins in food. Fifty-sixth report of the Joint FAO/WHO Expert Committee on Food Additives. World Health Organization - Technical Report Series 906

WHO (2006) Mycotoxins in African foods: Implication to food safety and health. AFRO Food Safety Newsletter, World Health Organisation Food Safety (FOS)

Wild CP, Pionneau FA, Montesano R et al (1987) Aflatoxin detected in human breast milk by immunoassay. Int J Cancer 40:328–333. https://doi.org/10.1002/ijc.2910400308

Zakaria A, Amin Y, Khalil O et al (2019) Rapid detection of aflatoxin M1 residues in market milk in Aswan Province, Egypt and effect of probiotics on its residues concentration. J Adv Vet Anim Res 6:197. https://doi.org/10.5455/javar.2019.f332

Zarba A, Wild CP, Hall AJ et al (1992) Aflatoxrn M 1 in human breast milk from The Gambia, West Africa, quantified by combined monoclonal antibody immunoaffinity chromatography and HPLC. Carcinogenesis 13:891–894. https://doi.org/10.1093/carcin/13.5.891

Zinedine A, Mañes J (2009) Occurrence and legislation of mycotoxins in food and feed from Morocco. Food Control 20:334–344. https://doi.org/10.1016/j.foodcont.2008.07.002

Zinedine A, González-Osnaya L, Soriano JM et al (2007) Presence of aflatoxin M1 in pasteurized milk from Morocco. Int J Food Microbiol 114:25–29. https://doi.org/10.1016/j.ijfoodmicro. 2006.11.001

In Vivo and In Vitro Toxicity Testing of Cyanobacterial Toxins: A Mini-Review

Samaneh J. Porzani, Stella T. Lima, James S. Metcalf, and Bahareh Nowruzi

Contents

S. J. Porzani · B. Nowruzi (✉)
Department of Biology, Science and Research Branch, Islamic Azad University, Tehran, Iran
e-mail: samanehbiology@gmail.com; bahareh.nowruzi@srbiau.ac.ir

S. T. Lima
Center for Nuclear Energy in Agriculture, University of Sao Paulo, Piracicaba, Brazil
e-mail: sdelimacamargo@gmail.com

J. S. Metcalf
Brain Chemistry Labs, Jackson, WY, USA
e-mail: james@ethnomedicine.org

P. de Voogt (ed.), *Reviews of Environmental Contamination and Toxicology Volume 258*,
Reviews of Environmental Contamination and Toxicology 258,
https://doi.org/10.1007/398_2021_74

Abstract Harmful cyanobacterial blooms are increasing and becoming a worldwide concern as many bloom-forming cyanobacterial species can produce toxic metabolites named cyanotoxins. These include microcystins, saxitoxins, anatoxins, nodularins, and cylindrospermopsins, which can adversely affect humans, animals, and the environment. Different methods to assess these classes of compounds in vitro and in vivo include biological, biochemical, molecular, and physicochemical techniques. Furthermore, toxic effects not attributable to known cyanotoxins can be observed when assessing bloom material. In order to determine exposures to cyanotoxins and to monitor compliance with drinking and bathing water guidelines, it is necessary to have reliable and effective methods for the analysis of these compounds. Many relatively simple low-cost methods can be employed to rapidly evaluate the potential hazard. The main objective of this mini-review is to describe the assessment of toxic cyanobacterial samples using in vitro and in vivo bioassays. Newly emerging cyanotoxins, the toxicity of analogs, or the interaction of cyanobacteria and cyanotoxins with other toxicants, among others, still requires bioassay assessment. This review focuses on some biological and biochemical assays (MTT assay, Immunohistochemistry, Micronucleus Assay, *Artemia salina* assay, *Daphnia magna* test, Radionuclide recovery, Neutral red cytotoxicity and Comet assay, Enzyme-Linked Immunosorbent Assay (ELISA), Annexin V-FITC assay and Protein Phosphatase Inhibition Assay (PPIA)) for the detection and measurement of cyanotoxins including microcystins, cylindrospermopsins, anatoxin-a, saxitoxins, and nodularins. Although most bioassay analyses often confirm the presence of cyanotoxins at low concentrations, such bioassays can be used to determine whether some strains or blooms of cyanobacteria may produce other, as yet unknown toxic metabolites. This review also aims to identify research needs and data gaps concerning the toxicity assessment of cyanobacteria.

Keywords Biological and biochemical methods · Cyanobacteria · Cyanotoxins · Toxicity tests · Toxins

Abbreviations

ATX	Anatoxin-a
BMAA	Beta-*N*-methylamino-l-alanine
BNCs	Binucleated cells
CBPI	Cytokinesis-block proliferation index
DAB	2,4-diaminobutyric acid
FISH	Fluorescence in situ hybridization
HBE	Human bronchial epithelial cells
HPLC	High performance liquid chromatography
i.p.	Intraperitoneal injection
i.v.	Intravenous injection
IHC	Immunohistochemical techniques
LC/MS/MS	Liquid chromatography tandem mass spectrometry
LOEC	Lowest observed effect concentration
MC-YR, LR, RR	Microcystins-YR, -LR and -RR
MNBC	Micronucleated binucleated cells
MNi	Micronuclei
MTT assay	3-(4,5-dimethylthiazol-2-yl)-2,5-diphenyltetrazolium bromide
NODs	Nodularins
NR	Neutral Red
PP2A	Protein phosphatase 2A
PPIA	Protein phosphatase inhibition assay
ROS	Reactive oxygen species
SCGE	Single-cell gel electrophoresis
SRB	Sulforhodamine B
STX	Saxitoxin

1 Introduction

Cyanobacteria can be characterized through their biodiversity, distribution, and ability to produce a wide array of secondary metabolites, often with bioactivity in certain test systems (Nowruzi et al. 2018b, 2019, 2020b). A wide range of cyanobacteria are natural toxin producers, which can adversely affect human and animal health, in addition to potentially affecting the environment (Nowruzi et al. 2012, 2013; Valério et al. 2010). Furthermore, with the distribution of cyanobacteria in coastal waters and the discovery of new marine species, cyanotoxins produced by these organisms may accumulate in or contaminate zooplankton and coastal edible aquatic vertebrates and invertebrates such as fish, mussels, and other crustaceans that may not be routinely tested for these compounds (Buratti et al. 2017; Nowruzi and Porzani 2021).

Numerous cases of lethal poisonings have been associated with cyanotoxin ingestion in wild animals and livestock exposed to toxic cyanobacterial blooms, especially *via* drinking water containing these cyanotoxins (Buratti et al. 2017). Although repeated/chronic exposure to low cyanotoxin concentrations still requires further attention and is a critical issue, a few episodes of acute human poisonings have been recorded (Buratti et al. 2017). Humans may be exposed orally to cyanotoxins through contaminated drinking water, consumption of cyanotoxin-containing freshwater fish, crops, vegetables, and food supplements, or by ingesting water during recreational activities (Funari and Testai 2008). Dermal and inhalation exposure may also occur during recreational, athletic, or professional activities (i.e., wind surfing) in cyanotoxin contaminated water bodies (Fig. 1).

Global change due to anthropological impacts, such as climate change, nutrient over-enrichment, and hydrological alteration of global water paths, may lead to the proliferation and persistence of cyanobacterial harmful blooms (Burford et al. 2020). Microcystin-producing *Microcystis* have spread into several water bodies, in the USA and in China, where many aquaculture plants are located with possible negative consequences on economic activities, such as fisheries production (Funari et al. 2012; Manganelli et al. 2012; Preece et al. 2015). In many countries, citizens are reliant on stored drinking water (sometimes desalinated water) in reservoirs. Such water bodies, in addition to recreational areas, drinking water supplies, irrigation water in dams and tanks, groundwater wells, hot springs, and rainwater pools, can be consistent hot spots for toxic cyanobacteria, and is a potential critical issue for these areas (Buratti et al. 2017; Chatziefthimiou et al. 2014, 2016; Costa et al. 2006; Mohamed 2008; Mohamed and Al Shehri 2007, 2009; Mohamed et al. 2016). In fact, extended dry periods with evaporation exceeding precipitation can favor the accumulation of brackish groundwater and increase lake salinity, potentially favoring the proliferation of toxic cyanobacteria, as evidenced in subtropical Australia with *Nodularia spumigena* (McGregor et al. 2012). The recent colonization of many subalpine Italian lakes by *Dolichospermum lemmermannii* is another interesting example of the effects of complex relations between environmental factors affected by climate change (Callieri et al. 2014). Some episodes of acute/short-term human intoxications due to consumption of contaminated water have been reported in some countries as a consequence of failure or inefficiency of water treatment management (Funari and Testai 2008) and recently in the USA after recreational activities (Buratti et al. 2017; Hilborn et al. 2014).

Recent studies have reported that global climate change may also favor hazardous cyanobacterial species, by increasing their growth rate (Malik et al. 2020), in addition to other factors that affect toxin production in cyanobacteria, including growth phase, high nutritional inputs such as nitrogen (N) and phosphorus (P), and environmental factors such as pH, light, and temperature (Paerl et al. 2001). Cyanotoxins are an important group of chemical compounds, from the perspective of structural diversity, ecotoxicology, toxicology, and environmental chemistry (Cullen et al. 2019). These molecules can be categorized into two main criteria: their mechanism of action, especially in mammals, and chemical structure. For the first, they can be divided into three principal classes: hepatotoxins, neurotoxins, and dermatotoxins (Woese 1987). For the second, structurally they can be classified as cyclic peptides, alkaloids, lipopolysaccharides (LPS), or organophosphates (Cohen

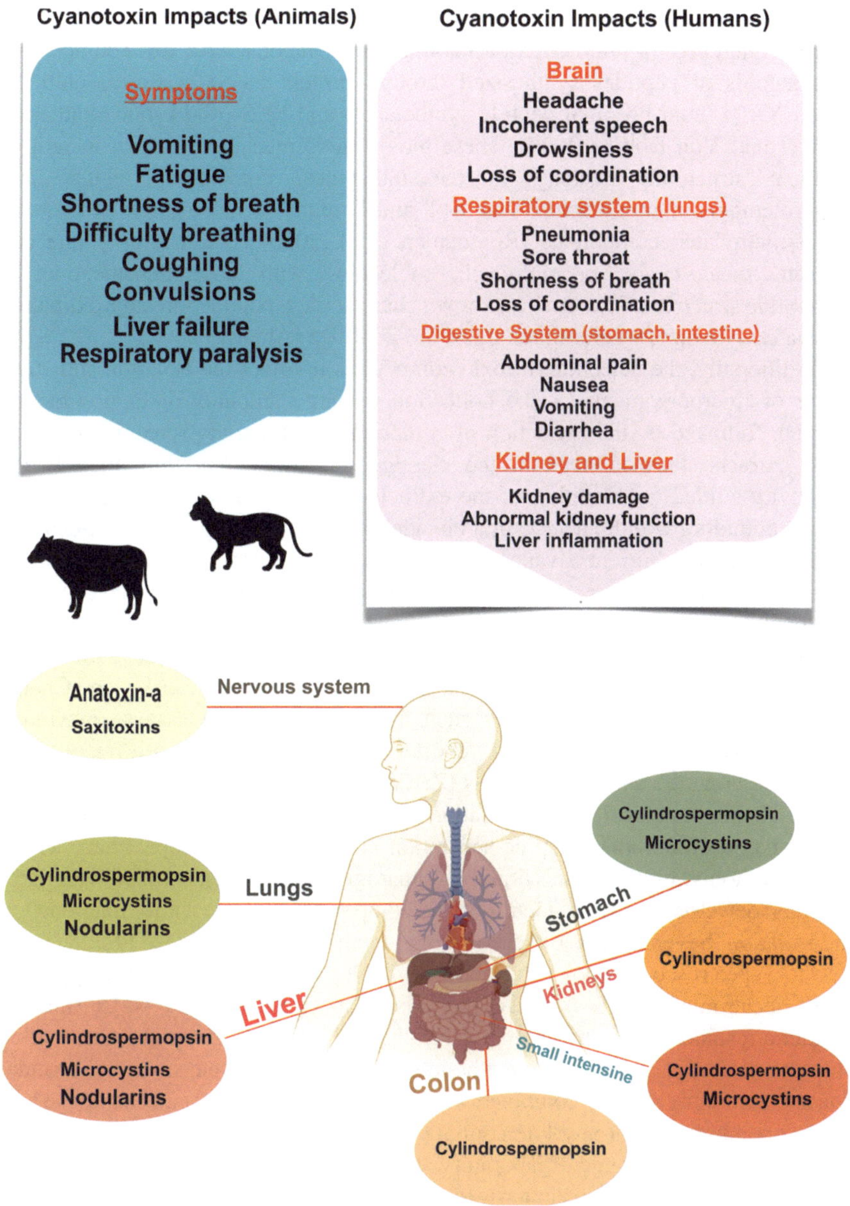

Fig. 1 Health impacts of cyanobacterial toxins on animal and human (tinnitus, headache, eyepain, vertigo, deafness, blindness, and convulsions) as well as different target of cyanotoxins on human organs such as lung, liver, stomach, kidney, intestine, and nervous system

and Gurevitz 2006; Ferrão-Filho and Kozlowsky-Suzuki 2011; Fiore et al. 2020; Zanchett and Oliveira-Filho 2013). Predominantly, cyanobacterial toxic compounds are alkaloids or peptides synthesized through protein based carrier assemblies, named NRPs (nonribosomal peptide synthetases) and PKS (polyketide synthases) (Welker and Von Döhren 2006). These biosynthetic machines give an incredible chemical structural diversity to the molecules, comprising regular and non-proteinogenic amino acids, such as β and D amino acids, as well as hydroxy and N-methylated compounds (Rouhiainen et al. 2010). The assembly line can generate linear, cyclic, or multicyclic molecules, with basic, depsipeptide, or lipopeptide structures. The smallest known characterized peptides are approximately 300 Da (Da) in size, whereas the largest are 2000 Da (Almesjö 2007).

Traditionally, the detection of toxic compounds involves the isolation and cultivation of microorganisms or the collection of large amounts of environmental material, followed by the extraction of compounds and activity screening of these crude extracts. Further steps in the classical discovery workflow include the bioactivity-guided fractionation of the extracts and the subsequent purification of the compound(s), with final structural elucidation and characterization (Potterat and Hamburger 2006). Current advances in technology have led to the implementation of high-throughput screening, where entire crude extract libraries are screened and fractionation is automated (Potterat and Hamburger 2006). In order to consider cyanobacteria, however, as a source of high-throughput drug leads, the initial toxicity screen of cyanobacterial extracts needs to be performed (Zerikly and Challis 2009) and standardized techniques such as cellular imaging assays, enzymatic inhibition assays, other cell-based reporter assays, as well as cell-free assays, could accelerate the discovery process (Zerikly and Challis 2009).

The chemo-analytical characterization of novel molecules, including toxins, is mainly based on chromatographic separation and their spectroscopic characterization. Commonly, high pressure liquid chromatography (HPLC), coupled with spectroscopic detectors, is used to identify toxic compounds of interest after fractionation and isolation. Several forms of spectroscopic methods, including DAD (diode array detector) spectroscopy, UV-vis (ultraviolet-visible), IR (infrared spectroscopy), MS (mass spectrometry), CD (circular dichroism) spectroscopy, and NMR (nuclear magnetic resonance) spectroscopy, enable the isolation and broad characterization of toxic molecules. The last two methodologies enable complete structural elucidation and configuration, and combined with X-ray crystallography can enable the best optimal combination for absolute configuration (Potterat and Hamburger 2006).

There is a diverse range of laboratory strategies for the detection and determination of cyanobacterial toxins in terrestrial and aquatic samples in the laboratory and in the field (Nowruzi et al. 2018a). These strategies can differ significantly in terms of the complexity of the data they supply. Relatively easy and inexpensive ways can be used for rapid risk assessment and permit the selection of management actions (Nowruzi et al. 2020a; Safavi et al. 2019). If necessary, extremely specific and sensitive analytical techniques are then applied to verify and quantify the cyanotoxins present. Procedures are selected, based on the facilities and experience available, in addition to the kind of information required. As an example, for recreational activities, there may be an urgent need to detect potentially harmful cyanobacteria to understand the risk of the presence of likely cyanotoxins and potential adverse effects to human and animal life (Christensen and Khan 2020). Other factors that are important include selectivity and sensitivity, as these may affect what strategies are selected. Hence, the main purpose of this review is the application and importance of different assays for the assessment of the toxicity of cyanobacterial toxins (Rajabpour et al. 2019).

The methods covered in this review include immunohistochemistry, MTT assay ((3-(4,5-dimethylthiazol-2-yl)-2,5-diphenyltetrazolium bromide), micronucleus assay (Micronuclei (MNi) assay), *A. salina* assay, *Daphnia magna* test, neutral red cytotoxicity, Comet assay, ELISA and PPIA assay, which are categorized into biological, biochemical, and physicochemical screening assays (Table 1). Generally, these biological assays differ in terms of the basis of detection, information regarding the toxicity of the sample and the assay simplicity/complexity (Table 2).

2 Radionuclide Recovery In Vivo

Techniques using radioisotopes often use competition in tests for evaluating toxins. For example, the radiolabeled protein phosphatase inhibition assay has been applied in water monitoring with a detection limit <0.1 μg/L (Xu et al. 2000). The drawbacks of the method are related to the need to produce labeled compounds with ^{32}P and its short half-life (Nicholson and Burch 2001), the cost of radioactive ATP and enzymes used, difficulties in commercially obtaining the ^{32}P isotope and strict regulations for laboratories to work with radioactive substances, although colorimetric reagents have removed many of the issues concerned with radioactivity. However, such radioactive reagents may be applied to locate where the toxins can be adsorbed (Koreivienė and Belous 2012). Although this method provides accurate quantitative results, some microcystin variants do not react with protein phosphatase enzymes to a similar extent and may affect quantification (Corbel et al. 2016; D'Anglada et al. 2015; Hyenstrand et al. 2003).

2.1 *Toxicity of Microcystin-LR In Vivo*

Microcystins (MC) are a class of cyclic peptides that typically contain seven amino acids. These heptapeptides are serine/threonine protein phosphatase inhibitors, which are synthesized by nonribosomal multidomain enzymatic complexes encoded by the microcystin biosynthetic gene cluster, *mcy* (Börner and Dittmann 2005). In rats and mice, the hepatotoxic microcystin causes rapid centrilobular hepatocyte necrosis, with breakdown of the sinusoidal endothelium, intrahepatic hemorrhage, and death. Robinson and coworkers (Robinson et al. 1990) found that [^{3}H] microcystin-LR is an excellent tool to investigate the toxicology of microcystins.

Table 1 Examples of biochemical assays used for the assessment of cyanotoxins from cyanobacteria

Cyanobacterial species	Method	Toxin	Outcome	References
Microcystis aeruginosa	Radionuclide recovery	Radiolabeled dihydro-MC-LR	Adsorption of toxin in cytosolic fraction, plasma membrane and rat livers and bile	Brooks and Codd (1987); Hooser et al. (1991); Pace et al. (1991)
Microcystis strain PCC-7820		^{14}C-labeled microcystin	Presence of toxin in liver of mice	Robinson et al. (1990)
Cylindrospermopsis raciborskii		^{14}C-labeled cylindrospermopsin	Existence of radiolabel in the solvent column	Norris et al. (2001)
Microcystis aeruginosa	Immunohistochemistry	Microcystin-LR	Staining intensity in the center-lobular regions due to hemorrhage and apoptosis; Accumulation of toxin in macrophages, livers of *Oncorhynchus mykiss*	Djediat et al. (2010); Yoshida et al. (1998)
Aphanizomenon ovalisporum		Cylindrospermopsin	Liver, bladder, intestine, and gills are the most immunoreactive organ	Guzmán-Guillén et al. (2014)
Anabaena spiroides, Microcystis aeruginosa and *Anabaena flos-aquae*	*Daphnia magna* test	Guanitoxin	Suppressed feeding levels in *Daphnia magna*	Costa (2005); Freitas et al. (2014); Ghadouani et al. (2004)
Microcystis aeruginosa		Microcystin	Producing reactive oxygen species (ROS) and induces oxidative stress enzymes and subsequently death	DeMott et al. (1991); Jungmann and Benndorf (1994); Okumura et al. (2007); Takenaka (2007); Wiegand et al. (2002)
Cylindrospermopsis raciborskii	*Artemia salina* assay	Cylindrospermopsin	Toxic to *A. salina*	Hawkins et al. (1997); Metcalf et al. (2002)
Anabaena sp.		Anatoxin-a	*Artemia* larvae mortality	Lahti et al. (1995)
Microcystis sp., *Anabaena* sp.		Microcystin	*Artemia* larvae mortality	Campbell et al. (1994)

Microcystis aeruginosa	Micronucleus assay	Microcystin-LR	Carcinogenic effect of microcystin	Abramsson-Zetterberg et al. (2010); Dias et al. (2014)
Cylindrospermopsis raciborskii		Cylindrospermopsin	Infliction of cytogenetic harm	Humpage et al. (2000)
Cylindrospermopsis raciborskii		Cylindrospermopsin	Metabolic activation in mediating its toxicity and indicates proteotoxicity of Cylindrospermopsin	Bazin et al. (2010)
Nodularia spumigena		Nodularin	Induction of oxidative DNA damage	Cheng et al. (1998); Lankoff et al. (2003); Lankoff et al. (2006a, b, c); Minshull et al. (1996); Sontag (2001); Žegura et al. (2011a, b)
Microcystis aeruginosa	Cytotoxicity (neutral red) assay	Microcystin	Toxicity assessment	Puerto et al. (2009)
Cylindrospermopsis raciborskii		Cylindrospermopsin	Decrease in protein content of CYN	Gutiérrez-Praena et al. (2011)
Microcystis aeruginosa	Comet assay	Microcystin	DNA-damage induced in the liver, kidney, intestine, and colon	Mankiewicz et al. (2002); Tice et al. (2000); Žegura et al. (2008)
Cylindrospermopsis raciborskii		Cylindrospermopsin	Induction of genotoxic effects	Humpage et al. (2005)
Cylindrospermopsis raciborskii	Annexin V-FITC assay	Cylindrospermopsin	CYN is not clastogenic in CHO-K1 cells and decreases the frequencies of mitotic indices, decreasing proliferation irrespective of metabolic activation system	Lankoff et al. (2007)
Microcystis aeruginosa, *Anabaena* sp.		Microcystin and Anatoxin-a	Apoptotic and necrotic cells assessment	Teneva et al. (2005)
Microcystis aeruginosa		Microcystin	Inhibition of cell viability and evoking HBE cell apoptosis	Li et al. (2016)
Nodularia spumigena		Nodularin	Genetic alterations and carcinogenesis	Lankoff et al. (2006a, b, c)

(continued)

Table 1 (continued)

Cyanobacterial species	Method	Toxin	Outcome	References
Cylindrospermopsis raciborskii	MTT assay	Cylindrospermopsin	Inhibition of progesterone production	Young et al. (2008)
Cylindrospermopsis raciborskii		Cylindrospermopsin	Toxicity of C3A and HepG2 cell lines	Froscio et al. (2009)
Microcystis aeruginosa, Cylindrospermopsis raciborskii		Cylindrospermopsin, microcystin, Anatoxin-a and beta-N-methylamino-l-alanine	Inflammation and apoptosis in immune and brain cells at low doses	Takser et al. (2016)
Microcystis aeruginosa		Microcystin-LR, −RR, and -LF	Cytotoxic effects on kidney cells, apoptosis	Piyathilaka et al.(2015)
Nodularia spumigena		Nodularin	Moderate cytotoxic effects on kidney cells, apoptosis	Manage (2019); Piyathilaka et al. (2015)
Microcystis aeruginosa, Nodularia spumigena	ELISA	Microcystins, nodularin	Toxin detection	Fischer et al. (2001); Spoof et al. (2003)
Microcystis aeruginosa		Microcystin	Toxin detection	Preece et al. (2015)
Aphanizomenon spp., Anabaena circinalis, Cylindrospermopsis raciborskii, Planktothrix spp., and *Lyngbya wollei*		Saxitoxins	Toxin detection	Kleinteich et al. (2013)
Microcystis aeruginosa	PPIA	Microcystin- LR	Toxin detection	Heresztyn and Nicholson (2001)
Microcystis aeruginosa		Microcystin	Screening hepatotoxins	Kaloudis et al. (2013)
Microcystis aeruginosa, Nodularia spumigena		Microcystin and nodularin	Toxin detection	Triantis et al. (2010); Wharton et al. (2019)

Table 2 Uses of bioassays for the detection of cyanotoxins *in vitro* and *in vivo*

Cyanotoxins	Anatoxins	Cylindrospermopsins	Microcystins	Saxitoxins	Nodularins
Biological assays					
Artemia salina assay	✓ In vivo	✓ In vivo	✓ In vivo	✓ In vivo	
Protein phosphatase inhibition assays (PPIA)			✓ In vitro		✓ In vitro
Micronucleus assay		✓ In vitro	✓ In vivo		✓ In vivo
Enzyme-linked immunosorbent assays (ELISA)			✓ In vivo ✓ In vitro	✓ In vitro	✓ In vivo ✓ In vitro
MTT assay	✓ In vitro	✓ In vivo ✓ In vitro	✓ In vitro		
Immunohistochemistry		✓ In vivo	✓ In vivo		
Daphnia magna test	✓ In vivo ✓ In vitro		✓ In vivo ✓ In vivo		
Radionuclide recovery		✓ In vivo	✓ In vivo		
Neutral red cytotoxicity		✓ In vivo	✓ In vitro		
Comet assay		✓ In vitro	✓ In vivo		
Annexin V-FITC assay	✓ In vitro	✓ In vitro	✓ In vitro		✓ In vitro

They measured the distribution, excretion and hepatic metabolism of [^{3}H] microcystin-LR (sublethal i.v.) in mice and revealed that all of the radiolabel was associated with cytosolic components bound covalently within 1 day (83 ± 5%), although by day 6 the amount of covalently bound isotope had decreased (42 ± 11%) (Robinson et al. 1990). Studies using radiolabeled microcystins have shown that it is rapidly taken up by the liver. When [^{125}I]-labeled microcystin-YM was administrated to rats by intravenous injection (i.v.), 20% of the radioactivity was found in the liver within 30 min, with no more than 6% located in any other organ and confirmed the studies on rapid adsorption of radiolabeled dihydro-microcystin-LR in fresh hepatocytes from perfused liver under normal conditions (Falconer 1986).

Studies of Hooser and collaborators (Hooser et al. 1991) with purified MC-LR in rats and mice showed that it is specifically toxic to the liver and this toxicosis is characterized by rapid, progressive necrosis. Moreover, studies in perfused rat liver (Pace et al. 1991) showed that bile formation decreased shortly after exposure of the liver to a toxic dose of microcystin and intraperitoneal injection (i.p.) of [^{14}C]-microcystins to mice showed that 70% of the radioactivity was localized in the liver after 1 min, and 90% after 180 min (Brooks and Codd 1987).

2.2 *Toxicity of Cylindrospermopsins In Vivo*

Cylindrospermopsins (CYN) are a class of hepatotoxins produced by different cyanobacterial species, including *Cylindrospermopsis raciborskii*, *Umezakia natans,* and *Aphanizomenon ovalisporum.* Hawkins and collaborators demonstrated the toxicity of an intraperitoneal injection of an aqueous sodium chloride extract of C. *raciborskii* (Hawkins et al. 1985), and further toxicological studies in mice were performed with oral administration of culture suspensions (Seawright et al. 1999). The toxicity of pure CYN administered i.p. was reported by Terao and co-workers (Terao et al. 1994) and the average lethal dose of CYN administrated by i.p. in mice was approximately 0.2 mg/kg (Ohtani et al. 1992). All of these studies confirmed that the liver was the principal organ affected. The liver became pale and swollen, and hepatocellular cytoplasmic vacuolation, typical of fat shift and varying degrees of periacinal necrosis were observed by histology (Seawright et al. 1999). Norris and collaborators (Norris et al. 2001) studied the dissemination or toxin depletion in vivo in Quackenbush male mice utilizing radiolabeled [^{14}C]-cylindrospermopsins. They found that the excretion patterns showed substantial interindividual variability between predominantly fecal or urinary excretion and were not related to toxicity.

3 Immunochemistry

Immunochemical strategies are often helpful for the detection of cyanotoxins. Antibodies are currently the foremost used reagents, largely due to the high sensitivity of these components in assays for cyanotoxin analysis. However, before their use, the antibodies used in a particular immunochemical assay system need to be defined and tested with different samples and toxins (Weller 2013).

3.1 Toxicity of Microcystin In Vivo

Correlating the occurrence of intralobular sites and the distribution of microcystin-LR (MC-LR) was evaluated via an immunohistochemical technique using a monoclonal antibody against MC-LR in mouse liver. Results showed a perceptible staining intensity in the centrilobular regions, which could occur with hemorrhage and apoptosis. Liver research indicated, however, that in the MC-LR-PP1/PP2A adducts free MC-LR was below detection limits. Such results suggested that the immunohistochemical localization of MC-LR in centrilobular liver cells is strongly related to the incidence of hemorrhage and cell death, and subjected to adduct development, notwithstanding the occurrence of coagulative necrosis that can often be linked with different causes, such as ischemic hypoxia (Yoshida et al. 1998). Chakib Djediat et al. (2010) reported the analysis of the presence of MC-LR in contaminated fish tissues using immunohistochemical methods. Immunohistochemistry of contaminated medaka fish showed that the animals had been severely affected by MC-LR. In this scenario, liver necrosis was undoubtedly the primary cause of death. Accumulation of MC-LR was also observed in the spleen in macrophages (Djediat et al. 2010). In another study, the effect of toxin exposure and MC-LR pathology was evaluated in the livers of yearling rainbow trout with MC-gavage (*Oncorhynchus mykiss*), through immunohistochemical identification and MC-specific antibodies (Tencalla and Dietrich 1997). The immunohistochemical analysis indicated MC accumulation and subsequent changes in cell morphology. Protein phosphatase inhibition and hepatocyte necrosis were significant points in the microcystin-induced hepatotoxicity and appeared to be related to non-covalent binding of MCs with protein phosphatases (Fischer et al. 2000; Tencalla and Dietrich 1997).

3.2 Toxicity of Cylindrospermopsins In Vivo

An investigation and comparison of the distribution of CYN (isolated from *Aphanizomenon ovalisporum*) in various organs of tilapia (*Oreochromis niloticus*) was carried out exploring, in this case, two exposure routes – administration of CYN via intraperitoneal injection or via the oral route. For the second route, fish were exposed by immersion in lyophilized CYN-producing *A. ovalisporum* cells, using different exposure times and concentrations. The results showed that the liver, bladder, intestine, and gills were the most immunoreactive organs. The results indicate that IHC procedures are a powerful tool for researching CYN distribution (Guzmán-Guillén et al. 2014).

4 Daphnia Magna Test

Daphnids are excellent representatives of the Cladocera, an essential and keystone group of organisms in freshwater systems and aquatic foodwebs. These organisms are also commonly used in toxicity assays (Okumura et al. 2007) and are a model organism when considered as an ecotoxicology tool for aquatic environments, allowing the analysis of potentially toxic secondary metabolites, including cyanotoxins (Ferrão-Filho et al. 2009; Logan 1997). This ecotoxicological test is stated in guidelines by water management agencies worldwide for biomonitoring and determining the impacts of toxic compounds on freshwaters, often using *Daphnia* species with pure toxins or extracts.

4.1 Toxicity of Microcystin In Vivo

In acute toxicity tests, Okumura and coworkers used a *Microcystis aeruginosa* (NPLJ-4) crude extract, finding an LC_{50} of 230 mg L^{-1} after 48 h for *Daphnia similis* (Okumura et al. 2007). Microcystins, including MC - LR, can have a remarkable effect on producing reactive oxygen species (ROS) and inducing oxidative stress enzymes, ultimately resulting in death of D. *magna* (Wiegand et al. 2002). In another study, after 4 h, LC_{50} values for MC-LR ranged from 9,600 to 21,400 mg MC-LR L^{-1} for some *Daphnia* species (DeMott et al. 1991; Jungmann and Benndorf 1994). Increasing the dosage of MC-LR from 1286.9 to 1463.3 mg MC-LR g^{-1} decreased the probability of *Daphnia magna* longevity by 50%, and sublethal concentrations above 62.5 mg L^{-1} or 375.6 mg MC-LR g^{-1}, the Lowest Observed Effect Concentration Value (LOEC), decreased the feeding frequency (Okumura et al. 2007; Takenaka 2007). The adsorption of microcystins via the gut and their detrimental consequences on *Daphnia galeata* have been also studied. Several characteristics such as gut shape, blood microcystin concentration and the response of *Daphnia* fed with *Microcystis* PCC 7806 or *Microcystis* PCC 7806 microcystin-knock out mutant have been studied. Microcystins rapidly spread into the blood after rupture of the gut epithelial tissue, irritating the midgut muscles and decreasing foregut function, with low amounts of microcystin (10.2–18.3 ng) being sufficient to kill *D. galeata* within 2 days (Rohrlack et al. 2005). DeMott et al. (1991) showed a decrease in clearance rates of *Microcystis aeruginosa* and *Anabaena flos-aquae* for two species of *Daphnia*. Ghadouani et al. (2004) also noticed an irreversible decline in the level of appendage beat and mandibular motion, and stimulation of the rate of labral rejection when D. *pulicaria* was exposed to purified MC-LR and a non-toxic alga (Freitas et al. 2014; Ghadouani et al. 2004).

4.2 Toxicity of Anatoxin-a In Vivo

Anatoxin-a (ATX) is a neurotoxin and neuromuscular blocking agent that acts on the cholinergic system, and at high concentrations, can cause fatal asphyxia in animals after injection or ingestion, inducing oxidative stress, and inhibiting the growth of some aquatic macrophytes and phytoplankton (Chia et al. 2019). Currently, few experiments have been carried out on the impact of anatoxin-a on zooplanktonic species. Costa et al. (2006) performed acute toxicity assays for anatoxin-a isolated from *Anabaena spiroides* on *D. magna*. They observed that extracts obtained from freeze-dried cells had slight effects on the Cladocerans *Daphnia gessneri*, *Daphnia pulex,* and *Moina micrura.* Over the 24-h period of exposure, the neurotoxic extracts suppressed feeding levels of *Daphnia magna* by nearly 50% (Costa et al. 2006).

5 Artemia Salina Assay

The *A. salina* bioassay is an attractive and commonly used invertebrate biotest. As a research organism, one benefit of *A. salina* is that it is easy to hatch nauplii and mature *Artemia* from dried eggs commercially available at aquarium stores. The eggs are large enough to be seen with the naked eye and tiny enough to be stored in a limited volume of liquid (Eppley 1974; Kiviranta et al. 1991). The standard *A. salina* bioassay is a reliable toxicity-based cyanotoxin detection system and may be an alternative for the mouse bioassay due to ethical concerns. In the case of CYN, use of the *A. salina* bioassay, which is often more sensitive than the mouse bioassay, could lead to lower experimental costs since the unit costs is less than an in vitro protein synthesis assay (Eaglesham et al. 1999; Metcalf et al. 2002). By comparing this bioassay with responses to purified toxins and sufficient physicochemical methods can provide clear identification and toxicity evaluation.

5.1 Toxicity of Cylindrospermopsins In Vivo

Metcalf et al. (2002) investigated the toxicity of Cylindrospermopsin, microcystin-LR, and *Microcystis* extracts on *A. salina* populations. Classical protein synthesis inhibitors and *C. raciborskii* isolates containing different amounts of Cylindrospermopsins were also assayed. The results showed that the toxic effects of the C. *raciborskii* cultures to A. *salina* can be inferred by this toxin, although additional toxic factors may be present in some cyanobacterial extracts. This possibility arises since protracted toxic effects in mice have been observed when dosed with *Aphanizomenon* sp. and *Cylindrospermopsis* sp. (Hawkins et al. 1997). Metcalf et al. (2002) found that Cylindrospermopsin was slightly less toxic than microcystin-LR but more toxic to *A. salina* than the inhibitors of protein synthesis cycloheximide, chloramphenicol, and tetracycline. Morover, the LC_{50} concentration for *Cylindrospermopsis raciborskii* over time was greater than the LC_{50} for purified cylindrospermopsin, with the exception of *C. raciborskii* strain CR1 (Metcalf et al. 2002).

5.2 Toxicity of Anatoxin-a In Vivo

For the assessment of the toxicity of cyanobacterial blooms in raw water and recreational water sources, quick, simple, and easy-to-use bioassays are needed. In order to detect the toxicity of cyanobacteria, Kiviranta and collaborators (Kiviranta et al. 1991) used a basic *A. salina* bioassay. *Artemia* brine shrimp larvae were tested with hepatotoxic and neurotoxic bloom samples and isolates of cyanobacteria. Several laboratory isolates caused mortality to *A. salina* even though they were non-toxic to mice (Kiviranta et al. 1991). Easy-to-use and straightforward tests were performed with *A. salina* (brine shrimp) larvae, luminescent bacteria and *Pseudomonas putida* by Lahti et al. (1995) to investigate and compare the toxicity of cyanotoxins. The *Artemia* assay appeared to be very sensitive to cyanobacterial hepatotoxins and was shown to be effective for the identification of anatoxin-a-induced toxicity. Although the fractionated extract of anatoxin-a was not lethal to *Artemia*, the neurotoxin could disrupt the larvae's ability to move further and filter cyanobacteria cultures resulting in anatoxin-a triggered *Artemia* larvae mortality (Lahti et al. 1995).

5.3 Toxicity of Microcystin In Vivo

Use of the *A. salina* bioassay has been advocated for the initial screening of cyanobacterial blooms for microcystin hepatotoxins, based on brine shrimp mortality determined with microcystin-containing and non-microcystin-containing cyanobacteria. Campbell et al. (1994) analyzed the response of the brine shrimp assay to extracts of MC-LR-containing cyanobacterial isolates, bloom samples of *Microcystis*, extracts of *Anabaena* bloom material, and a laboratory isolate of *Microcystis elabens* M-177 which does not produce microcystins. In all cases, the extracts were fractionated before testing so that responses to the fractions could be tentatively assigned to the presence of microcystins. This approach clearly showed that the brine shrimp assay responded to fractions from microcystin-containing and non-microcystin-containing cyanobacteria (Campbell et al. 1994).

5.4 *Toxicity of Saxitoxin In Vivo*

Paralytic shellfish toxins (PSTs), such as saxitoxin (STX), are molecules considered natural products produced by freshwater cyanobacteria and marine dinoflagellates, which have a high affinity for voltage-gated sodium channels (VGSCs; Lukowski et al. 2020). Saxitoxin-producing cyanobacterial genera include *Aphanizomenon flos-aquae*, *Lyngbya wollei,* and *Cylindrospermopsis raciborskii* (Nicholson and Burch 2001). The saxitoxins are a relatively complex toxin class with widely differing levels of toxicity among the variants. They can be divided into three groups based on the net charge of the molecule under acidic conditions: saxitoxin (STX), neosaxitoxin (neoSTX) including decarbamoyl derivatives (charge +2;), the gonyautoxins (GTXs) including decarbamoyl derivatives (charge +1) and C toxins (charge 0). Park et al. (1986) exposed *A. salina* larvae to STX. Approximately 30 larvae were exposed to different concentrations of STX, resulting in dead or moribund shrimp once administrated (Park et al. 1986). The results showed that *A. salina* was sensitive to saxitoxin at doses of 5 μg after 48 h (Park et al. 1986).

6 Micronucleus Assay

Micronuclei represent fragments of chromosomes or whole chromosomes that at mitosis are not integrated into the main nuclei and thus occur only in cells that have undergone a nuclear division. The simplicity of scoring and the wide applicability of the in vitro micronucleus test in different cell types make it an attractive tool to assess cytogenetic abnormality. However, the limited understanding of the induction mechanisms of micronuclei and the absence of adequate methods to assess concurrently cell proliferation and micronucleus induction delayed the validation of the in vitro micronucleus test for in vitro genotoxicity testing (Fenech and Morley 1986). The method applied to cyanobacteria is statistically more precise than the conventional micronucleus method, since for a particular number of cells scored, number of micronuclei observed was twice than observed in daughter nucleus cells (Fenech and Morley 1986).

In order to contribute to the database concerning the suggested guideline concentration for MC-LR in drinking water, Lilianne Abramsson-Zetterberg and coworkers (Abramsson-Zetterberg et al. 2010) reported the results from some short-term tests using the micronucleus assay in vivo and in vitro. The flow cytometry-based micronucleus assay in the mouse was used to increase the sensitivity (Abramsson-Zetterberg et al. 2010).

Both cyanobacterial extracts originating from four different lakes in Sweden and pure MC-LR were evaluated. Although, doses up to near lethality were used and an average of 200,000 young erythrocytes, polychromatic erythrocytes, were analyzed from each male CBA mice. In in vitro micronucleus analysis using human lymphocytes, no genotoxic effect was detected, nor could any effect be demonstrated

(Abramsson-Zetterberg et al. 2010). These results showed that low concentrations of microcystins in drinking water did not increase the cancer risk through chromosome breaks or mal-distribution of chromosomes. This technique commonly uses pure toxins applied to animal cells but can also be an efficient method for toxin detection in freshwater samples.

6.1 Toxicity of Microcystin-LR In Vivo

Many reports have shown that MC-LR induces DNA damage in liver cells in vivo and in distinct cells in vitro (Dias et al. 2014; Ding et al. 1999; Lankoff et al. 2004; Mankiewicz et al. 2002; Nong et al. 2007; Rao et al. 1998; Žegura et al. 2003, 2004, 2006, 2008). Nevertheless, the process behind the DNA breakage identified is not reliable and appears to be dose- and cell-type based. In addition, experiments showing activation of micronuclei (MNi) in vivo (Ding et al. 1999) or in vitro (Zhan et al. 2004) have suggested a lasting chromosome damaging effect for microcystins. In contrast with that, in several cell models (Abramsson-Zetterberg et al. 2010; Lankoff et al. 2006b; Žegura et al. 2011a), some authors recorded no effect of MC-LR on the micronucleus size, in line with the negative findings of chromosome abnormalities (Dias et al. 2014). Lankoff and collaborators (Lankoff et al. 2006a, b, c) were the first to show that MC-LR attenuates the nucleotide excision repair (NER) system. They pre-treated CHO-K1 (Chinese hamster ovary) cells with MC-LR (1, 10, and 20 mg/ml) for 6 h, followed by exposure to UV (25 J/m^2) radiation (Lankoff et al. 2006b). Assessing the DNA repair kinetics with the alkaline comet assay after 0, 30, 60, and 120 min of recovery, they confirmed that MC-LR did not induce DNA damage. In cells treated with 10 and 20 mg/mL MC-LR prior to UV-irradiation, the level of residual DNA damage was after 120 min of recovery, significantly higher when compared to cells exposed to UV only. These results suggested that MC-LR inhibited the incision/excision phase as well as the rejoining phase of NER (Lankoff et al. 2006b).

Evaluation of MC-LR genotoxicity in cytokinesis-blocked Vero-E6 and HepG2 cells was conducted with the micronucleus test. The two tested concentrations of MC-LR have been able to considerably increase the frequency of micronucleated binucleated (MNBC) cells in both cell lines. For Vero-E6 cells a substantial 1.8-fold increase was observed for the maximum toxin concentration, and a 2.1-fold increase for HepG2 cells was also observed. The proportion of mono-, bi-, correlated multinucleated cells was measured for each treatment status in order to see the cytokinesis-block proliferation index (CBPI), which provides an indication of toxicity. The proportion of binuclear cells in Vero-E6 and HepG2 colonies ranged from 45–56% to 64%, respectively. No difference was observed between MC-LR-treated and untreated cells, regardless of the dosage (Fenech 2000; Parry et al. 2002). The calculable cytokinesis-block proliferation index (CBPI) was 1.5–1.6 for Vero-E6 cells and 1.6–1.7 for HepG2 cells, and no change was observed with the MC-LR treatment, proposing that the levels of toxins evaluated do not interfere with the cell cycle's conventional progression, resulting in no impact on cell viability (Dias et al. 2014; Rosefort et al. 2004; Zegura et al. 2009).

From these studies we can conclude that one pathway of MCs genotoxic activity is mediated by induction of ROS formation, which causes DNA strand breaks and mutagenic oxidative DNA lesions. In addition, MC-LR has been shown to inhibit two pathways of DNA repair, NER and the DSB repair by NHEJ, which is in line with other experiments that demonstrated that PP1 and PP2A inhibition significantly decreases the activity of the DNA repair systems (Ariza et al. 1996; Khodarev et al. 2001). The impairment of DNA repair is along with DNA damage a very important factor involved in carcinogenesis. If the repair process is compromised or the lesion is repaired by some other error-prone mechanism, damaged DNA can cause mutation and ultimately contributes to neoplastic transformation (Žegura et al. 2011b).

6.2 Toxicity of Cylindrospermopsins In Vitro

The calculation of micronuclei frequency next to the detection of apoptotic cells solely through morphological analysis is becoming more and more accepted. This provides a lot of accurate cellular susceptibility recognition for genotoxins (Abend et al. 1995; Fenech 1997) and may result in some of the benefits that the MN assay provides to the area, such as the wide number of analyzable cells, the flexibility of the procedure, the possible precision, and even the opportunity to discover aneugens (Llana-Ruiz-Cabello et al. 2015; Nesslany and Marzin 1999). Humpage et al. (2000) used the WIL2-NS lymphoblastoid cell line in a well-validated cytokinesis-block micronucleus study to assess CYN's carcinogenic activity (Humpage et al. 2000). The findings showed that CYN caused a substantial increase in micronuclei in binucleated cells (BNCs) subjected to different concentrations, and a significant increase in centromere-positive micronuclei at all concentrations studied for CYN. The results indicated that CYN works to inflict cytogenetic harm by two pathways, one at the DNA level to trigger strand splits, and the other at the kinetochore/spindle level to generate a loss of whole chromosomes (aneuploidy). As *C. raciborskii* occurs worldwide in several human drinking water sources, these research results may have significant public health consequences (Humpage et al. 2000).

6.3 Toxicity of Nodularin In Vivo

The hepatotoxin nodularin (NOD) is commonly detected in cyanobacterial blooms containing *Nodularia spumigena* and isolates of this species from blooms (Baker and Humpage 1994; Bolch et al. 1999; Sivonen et al. 1989). The toxicity of *N. spumigena* to animals and humans is a growing concern as the occurrence of harmful cyanobacterial blooms increases. Although nodularin could be a liver carcinogen, possessing both initiating and tumor-promoting activities, the mechanisms by which this toxin damages DNA and induces liver cancer are still unknown (Meili et al. 2016). The clastogenic effects of NOD were investigated in CHO-K1 cells and HepG2 cells by Lankoff and coworkers (Lankoff et al. 2006a). An accumulated frequency of micronuclei was found in CHO-K1 cells incubated with NOD, whereas MC-LR induced solely a small impact. In HepG2 cells, no increase in micronuclei formation was ascertained and the accumulated frequency of micronuclei was conjointly detected at concentrations >2.5 mg/mL. These findings revealed that NOD possesses excessive robust clastogenic behavior that causes aneuploidy relative to MCs. When the clastogenicity of NOD related to MCs was investigated, it should be considered that MCs have not yet been tested for their aneugenic behavior (Cheng et al. 1998; Lankoff et al. 2003, 2006c; Minshull et al. 1996; Sontag 2001; Žegura et al. 2011b).

7 Cytotoxicity (Neutral Red) Assay

The use of neutral red as a cytotoxic indicator in assays has developed from an infectious agent cytopathogenicity test (Borenfreund and Puerner 1985). Although neutral red (NR) may be a poor cationic dye intended to reach the cell by non-ionic diffusion across the cell membrane, the pigment then accumulates in the living cells' lysosomes (Borenfreund and Puerner 1985). The test consists of two parts: (a) screening of cells under an inverted microscope with phase optics, to detect morphological alterations; and (b) a quantitative spectrophotometric procedure, based on the incorporation of neutral red into the lysosomes of viable cells (Borenfreund and Puerner 1985). This simple and sensitive procedure, based on both morphologic and spectrophotometric criteria, has been developed to facilitate the determination of surviving viable cells upon termination of experimental procedures. The particular advantage of this assay is the application to monolayer cultures, thus circumventing the need for trypsinization and use of the trypan blue exclusion test. This assay, suitable for screening cytotoxic agents, has been standardized and adapted for use with an automated spectrophotometric microplate reader. This technique is commonly used with pure toxins exposed to cells.

7.1 Toxicity of Microcystin In Vitro

The impact of MC-LR on the Caco-2 cell line was studied with differentiated and non-differentiated cells lines (Puerto et al. 2010). No increased sensitivity was detectable with the differentiated Caco-2 cell line, although the impact after 48 h of exposure compared to 24 h was apparent in undifferentiated cells. Similarly, EC_{50} values for neutral red uptake in differentiated cells were not discernibly different from those obtained in undifferentiated Caco-2 cells. Still, the results found for neutral red uptake after 48 h in both types of cells were more pronounced (Puerto et al. 2010). The toxicity of one of the most common MC variants, MC-LR, is well known, although studies dealing with MC-RR and MC-YR are less reported. Thus, toxic effects of MC-RR and MC-YR at various concentrations were examined in either undifferentiated or differentiated cells by Puerto and collaborators in a human colon cancer cell line Caco-2 (Puerto et al. 2009). Results revealed that protein content was the most sensitive endpoint for MC-RR, with reductions of 45% in differentiated cells after 48 h of exposure to 200 mM MC-RR. Regarding MC-YR, in undifferentiated cells, inhibition of neutral red uptake was observed with reductions of more than 80% at 100 mM. Moreover, the morphological studies indicated changes within the cells, especially at high concentrations, experiencing a general decrease in cell diversity and hydropic degeneration. Over the course of the treatment, the sensitivity of cultures to these contaminants was exceptionally low and, to a lesser degree, MC-YR demonstrated greater toxicity than MC-RR (Puerto et al. 2009).

7.2 *Toxicity Detection of Cylindrospermopsins In Vivo and In Vitro*

Both in vivo and in vitro studies have been performed with CYN to elucidate the toxic effects of this cyanotoxin. In vivo studies in mice suggest that the liver is a major target organ, and CYN has traditionally been classified as a hepatotoxic cyanotoxin (De Figueiredo et al. 2004). However, the heart, thymus, spleen, lungs, kidneys, and ovaries can also be affected (Young et al. 2008). Chong and coworkers (Chong et al. 2002) demonstrated that the uptake of CYN into hepatocytes may involve more than one mechanism (Chong et al. 2002). Cell culture-based toxicity tests have the potential to screen samples for a biochemical response characteristic to the toxin class, while retaining the ability to detect more general cytotoxicity endpoints (Froscio et al. 2009). The toxicity of CYN can be detected via cell-based assays known to produce damage in a range of different cell types.

Freshly isolated hepatocytes were used as a sensitive in vitro model for CYN toxicity (Humpage et al. 2005; Pegram et al. 2008) with concentrations below micromolar range producing disruption of the cell membrane within 18–24 h. While the sensitivity of primary hepatocytes was hypothesized to be through CYP450 activation of CYN to toxic metabolites, CYN was also shown to display effects in cell lines considered to be less metabolically competent. In this sense, the cell line chosen by Humpage and collaborators (Humpage et al. 2005), PLHC-1, a hepatocellular carcinoma of the cyprinid fish *Poeciliopsis lucida*, has a hepatocellular origin. This permanent cell line, very significant in the pathogenicity of CYN, was also expected to have metabolic activity response and uptake capabilities, in addition to an inducible CYP450 activity, which can be inhibited (Humpage et al. 2005); (Babich et al. 1991). Moreover, PHLC-1 cells can be used in an in vitro system to screen environmentally relevant stressors using a combined stress protein and cytotoxicity assay (Ryan and Hightower 1994). Retention of hepatocyte properties has made this cell line ideal for in vitro toxicology assays.

PHLC-1 cells have been well characterized by environmental toxicologists, reporting that in vitro cytotoxicity experiments of some chemicals are positively correlated with the acute toxicity in vivo, and therefore cytotoxicity assays could be applied as an alternative for acute fish toxicity testing (Fent 2001). However, the scientific community does not have much knowledge about the toxic effects induced by the cyanotoxin CYN on fish. In this sense, this experimental model makes the PLHC-1 cell line suitable for studying the toxic effects caused by cyanobacteria, not only because they can be in contact with water, but also because cyanobacteria are an essential nutritional component of fish cichlids and cyprinids (Prieto et al. 2007; Zurawell et al. 2005). Another line of research for cytotoxicity assays are the inhibitory effect on protein and glutathione synthesis reported in vitro by Daniel Gutiérrez-Praena and coworkers (Gutiérrez-Praena et al. 2011). Research on the toxicity of CYN to fish, both in vivo and in vitro, has been carried out to determine whether CYN can be toxic to fish, causing damage to cells and oxidative stress, which can lead to pathogenicity. PLCH-1 cells from fish liver were exposed to different levels of CYN over time by Gutiérrez-Praena and collaborators. The results showed that the PLHC-1 cells experienced a concentration and time-dependent decrease in protein content (Gutiérrez-Praena et al. 2011).

The toxicity caused by CYN to aquatic organisms remains an area that still needs further research. Thus, cytotoxic assays using well-known cells lines could be a promising alternative to fill this important gap, especially as cyanobacterial harmful blooms may increase due to global climate change that can adversely affect human and animal health, in addition to the environment (Christensen and Khan 2020).

8 Comet Assay

A more useful approach for assessing DNA damage is the single-cell gel (SCG) or Comet assay. Since the introduction of the alkaline ($pH > 13$) Comet assay in 1988 (Tice et al. 2000), the number of applications and investigators using this technique has increased exponentially. Compared with other genotoxicity assays, the advantages of this technique include: (1) sensitivity for detecting low levels of DNA damage; (2) requirement for small numbers of cells per sample; (3) flexibility; (4) low cost; (5) ease of application; (6) ability to conduct studies using relatively small amounts of a tested substance; and (7) the relatively short time period (a few days) needed to complete an experiment (Tice et al. 2000). Attractive uses of this assay in genetic toxicology include: (a) a potentially high-throughput screening assay; (b) mechanism to distinguish genotoxicity and cytotoxicity induced chromosomal damage; (c) a mechanism in vivo to differentiate genotoxic and non-genotoxic carcinogens; and (d) the potential to be part of a set of in vitro/in vivo assays used for regulatory submissions. This assay has grown into a fundamental method for researchers involved in research areas ranging from human and environmental monitoring, looking for genetic toxicology and/or DNA repair processes. This technique typically relies on pure toxins and their application to cells.

8.1 *Toxicity of Microcystin In Vivo*

Using the comet assay Tice et al. (2000) observed that after i.p. administration of MC-LR, DNA lesions were mainly induced in the liver, kidney, intestine, and colon. The responsiveness of the i.p. route in contrast with the oral route revealed a difference in the toxin's bioavailability, with MC-LR causing DNA damage irrespective of the path of administration (Gaudin et al. 2008).

The alkaline version of the comet assay developed by Singh and coworkers (Singh et al. 1988) recognized genotoxic agents by detecting multiple DNA damage along with single-strand breaks, double-strand breaks (DSB), alkali-labile sites, cross-links of DNA–DNA and DNA–protein, and single-strand breaks linked to inadequate excision-repair locations (Singh et al. 1988). The comet assay's advantages included an ability to identify low rates of DNA injury, a limited number of cells needed per sample and the low cost. In addition, the in vivo procedure could be applied to any tissue of interest enabling identification of possible target organs (Tice et al. 2000).

Mankiewicz et al. (2002) compared the toxicity of cyanobacterial compounds with microcystins (mainly from the genus *Microcystis*) through the SOS chromotest with prokaryotic cells (*Escherichia coli*) and comet assay with human lymphocytes, all illustrating the possible genotoxic effect of cyanobacterial extracts with microcystins (CEMs). Extracts of cyanobacteria triggered in vitro DNA damage in human lymphocytes. The findings showed that CEMs from Polish water reservoirs were cytotoxic, once it was expressed in bacterial cells by the stimulation of the SOS repair system and in human lymphocytes triggered by DNA damage (Mankiewicz et al. 2002). Filipič and collaborators (Filipič et al. 2007) conducted lymphocyte, liver, kidney (cortex and medulla), heart, spleen, and brain cell destruction of male Fisher F344 rats exposed to a sublethal MC-YR dose, utilizing the comet assay. Subchronic exposure to sublethal levels of MCs was reported to induce systemic hepatotoxicity in mammals, impacting not just the liver but also other vital organs (Filipič et al. 2007).

The genotoxic results of cyanobacterial extracts (including *Leptolyngbya* sp., *Oscillatoria earlei*, *Phormidium* sp. and uncultured cyanobacterial strains) were established by Comet Assay with human lymphocyte cells. Interestingly, the lymphocyte cells treated with cyanobacterial extracts were around 8 or 19-fold less toxic to DNA than the ones treated with H_2O_2 (Andeden et al. 2018). If the ratio of tail length to head length is <5%, comet test findings indicated that there is no DNA disruption (Yen et al. 2001). This shows that cyanobacterial extracts do not cause major DNA damage when compared to controls treated with H_2O_2 (Andeden et al. 2018). Furthermore, the effect of non-cytotoxic concentrations of microcystin-LR on the production of intracellular reactive oxygen species has been assessed, examining

DNA damage in human colon adenocarcinoma CaCo-2, human astrocytoma IPDDC-A2 as well as human B-lymphoblastoid NCNC lines. Using the comet assay it was shown that MC-LR caused a time- and dose-dependent rise in DNA damage in CaCo-2 cells at non-cytotoxic concentrations, but not substantially in IPDDC-2A and NCNC cells. These findings showed that colon cells should also be seen as a priority for assessing microcystin toxicity, in addition to liver cells, hypothesizing that exposure to low concentrations of microcystins would impact the intestinal tissue as well (Žegura et al. 2008).

In a study by Ding and collaborators (Ding et al. 1999), the genotoxic potential of a microcystin extract from cyanobacteria isolated from a field sample (1.25, 12.5 and 125 mg lyophilized algae cells/ml) and pure MC-LR (1 mg/mL) was demonstrated in isolated primary rat hepatocytes (Ding et al. 1999). The extracts contained MC-LR at a concentration of 0.45 μg per mg of lyophilized cells, besides several other MC variants (Ding et al. 1999). Through the comet assay, after 4 h of exposure, a dose-dependent increase in DNA damage in cells exposed to the *Microcystis* extract and pure MC-LR was observed. However, the investigation of morphological changes to hepatocytes treated with the same concentrations of *Microcystis* extract or MC-LR revealed condensed and segregated nuclei, and most of the cells had shrunk, indicating severe cytotoxicity (Ding et al. 1999). As dead or dying cells can undergo rapid DNA fragmentation, which increases DNA migration in the comet assay, the observed DNA damage could be the consequence of cytotoxicity rather than genotoxicity. On the other hand, in the same study, in vivo assessment of the *Microcystis* extract in mice induced micronuclei formation in polychromatic erythrocytes (PCE) (Ding et al. 1999). Although pure MC-LR was not tested in vivo, the induction of free radical formation and mitochondrial alterations associated with MCs hepatotoxicity was shown (Ding et al. 1999).

In a human hepatoma cell line (HepG2 cells), MC-LR induced DNA strand breaks after exposure to 0.01, 0.1, and 1 mg/mL in a time- and dose-dependent manner (Žegura et al. 2003). The level of DNA strand breaks increased according to the time of incubation and reached a maximum after 4 h of exposure, and then, with further exposure gradually declined. Using the DNA repair inhibitors (cytosine beta-D-arabinofuranoside hydrochloride and hydroxyurea) it was shown that the MC-LR-induced DNA strand breaks were predominantly intermediate as a result of the cellular repair of DNA lesions (Žegura et al. 2003). Recently, it was shown that MC-LR, at non-cytotoxic concentrations (0.01, 0.1, 1, and 10 mg/mL) induced increasing DNA damage, in a time- and dose-dependent manner, as well as increased the formation of reactive oxygen species (ROS) in human colon adenocarcinoma (CaCo-2 cells), but not in the human B-lymphoblastoid cell line (NCNC cells) (Žegura et al. 2003).

In human lymphocytes, cyanobacterial extracts collected from the Sulejów water reservoir, in Poland, containing MCs (0.25–1 mM) and pure MC-LR (0.25–4 mM) induced DNA damage, with the maximum rate reached after 12 h of incubation (Mankiewicz et al. 2002). Pure MC-LR induced less DNA strand breaks than the extracts (Mankiewicz et al. 2002). Palus et al. (2007) obtained similar results when studying the toxic and genotoxic potential of cyanobacterial extracts collected in the

Sulejów reservoir during summer of 2004 (Palus et al. 2007). Higher cytotoxicity and genotoxicity of the extract was correlated with higher content of MCs within the cells (Palus et al. 2007). In another study with human lymphocytes, MC-LR (1, 10, and 25 mg/mL) induced DNA strand breaks, which reached the maximum rate after 18 h of exposure (Lankoff et al. 2004). The authors compared the extent of DNA damage examined by the comet assay to the frequency of apoptotic cells determined by the TUNEL method and found a significant positive correlation (Lankoff et al. 2004). Based on these findings, they concluded that the induced DNA damage caused by MC-LR observed in the comet assay may be related to the early stages of apoptosis, due to cytotoxicity but not genotoxicity (Lankoff et al. 2004). In the human hepatoma cell line (HepG2 cells), MC-LR (0.01, 0.1, and 1 mg/mL) induced the oxidation of pyrimidines and purines (Žegura et al. 2003, 2004). The oxidized bases were detected with the modified comet assay using the bacterial enzymes endonuclease III (endo III) and formamidopyrimidine DNA glycosylase (fpg), which catalyze the excision of oxidized pyrimidines and oxidized purines, respectively. DNA strand breaks and oxidized pyrimidines were repaired within a short time of exposure to MC-LR (8 h), while oxidized purines remained un-repaired and accumulated in the DNA (Žegura et al. 2004).

8.2 *Toxicity of Cylindrospermopsins In Vitro and In Vivo*

While CYN itself is a potent protein synthesis inhibitor, its acute toxicity appears to be mediated by cytochrome P450 (CYP450) generated metabolites. CYN also induces genotoxic effects both in vitro and in vivo, and preliminary evidence suggests that tumors are generated after oral exposure to CYN. To better understand the role of CYP450-activated CYN metabolites in in vitro genotoxicity, Humpage et al. (2005) quantified the process in primary mouse hepatocytes using the comet assay, both in the presence and absence of CYP450 inhibitors known to block acute CYN cytotoxicity (Humpage et al. 2005). At concentrations above 0.1 μM (EC_{50} = 0.5 μM), CYN was cytotoxic and resulted in substantial increases in Comet tail length, area and tail moment at 0.05 μM and above, concluding that genotoxicity due to toxicity is unlikely to be secondary to metabolic disruption. The CYP450 inhibitors omeprazole (100 μM) and SKF525A (50 μM) completely inhibited the genotoxicity induced by CYN. The cyanotoxin also inhibited production of glutathione (GSH) which could potentiate cytotoxicity, and by implication genotoxicity, via reduced reactive oxygen species (ROS) quenching (Humpage et al. 2005). The lipid peroxidation marker, malondialdehyde (MDA) was quantified in CYN-treated cells, and the effect of the reduced glutathione (GSSG) reductase (GSSG-rd.) inhibitor 1,3-bis(chloroethyl)-l-nitrosourea (BCNU) on both MDA production and lactate dehydrogenase (LDH) leakage was examined. MDA levels were not elevated by CYN treatment, and blockage of GSH regeneration by BCNU did not affect lipid peroxidation or cytotoxicity. It therefore seems likely that CYP450-

derived metabolites are responsible for both the acute cytotoxicity and genotoxicity induced by CYN (Humpage et al. 2005).

9 Annexin V-FITC Assay (Fluorescein Isothiocyanate Labeled Annexin V)

In the early stages of apoptosis changes occur at the cell surface, which until now have remained difficult to recognize. One of these plasma membrane alterations is the translocation of phosphatidylserine (PS) from the inner side of the plasma membrane to the outer layer, whereby PS becomes exposed at the external surface of the cell. Annexin V is a Ca^{2+}-dependent phospholipid-binding protein with high affinity for PS and can be used as a sensitive probe for PS exposure upon the cell membrane. The Annexin V assay offers the possibility of detecting early phases of apoptosis before damage to cell membrane integrity, allowing measurement of the kinetics of apoptotic death in relation to the cell cycle. However, this method is subjective and offers only a numerical impression of the occurrence of apoptosis in cytological preparations. Therefore, it is not suitable for kinetic studies or statistical analysis. This type of assay measures only cells which have already reached the stage where a substantial amount of the DNA has been fragmented and leaked from the cell or has been altered and lost staining capacity. Thus, the annexin V-FITC experiment does not provide information about the number of apoptotic cells which are in the initial phase of the process (Vermes et al. 1995), as plasma membrane disruption will be accrued during cell processing.

9.1 Toxicity of Cylindrospermopsins In Vitro

CYN is considered to be responsible for tumor initiation in animals, despite the fact that mechanisms related to CYN-induced carcinogenesis are still scarce and equivocal. Lankoff et al. (2007) investigated the impact of metabolic activation on CYN-induced DNA damage (Lankoff et al. 2007). The effect of different doses of CYN (0.05–2 mg/mL) on DNA damage was determined in CHO-K1 cells after 3, 16, and 21 h of treatment. The chromosome aberration assay with and without metabolic activation was applied to evaluate the clastogenic activity of CYN and its metabolite(s). In addition, the occurrence of apoptosis and necrosis was estimated by the annexin method using flow cytometry. The results revealed that CYN is not clastogenic in CHO-K1 cells irrespective of S9 fraction-induced metabolic activation. However, CYN significantly decreased the frequencies of mitotic ratio and decreased the unattached proliferation of the metabolic activation system. CYN increased the frequency of necrotic cells in a dose- and time-dependent manner, whereas it had a very slight impact on apoptosis. Moreover, the presence of metabolic activation induced susceptibility to necrotic but not apoptotic cell death (Lankoff et al. 2007).

9.2 *Toxicity of Microcystin and Anatoxin-a In Vitro*

Double staining with annexin V-FITC, through binding to an early apoptotic precursor called phosphatidylserine, and propidium iodide, through binding to nuclear DNA following plasma membrane degradation, was used to evaluate the specificity and apoptotic effect of microcystin (MC) and anatoxin-a (ATX) in T cells and B cells. This method can be an efficient way for distinguishing apoptotic and necrotic cells. The results showed that MC-LR only induced apoptosis in B cells, possibly through B-cell receptor-specific activation of kinase pathways, whereas the T cells were not influenced. Mouse splenocyte sensitivity to anatoxin-a in comparison with MC-LR suggests that most cells are in the late apoptotic or secondary necrotic process (Annexin-V+/PI+). In viable subpopulations of T and B cells, staining of Annexin-V was not found. The cytotoxic effects of MC-LR and ATX on lymphocytes therefore operate through various mechanisms (Teneva et al. 2005). Using human bronchial epithelial (HBE) cells, Li et al. (2016) assessed MC-LR-induced toxicity and its potential mechanisms (Li et al. 2016). Cell viability, ROS, mitochondrial membrane potential (MMP), apoptosis rate, and protein expression levels of caspase-3, caspase-9, cytochrome *c* (Cyt *c*), Bax and Bcl-2, were determined to investigate MC-LR toxicity, and to explore the role of the mitochondrial pathway in MC-LR-induced apoptosis of human bronchial epithelial cells. Human bronchial epithelial cells were treated with differing amounts of MC-LR and Z-VAD-FMK, a pan-caspase inhibitor. As a result, MC-LR surprisingly inhibited cell viability and induced human bronchial epithelial cells apoptosis in a dose-dependent way, through Annexin V/propidium iodide assay. Consequently, MC-LR may contribute to cell apoptosis, excessive development of reactive oxygen species, possible failure of the mitochondrial membrane, upregulated Bax expression, as well as regulated B-cell lymphoma-2 expression in human bronchial epithelial cells (Li et al. 2016).

9.3 *Toxicity of Nodularin In Vitro*

It has been shown that nodularin, a cyanotoxin with hepatotoxic activity, could result in carcinogenic events in the liver, even though the mechanisms of the induced DNA damage by this toxin remain unknown (Lankoff et al. 2006c). Lankoff et al. (2006c) explored the DNA destroying effects of nodularin with assays using HepG2 cells. The impacts of multiple doses of nodularin on DNA damage were observed over time. The modified comet assay was analyzed according to both the analysis of DNA disruption induced by ROS and the FISH-micronucleus assay (clastogenic or aneugenic nodularin activity). Furthermore, the frequency of apoptosis was determined through a morphological study of chromatin condensation. It was observed that nodularin caused oxidative DNA damage by purine oxidation, and due to aneugenic behavior increased the development of positive centromere micronuclei. In addition to genotoxicity, nodularin confers cytotoxic activity by activating apoptosis in HepG2 cells. These results suggest that nodularin has a negative function in the process leading to genetic alterations which would include carcinogenesis (Lankoff et al. 2006c).

10 MTT Assay

The general purpose of the MTT (3-[4,5-dimethylthiazol-2-yl]-2,5 diphenyl tetrazolium bromide) assay is that of a relatively high-throughput assay to measure viable cells without the necessity of elaborate cell counting. Therefore, the most common use is to determine cytotoxicity of several drugs at different concentrations. The MTT assay is suitable for the measurement of drug sensitivity in established cell lines as well as primary cells. Lately, experimental data have suggested that MTT and other positively charged tetrazolium salts are primarily reduced not by the traditionally accepted succinate dehydrogenase but by oxidoreductase enzymes (Porzani et al. 2021). The majority of the assays utilize the reduced pyridine nucleotide NADH (Li and Song 2007). This technique is usually based on pure toxins and exposure of the cells to them.

10.1 Toxicity of Cylindrospermopsins In Vivo

The application of the tetrazolium salts method in cyanobacteria is rarely reported (Teneva et al. 2013). The MTT test measured cytotoxicity and a radioimmunoassay quantified progesterone or estrogen to assess the impact of CYN on primary human IVF-derived granulosa cells. Young et al. (2008) studied the effects of CYN on human IVF-derived granulosa cells and showed that CYN was hepatotoxic when administered orally to mice, and cytotoxic and genotoxic to human cell lines (Young et al. 2008). To determine the effects of CYN on primary human IVF-derived granulosa cells, 0–1 μg/ml CYN was added to cells for 2, 4, or 6 h ± hCG ($n = 6$), and for 24, 48, and 72 h ($n = 6$). Cytotoxicity was evaluated by MTT assay, and secreted progesterone or estrogen quantified by radioimmunoassay. After 24 h of exposure, 1 μg/ml CYN was cytotoxic ($p < 0.05$), whereas 0.0625 μg/mL CYN did not cause cytotoxicity or affect estrogen production, but inhibited basal progesterone production ($p < 0.01$). Similarly, 6 h of exposure to 1 μg/mL CYN did not affect cytotoxicity or hCG-stimulated estrogen production, but inhibited hCG-stimulated progesterone production ($p < 0.01$). In this in vitro assay, CYN inhibited progesterone production showing the potential to be an endocrine disrupter by changing the progesterone:estrogen ratio in women (Young et al. 2008). Another study explored the sensitivity of CYN (Froscio et al. 2009) and showed that the hepatic-derived cell lines, C3A and HepG2, were most at risk of CYN effects based on MTT assessment. Apparently, the sensitivity of the cell line decreased in cells derived from progressively more distal regions of the duct tract: gastric > duodenal > ileal > colonic. Nevertheless, any investigation would have been required to see whether or not this can be a development applicable to intestinal adsorption rates (Froscio et al. 2009).

10.2 Toxicity of Cylindrospermopsins, Microcystins, and Anatoxin-a In Vitro

The effect of several toxins has been studied in murine macrophage-like RAW264.7, microglial BV-2, and neuroblastoma N2a cell lines on cell viability, to evaluate how cyanotoxins can interfere in programmed cell death and inflammation events (Takser et al. 2016). CYN, MC-LR, and anatoxin-a were analyzed independently and in combination. The neurotoxins beta-N-methylamino-l-alanine (BMAA) and its isomer 2,4-diaminobutyric acid (DAB) were also studied. The MTT assay determined cellular viability. Results showed that significant low doses of cyanotoxins, frequently found as mixtures in drinking water, have the ability to cause inflammation and apoptosis in immune and brain cells. Further studies of the neuroinflammatory consequences of these compounds is needed in vivo to stipulate the protection limits for cyanotoxins in water supplies and, eventually, in food (Takser et al. 2016).

10.3 Toxicity of Microcystin-LR In Vitro

The cytotoxicity and possible apoptotic effects of MC-LR on human embryonic kidney cell line (HEK-293) and human kidney adenocarcinoma cell line (ACHN) were evaluated via MTT assay and sulforhodamine B (SRB) cell viability assays by Piyathilaka and collaborators (Piyathilaka et al. 2015). They showed that cyanotoxins could cause cytotoxic effects on kidney cells, with MC-LR having the most pronounced effect, whereas MC-LW was least toxic, on both cell types tested. Alternatively, MC-RR, MC-LF, and nodularin showed moderate toxicity on human nephritic cells. Thus, the research presents evidence that exposure to MC-LR may cause morphological changes linked to apoptosis, up-regulation of the expression of Bax and p53 genes, which may modulate the expression of the Survivin gene, and increase the function of caspase 3 and 9 in human embryonic kidney cells (HEK-293) and human kidney carcinoma cells. Bax, p53, Survivin, caspase 3 and 9 are, therefore, presumably concerned with cellular damage induced by MC-LR. Furthermore, this discovery tends to elucidate the toxicological process that underlies the effects of MC-LR on human kidneys (Manage 2019; Piyathilaka et al. 2015).

11 Enzyme-Linked Immunosorbent Assay (ELISA)

The Enzyme-Linked Immunosorbent Assay (ELISA) is an antibody-based test that may be either monoclonal, unique to a single molecule type, or polyclonal. (Metcalf et al. 2000; Zeck et al. 2001). ELISAs have several attributed characteristics, such as being fast, sensitive, and easy to perform. Analytical techniques based on ELISA have high sensitivity, although the cross-reactivity between cyanotoxin variants and between cyanotoxin classes can vary, such as for microcystin variants and nodularin (Metcalf et al. 2000). Therefore, an issue to be considered when such antibodies are applied to a sample containing an unknown or complex mixture of microcystins variants is the potential for poor reaction with some components. Depending on both the cross-reactivity characteristics and the toxicity of the microcystin variant, ELISAs can over- or underestimate the toxin concentration when assessed as microcystin-LR toxicity equivalents. In cases where the mixture of toxins is well characterized, application of ELISA for continued monitoring is quite acceptable (Sanseverino et al. 2017). Although quantification by ELISA may not be as reliable as for physicochemical methods, this technique can be an efficient method for cyanotoxin detection and monitoring.

11.1 Toxicity of Microcystins and Nodularin In Vivo and In Vitro

Berry et al. (2011) collected samples from Lago de Patzcuaro, Mexico, with the aim to explore the toxicity of microcystins in fish species ("charales," *Chirostoma* sp.; *Goodea* sp.; and "carp," *Cyprinus carpio*) (Berry et al. 2011). They performed several analytical strategies including ELISA and protein phosphatase inhibition assay. The lake where the samples were collected is a habitat for important fish species, especially in terms of human consumption and the local economy. The results showed that microcystin was accumulating in fish tissue as a result of a toxic *Microcystis* bloom (Berry et al. 2011). An indirect ELISA for microcystin and nodularin detection by Fischer et al. (2001) was used, based on detection of the Adda side chain. The assay had low detection limits and excellent cross-reactivity to cyclic cyanotoxins, including MC-LR, -RR, -YR, -LW, -LF, 3-desmethyl-MC-LR, 3-desmethyl-MC-RR, and nodularin (Fischer et al. 2001).

Evaluating the detection of microcystins and nodularin from Finnish water supplies revealed that the most commonly identified microcystins were MC -RR, -LR, and -YR with differing degrees of methylation (nonmethylated, mono methylated, or demethylated) showing good agreement when ELISA results were compared with LC-MS analysis (Spoof et al. 2003). Tillmanns et al. (2007) investigated the ELISA assay to select for the best normalizing parameter for microcystin concentrations, showing that microcystins were up to fourfold greater in lake samples. However, analysis of samples obtained by plankton tow routinely showed higher concentrations than complete water samples (Tillmanns et al. 2007). When combined with LC-MS/MS analysis, ELISA analysis of Puget Sound Washington region lakes, Washington, USA showed maximum concentrations of 2,700 μg/L in inland freshwater lakes, up to 0.34 μg/L in marine water, and in mussels up to 6.5 μg/kg, with the MC-LA variant being the major toxin (Preece et al. 2015).

Microcystins were analyzed in Sardinian ponds (Cabras and Tortolì), Italy, in the spring and summer season using ELISA (Sedda et al. 2016). The results obtained revealed a similarity in microcystin concentrations for Cabras pond in August, and Tortolì pond in September. Such information will help to enrich data about the environmental health of Sardinian ponds and their economically important products (Sedda et al. 2016). Analysis of the microcystin content in cyanobacterial isolates from the Ganga River, India (Dixit et al. 2017) examined intracellular content, whereas water samples were used to measure extracellular release. Water samples from the Nonindustrial area of Kanpur (MGK), Industrial area of Kanpur (SGK), and Gomti River near Nishatganj in Lucknow (GNL) displayed concentrations of 3, 2, and 0.8 ppb of MC, respectively. The existence of MC was not seen at numerous water sampling sites of the Sai River near Banthra in Lucknow (SBL) and the Sharda Canal in Unnao (SCU), although toxic cyanobacterial isolates *Oscillatoria* sp. RBD01 and *Leptolyngbya* sp. RBD05 were shown to contain MCs at 23 and 17 ppb of MC in cells, respectively (Dixit et al. 2017).

The Adda-ELISA is an indirect and efficient ELISA which detects the Adda moiety (novel amino acid found in all microcystins and nodularins). This assay will identify both MCs and nodularins and possible Adda degradation products. Analysis of 31 water bodies in Michigan, USA, between July and October, 2017 found MC-LA, -LR, -RR, D-Asp3-LR, and YR were the most commonly found microcystins variants (Birbeck et al. 2019), with D-Asp3-MC-LR being the only one not included in the US Environmental Protection Agency (US EPA) Method 544 (liquid chromatography tandem mass spectrometry (LC-MS/MS) and solid phase extraction (SPE)). The MC data were comparable in July and August but started to differ in September and October, when the Adda-ELISA results were higher than the LC-MS/MS data, as a result of the accumulation of biodegradation products of MCs in lakes, late in the season (Birbeck et al. 2019).

Cyanotoxins, including MCs and NODs have been discussed as a potential source of dolphin mortality (Brown et al. 2018). Due to the fact that estuarine dolphins occupy areas with repeated or persistent harmful cyanobacterial blooms, they are regarded as a fairly high-risk population regarding sensitivity to cyanotoxins in coastal waters. MMPB (2-methyl-3-methoxy-4-phenylbutyric acid) was used to assess all liver tissue concentrations of MCs and NOD (bounded and free). Extractions of free MCs/NODs were conducted and analyzed using ELISA to assess positive controls for MMPB. Analysis with MMPB resulted in the discovery of total low levels of MCs and NOD in certain specimens. The data indicated that these cyanotoxins were present in both estuarine and coastal dolphins, with potentially toxic and immune health impacts (Brown et al. 2018).

Cristiana Moreira and collaborators (Moreira et al. 2020) explored seven freshwater habitats in Portugal's Northern and Central areas. During April and

September, 2017, these places were examined for large cyanotoxin-containing blooms (microcystins, cylindrospermopsins, anatoxin-a, and saxitoxins) utilizing ELISA assays. Tests showed that 48% of the samples were above the guideline value for microcystin-LR in drinking water, and 33% of the remaining samples were over the guideline value for anatoxin-a, with one sample above the recommended value for saxitoxins, obtained from an area in the North. The research could improve Portugal's mitigation policies and encourage strategies for environmental protection and residues management (Moreira et al. 2020).

11.2 Toxicity of Saxitoxins In Vitro

Kleinteich et al. (2013) performed a study to screen for the presence of saxitoxins in five cyanobacterial arctic communities using an ELISA assay (Kleinteich et al. 2013). In particular, samples were screened to check the production of MC and STX based on the presence of the biosynthetic gene clusters, which confer the ability of cyanobacterial populations in the Arctic to synthesize these cyanotoxins. Detection of these substances in the cryosphere enhances the understanding of the biogeography and regional spread of harmful cyanobacterial blooms (Kleinteich et al. 2013).

12 Protein Phosphatase Inhibition Assay (PPIA)

Protein phosphatase inhibition assays are sensitive procedures for determining microcystins and nodularins produced by cyanobacteria. While the response and measurement of toxins in this assay is related to enzyme inhibition by microcystins, which although not related to the totality of the toxicity of microcystins, does provide an indication of toxicity. The current data suggest that such assays may give an overestimation of toxin concentration based on microcystin-LR toxicity equivalents. However, more accurate data, with a larger range of microcystin variants, need to be obtained in order to determine whether there is an underestimation of toxin concentrations for some variants. At this stage, these assays are restricted to use as screening tools (Nicholson and Burch 2001) and are commonly used to assess contaminated water bodies.

12.1 Identification of Microcystins and Nodularins Through PPIA In Vitro Assays

Since microcystins and nodularins are known to be effective protein phosphatase inhibitors, these toxins can be identified using PPIA, a simple, quick, responsive, and reproducible colorimetric assay (Liu et al. 2014). Heresztyn and Nicholson (2001) used PPIA and p-nitrophenyl phosphate as a substrate to determine two hepatotoxins in water samples without standards. The study concluded that this technique was better than radiolabeled substrates in terms of commercial availability and simplicity (Heresztyn and Nicholson 2001). Many studies have applied the PPIA assay to evaluate the presence of the cyanobacterial hepatotoxins, microcystins and nodularins, as well as their variants, in freshwater samples (Kaloudis et al. 2013; Triantis et al. 2010).

There are several ways in which microcystins can be identified, but the PPIA assay includes evidence on the biological behavior of MCs and NOD with the advantage of not requiring live whole animal specimens with associated ethical concerns. Wharton et al. (2019) improved the sensitivity and specificity of the traditional PPIA assay through the use of an immunocapture step. This procedure utilizes an adda-specific antibody to collect and accumulate MCs and NOD up to 10 times from, e.g., human urine, before tests of PPIA toxicity compared to inhibition by MC-LR. Results showed that the quantitative range of 0.050–0.500 ng/mL for MC-LR can be applied using this method, suggesting it to be a valuable method to track low cyanotoxin concentrations in humans (Wharton et al. 2019).

13 Conclusion

Cyanobacterial harmful blooms occur worldwide and have been increasing over the last decades due to global climate change and the consequences of intense anthropological activities. These changes might act synergistically to promote cyanobacterial dominance and persistence, adversely affecting human, animal, and ecosystem health globally. The development of multi-strategy testing is essential to achieve fast, accurate, and interdisciplinary results. Thus, water management agencies need to adopt the precautionary principle, to consider any cyanobacterial bloom as toxic until analyzed. The expansion of cyanobacterial harmful blooms may have serious consequences for human drinking water, irrigation water supplies, fisheries, and recreational sources. Implementation of initial screening methods in both freshwater and marine environments, to assess for cyanotoxins, should be a part of all management protocols. Therefore, this review sought to bring together the most conventional methods for the in vivo and in vitro assessment of cyanotoxins using biological and biochemical assays, as well as cytotoxic and genotoxic methods. This mini-review also showed the advantages and disadvantages of the various methods to allow scientists in biochemistry, chemistry, molecular biology, and environmental science to develop management and action plans to help control the risks posed by cyanobacterial toxins.

Author Contributions
Samaneh J. Porzani: writing – the original draft, the main manuscript text and graphical abstract.
Stella T. Lima: writing – review and editing, visualization.
James S. Metcalf: writing – review and editing, visualization.
Bahareh Nowruzi: writing – review, editing, and proof-read; supervised the manuscript.

Conflict of Interest The authors declare that they have no conflict of interest.

Informed Consent No human participants were involved in this study.

Funding No funding was received for this work.

References

Abend M, Rhein A, Gilbertz K-P, Blakely W, Van Beuningen D (1995) Correlation of micronucleus and apoptosis assays with reproductive cell death. Int J Radiat Biol 67:315–326

Abramsson-Zetterberg L, Sundh UB, Mattsson R (2010) Cyanobacterial extracts and microcystin-LR are inactive in the micronucleus assay in vivo and in vitro. Mutat Res 699:5–10

Almesjö L (2007) Filamentous cyanobacteria in the Baltic Sea-spatiotemporal patterns and nitrogen fixation. Systemekologiska institutionen, Stockholm, pp 1–32

Andeden EE, Ozturk S, Aslim B (2018) Antiproliferative, neurotoxic, genotoxic and mutagenic effects of toxic cyanobacterial extracts. Interdiscip Toxicol 11:267–274

Ariza RR, Keyse SM, Moggs JG, Wood RD (1996) Reversible protein phosphorylation modulates nucleotide excision repair of damaged DNA by human cell extracts. Nucleic Acids Res 24:433–440

Babich H, Rosenberg D, Borenfreund E (1991) In vitro cytotoxicity studies with the fish hepatoma cell line, PLHC-1 (Poeciliopsis lucida). Ecotox Environ Safe 21:327–336

Baker P, Humpage A (1994) Toxicity associated with commonly occurring cyanobacteria in surface waters of the Murray-Darling basin, Australia. Mar Freshw Res 45:773–786

Bazin E, Mourot A, Humpage AR, Fessard V (2010) Genotoxicity of a freshwater cyanotoxin, cylindrospermopsin, in two human cell lines: Caco-2 and HepaRG. Environ Mol Mutagen 51(3):251–259

Berry J, Lee E, Walton K, Wilson A, Bernal-Brooks F (2011) Microcystin production by a persistent cyanobacterial bloom in Lago de Pátzcuaro (Michoacán, Mexico), and apparent bioaccumulation of the toxin in small commercial catches of fish. Environ Toxicol Chem 30:1621–1628

Birbeck JA, Westrick JA, O'Neill GM, Spies B, Szlag DC (2019) Comparative analysis of microcystin prevalence in Michigan lakes by online concentration LC/MS/MS and ELISA. Toxins 11:13

Bolch CJ, Orr PT, Jones GJ, Blackburn SI (1999) Genetic, morphological, and toxicological variation among globally distributed strains of Nodularia (Cyanobacteria). J Phycol 35:339–355

Borenfreund E, Puerner JA (1985) A simple quantitative procedure using monolayer cultures for cytotoxicity assays (HTD/NR-90). J Tissue Cult Methods 9:7–9

Börner T, Dittmann E (2005) Molecular biology of cyanobacterial toxins. In: Harmful cyanobacteria. Springer, pp 25–40

Brooks W, Codd G (1987) Distribution of Microcystis aeruginosa peptide toxin and interactions with hepatic microsomes in mice. Pharmacol Toxicol 60:187–191

Brown A, Foss A, Miller MA, Gibson Q (2018) Detection of cyanotoxins (microcystins/nodularins) in livers from estuarine and coastal bottlenose dolphins (Tursiops truncatus) from Northeast Florida. Harmful Algae 76:22–34

Buratti FM, Manganelli M, Vichi S, Stefanelli M, Scardala S, Testai E, Funari E (2017) Cyanotoxins: producing organisms, occurrence, toxicity, mechanism of action and human health toxicological risk evaluation. Arch Toxicol 91:1049–1130

Burford MA, Carey CC, Hamilton DP, Huisman J, Paerl HW, Wood SA, Wulff A (2020) Perspective: advancing the research agenda for improving understanding of cyanobacteria in a future of global change. Harmful Algae 91:101601. https://doi.org/10.1016/j.hal.2019.04.004

Callieri C, Bertoni R, Contesini M, Bertoni F (2014) Lake level fluctuations boost toxic cyanobacterial "oligotrophic blooms". PLoS One 9:e109526

Campbell D, Lawton L, Beattie K, Codd G (1994) Comparative assessment of the specificity of the brine shrimp and microtox assays to hepatotoxic (microcystin-LR-containing) cyanobacteria. Environ Toxicol Water Qual 9:71–77

Chatziefthimiou AD, Richer R, Rowles H, Powell JT, Metcalf JS (2014) Cyanotoxins as a potential cause of dog poisonings in desert environments. Vet Rec 174:484–485

Chatziefthimiou AD, Metcalf JS, Glover WB, Banack SA, Dargham SR, Richer RA (2016) Cyanobacteria and cyanotoxins are present in drinking water impoundments and groundwater wells in desert environments. Toxicon 114:75–84

Cheng A, Balczon R, Zuo Z, Koons JS, Walsh AH, Honkanen RE (1998) Fostriecin-mediated G2-M-phase growth arrest correlates with abnormal centrosome replication, the formation of aberrant mitotic spindles, and the inhibition of serine/threonine protein phosphatase activity. Cancer Res 58:3611–3619

Chia MA, Kramer BJ, Jankowiak JG, Bittencourt-Oliveira MC, Gobler CJ (2019) The individual and combined effects of the cyanotoxins, anatoxin-a and microcystin-LR, on the growth, toxin production, and nitrogen fixation of prokaryotic and eukaryotic algae. Toxins 11:43

Chong M, Wong B, Lam P, Shaw G, Seawright A (2002) Toxicity and uptake mechanism of cylindrospermopsin and lophyrotomin in primary rat hepatocytes. Toxicon 40:205–211

Christensen VG, Khan E (2020) Freshwater neurotoxins and concerns for human, animal, and ecosystem health: a review of anatoxin-a and saxitoxin. Sci Total Environ 736:139515. https://doi.org/10.1016/j.scitotenv.2020.139515

Cohen Y, Gurevitz M (2006) The cyanobacteria—ecology, physiology and molecular genetics. The Prokaryotes 4:1074–1098

Corbel S, Mougin C, Nélieu S, Delarue G, Bouaïcha N (2016) Evaluation of the transfer and the accumulation of microcystins in tomato (Solanum lycopersicum cultivar MicroTom) tissues using a cyanobacterial extract containing microcystins and the radiolabeled microcystin-LR (14C-MC-LR). Sci Total Environ 541:1052–1058

Costa S (2005) Efeitos de saxitoxinas produzidas por Cylindrospermopsis raciborskii e de outras cianotoxinas sobre cladóceros (Branchiopoda) Programa de Pó s-graduaça o em Ciências Bioló gicas Universidade Federal do Rio de Janeiro, Rio de Janeiro, Brazil (PhD thesis)

Costa ID, Azevedo S, Senna P, Bernardo R, Costa S, Chellappa N (2006) Occurrence of toxin-producing cyanobacteria blooms in a Brazilian semiarid reservoir. Braz J Biol 66:211–219

Cullen A, Pearson LA, Mazmouz R, Liu T, Soeriyadi AH, Ongley SE, Neilan BA (2019) Heterologous expression and biochemical characterisation of cyanotoxin biosynthesis pathways. Nat Prod Rep 36:1117–1136

D'Anglada LV, Donohue JM, Strong J, Hawkins B (2015) Health effects support document for the cyanobacterial toxin cylindrospermopsin. US Environmental Protection Agency, Office of Water

De Figueiredo DR, Azeiteiro UM, Esteves SM, Gonçalves FJ, Pereira MJ (2004) Microcystin-producing blooms—a serious global public health issue. Ecotox Environ Safe 59:151–163

DeMott WR, Zhang QX, Carmichael WW (1991) Effects of toxic cyanobacteria and purified toxins on the survival and feeding of a copepod and three species of Daphnia. Limnol Oceanogr 36:1346–1357

Dias E, Louro H, Pinto M, Santos T, Antunes S, Pereira P, Silva MJ (2014) Genotoxicity of microcystin-LR in in vitro and in vivo experimental models. Biomed Res Int 2014:9

Ding W-X, Shen H-M, Zhu H-G, Lee B-L, Ong C-N (1999) Genotoxicity of microcystic cyanobacteria extract of a water source in China. Mutat Res 442:69–77

Dixit RB, Patel AK, Toppo K, Nayaka S (2017) Emergence of toxic cyanobacterial species in the Ganga River, India, due to excessive nutrient loading. Ecol Indic 72:420–427

Djediat C, Malécot M, de Luze A, Bernard C, Puiseux-Dao S, Edery M (2010) Localization of microcystin-LR in medaka fish tissues after cyanotoxin gavage. Toxicon 55:531–535

Eaglesham GK et al (1999) Use of HPLC-MS/MS to monitor cylindrospermopsin, a blue–green algal toxin, for public health purposes. Environ Toxicol 14:151–154

Eppley RM (1974) Sensitivity of brine shrimp (Artemia salina) to trichothecenes. J AOAC Int 57:618–620

Falconer KJ (1986) The geometry of fractal sets, vol 85. Cambridge university press

Fenech M (1997) The advantages and disadvantages of the cytokinesis-block micronucleus method. Mutat Res 392:11–18

Fenech M (2000) The in vitro micronucleus technique. Mutat Res 455:81–95

Fenech M, Morley AA (1986) Cytokinesis-block micronucleus method in human lymphocytes: effect of in vivo ageing and low dose X-irradiation. Mutat Res 161:193–198

Fent K (2001) Fish cell lines as versatile tools in ecotoxicology: assessment of cytotoxicity, cytochrome P4501A induction potential and estrogenic activity of chemicals and environmental samples. Toxicology In Vitro 15(4–5):477–488

Ferrão-Filho AS, Kozlowsky-Suzuki B (2011) Cyanotoxins: bioaccumulation and effects on aquatic animals. Mar Drugs 9:2729–2772

Ferrão-Filho AS, Soares MCS, de Freitas MV, Azevedo SM (2009) Biomonitoring of cyanotoxins in two tropical reservoirs by cladoceran toxicity bioassays. Ecotox Environ Safe 72:479–489

Filipič M, Žegura B, Sedmak B, Horvat-Žnidaršic I, Milutinovič A, Šuput D (2007) Subchronic exposure of rats to sublethal dose of microcystin-YR induces DNA damage in multiple organs. Radiol Oncol 41:15–22

Fiore MF, de Lima ST, Carmichael WW, McKinnie SM, Chekan JR, Moore BS (2020) Guanitoxin, re-naming a cyanobacterial organophosphate toxin. Harmful Algae 92:101737

Fischer WJ, Hitzfeld BC, Tencalla F, Eriksson JE, Mikhailov A, Dietrich DR (2000) Microcystin-LR toxicodynamics, induced pathology, and immunohistochemical localization in livers of blue-green algae exposed rainbow trout (Oncorhynchus mykiss). Toxicol Sci 54:365–373

Fischer WJ, Garthwaite I, Miles CO, Ross KM, Aggen JB, Chamberlin AR, Towers NR, Dietrich DR (2001) Congener-independent immunoassay for microcystins and nodularins. Environ Sci Technol 35:4849–4856

Freitas EC, Pinheiro C, Rocha O, Loureiro S (2014) Can mixtures of cyanotoxins represent a risk to the zooplankton? The case study of Daphnia magna Straus exposed to hepatotoxic and neurotoxic cyanobacterial extracts. Harmful Algae 31:143–152

Froscio SM, Fanok S, Humpage AR (2009) Cytotoxicity screening for the cyanobacterial toxin cylindrospermopsin. J Toxicol 72:345–349

Funari E, Testai E (2008) Human health risk assessment related to cyanotoxins exposure. Crit Rev Toxicol 38:97–125

Funari E, Manganelli M, Sinisi L (2012) Impact of climate change on waterborne diseases. Annali dell'Istituto superiore di sanita 48:473–487

Gaudin J, Huet S, Jarry G, Fessard V (2008) In vivo DNA damage induced by the cyanotoxin microcystin-LR: comparison of intra-peritoneal and oral administrations by use of the comet assay. Mutat Res 652:65–71

Ghadouani A, Pinel-Alloul B, Plath K, Codd GA, Lampert W (2004) Effects of Microcystis aeruginosa and purified microcystin-LR on the feeding behavior of Daphnia pulicaria. Limnol Oceanogr 49:666–679

Gutiérrez-Praena D, Pichardo S, Jos Á, Cameán AM (2011) Toxicity and glutathione implication in the effects observed by exposure of the liver fish cell line PLHC-1 to pure cylindrospermopsin. Ecotox Environ Safe 74:1567–1572

Guzmán-Guillén R, Gutiérrez-Praena D, De los Ángeles Risalde M, Moyano R, Prieto AI, Pichardo S, Jos Á, Vasconcelos V, Cameán AM (2014) Immunohistochemical approach to study cylindrospermopsin distribution in tilapia (Oreochromis niloticus) under different exposure conditions. Toxins 6:283–303

Hawkins PR, Runnegar MT, Jackson A, Falconer I (1985) Severe hepatotoxicity caused by the tropical cyanobacterium (blue-green alga) Cylindrospermopsis raciborskii (Woloszynska) Seenaya and Subba Raju isolated from a domestic water supply reservoir. Appl Environ Microbiol 50:1292–1295

Hawkins PR, Chandrasena NR, Jones GJ, Humpage AR, Falconer IR (1997) Isolation and toxicity of Cylindrospermopsis raciborskii from an ornamental lake. Toxicon 35:341–346

Heresztyn T, Nicholson BC (2001) Determination of cyanobacterial hepatotoxins directly in water using a protein phosphatase inhibition assay. Water Res 35:3049–3056

Hilborn ED, Roberts VA, Backer L, Deconno E, Egan JS, Hyde JB, Nicholas DC, Wiegert EJ, Billing LM, Diorio M, Mohr MC, Hardy JF, Wade TJ, Yoder JS, Hlavsa MC, Centers for Disease Control and Prevention (2014) Algal bloom-associated disease outbreaks among users of freshwater lakes—United States, 2009–2010. MMWR Morb Mortal Wkly Rep 63(1):11–15

Hooser SB, Kuhlenschmidt MS, Dahlem AM, Beasley VR, Carmichael WW, Haschek WM (1991) Uptake and subcellular localization of tritiated dihydro-microcystin-LR in rat liver. Toxicon 29:589–601

Humpage AR, Fenech M, Thomas P, Falconer IR (2000) Micronucleus induction and chromosome loss in transformed human white cells indicate clastogenic and aneugenic action of the cyanobacterial toxin, cylindrospermopsin. Mutat Res 472:155–161

Humpage AR, Fontaine F, Froscio S, Burcham P, Falconer IR (2005) Cylindrospermopsin genotoxicity and cytotoxicity: role of cytochrome P-450 and oxidative stress. J. Toxicol 68:739–753

Hyenstrand P, Rohrlack T, Beattie KA, Metcalf JS, Codd GA, Christoffersen K (2003) Laboratory studies of dissolved radiolabelled microcystin-LR in lake water. Water Res 37:3299–3306

Jungmann D, Benndorf J (1994) Toxicity to Daphnia of a compound extracted from laboratory and natural Microcystis spp., and the role of microcystins. Freshw Biol 32:13–20

Kaloudis T, Zervou SK, Tsimeli K, Triantis TM, Fotiou T, Hiskia A (2013) Determination of microcystins and nodularin (cyanobacterial toxins) in water by LC–MS/MS. Monitoring of Lake Marathonas, a water reservoir of Athens, Greece. J Hazard Mater 263:105–115

Khodarev NN, Park JO, Yu J, Gupta N, Nodzenski E, Roizman B, Weichselbaum RR (2001) Dose-dependent and independent temporal patterns of gene responses to ionizing radiation in normal and tumor cells and tumor xenografts. PNAS 98:12665–12670

Kiviranta J, Sivonen K, Niemelä S, Huovinen K (1991) Detection of toxicity of cyanobacteria by Artemia salina bioassay. Environ Toxicol Water Qual 6:423–436

Kleinteich J, Wood SA, Puddick J, Schleheck D, Küpper FC, Dietrich D (2013) Potent toxins in Arctic environments–presence of saxitoxins and an unusual microcystin variant in Arctic freshwater ecosystems. Chem Biol Interact 206:423–431

Koreivienė J, Belous O (2012) Methods for cyanotoxins detection. Botanica 18:58–65

Lahti K, Ahtiainen J, Rapala J, Sivonen K, Niemelä S (1995) Assessment of rapid bioassays for detecting cyanobacterial toxicity. Lett Appl Microbiol 21:109–114

Lankoff A, Banasik A, Obe G, Deperas M, Kuzminski K, Tarczynska M, Jurczak T, Wojcik A (2003) Effect of microcystin-LR and cyanobacterial extract from polish reservoir of drinking water on cell cycle progression, mitotic spindle, and apoptosis in CHO-K1 cells. Toxicol Appl Pharmacol 189:204–213

Lankoff A, Krzowski Ł, Głąb J, Banasik A, Lisowska H, Kuszewski T, Góźdź S, Wójcik A (2004) DNA damage and repair in human peripheral blood lymphocytes following treatment with microcystin-LR. Mutat Res 559:131–142

Lankoff A, Banasik A, Duma A, Ochniak E, Lisowska H, Kuszewski T, Góźdź S, Wojcik A (2006a) A comet assay study reveals that aluminium induces DNA damage and inhibits the repair of radiation-induced lesions in human peripheral blood lymphocytes. Toxicol Lett 161:27–36

Lankoff A, Bialczyk J, Dziga D, Carmichael W, Lisowska H, Wojcik A (2006b) Inhibition of nucleotide excision repair (NER) by microcystin-LR in CHO-K1 cells. Toxicon 48:957–965

Lankoff A, Wojcik A, Fessard V, Meriluoto J (2006c) Nodularin-induced genotoxicity following oxidative DNA damage and aneuploidy in HepG2 cells. Toxicol Lett 164:239–248

Lankoff A, Wojcik A, Lisowska H, Bialczyk J, Dziga D, Carmichael W (2007) No induction of structural chromosomal aberrations in cylindrospermopsin-treated CHO-K1 cells without and with metabolic activation. Toxicon 50:1105–1115

Li J, Song L (2007) Applicability of the MTT assay for measuring viability of cyanobacteria and algae, specifically for Microcystis aeruginosa (Chroococcales, Cyanobacteria). Phycologia 46:593–599

Li Y, Li J, Huang H, Yang M, Zhuang D, Cheng X, Zhang H, Fu X (2016) Microcystin-LR induces mitochondria-mediated apoptosis in human bronchial epithelial cells. Exp Ther Med 12:633–640

Liu L, Jokela J, Wahlsten M, Nowruzi B, Permi P, Zhang YZ, Xhaard H, Fewer DP, Sivonen K (2014) Nostosins, trypsin inhibitors isolated from the terrestrial cyanobacterium Nostoc sp. strain FSN. J Nat Prod 77:1784–1790

Llana-Ruiz-Cabello M, Pichardo S, Maisanaba S, Puerto M, Prieto AI, Gutierrez-Praena D, Jos A, Cameán AM (2015) In vitro toxicological evaluation of essential oils and their main compounds used in active food packaging: a review. Food Chem Toxicol 81:9–27

Logan TJ (1997) Critical reviews in environmental science and technology. CRC Press

Lukowski AL, Mallik L, Hinze ME, Carlson BM, Ellinwood DC, Pyser JB, Koutmos M, Narayan AR (2020) Substrate promiscuity of a paralytic shellfish toxin amidinotransferase. ACS Chem Biol 15:626–631

Malik JK, Bharti VK, Rahal A, Kumar D, Gupta RC (2020) Cyanobacterial (blue-green algae) toxins. In: Handbook of toxicology of chemical warfare agents. Elsevier, pp 467–478

Manage P (2019) Cyanotoxins: a hidden cause of chronic kidney disease of unknown etiology (CKDu) in Sri Lanka-a review. Sri Lanka J Aquat Sci 24. https://doi.org/10.4038/SLJAS.V24I1.7562

Manganelli M, Scardala S, Stefanelli M, Palazzo F, Funari E, Vichi S, Buratti FM, Testai E (2012) Emerging health issues of cyanobacterial blooms. Ann Ist Super Sanita 48:415–428

Mankiewicz J, Walter Z, Tarczynska M, Palyvoda O, Wojtysiak-Staniaszczyk M, Zalewski M (2002) Genotoxicity of cyanobacterial extracts containing microcystins from polish water reservoirs as determined by SOS chromotest and comet assay. Environ Toxicol 17:341–350

McGregor GB, Stewart I, Sendall BC, Sadler R, Reardon K, Carter S, Wruck D, Wickramasinghe W (2012) First report of a toxic Nodularia spumigena (Nostocales/Cyanobacteria) bloom in sub-tropical Australia. I. Phycological and public health investigations. Int J Environ Res Public Health 9:2396–2411

Meili N, Christen V, Fent K (2016) Nodularin induces tumor necrosis factor-alpha and mitogen-activated protein kinases (MAPK) and leads to induction of endoplasmic reticulum stress. Toxicol Appl Pharmacol 300:25–33

Metcalf J, Bell S, Codd G (2000) Production of novel polyclonal antibodies against the cyanobacterial toxin microcystin-LR and their application for the detection and quantification of microcystins and nodularin. Water Res 34:2761–2769

Metcalf J, Lindsay J, Beattie K, Birmingham S, Saker M, Törökné A, Codd G (2002) Toxicity of cylindrospermopsin to the brine shrimp Artemia salina: comparisons with protein synthesis inhibitors and microcystins. Toxicon 40:1115–1120

Minshull J, Straight A, Rudner AD, Dernburg AF, Belmont A, Murray AW (1996) Protein phosphatase 2A regulates MPF activity and sister chromatid cohesion in budding yeast. Curr Biol 6:1609–1620

Mohamed ZA (2008) Toxic cyanobacteria and cyanotoxins in public hot springs in Saudi Arabia. Toxicon 51:17–27

Mohamed ZA, Al Shehri AM (2007) Cyanobacteria and their toxins in treated-water storage reservoirs in Abha city, Saudi Arabia. Toxicon 50:75–84

Mohamed ZA, Al Shehri AM (2009) Microcystin-producing blooms of Anabaenopsis arnoldi in a potable mountain lake in Saudi Arabia. FEMS Microbiol Ecol 69:98–105

Mohamed ZA, Deyab MA, Abou-Dobara MI, El-Raghi WM (2016) Occurrence of toxic cyanobacteria and microcystin toxin in domestic water storage reservoirs. Egypt J Water Supply Res Technol Aqua 65:431–440

Moreira C, Gomes C, Vasconcelos V, Antunes A (2020) Cyanotoxins occurrence in Portugal: a new report on their recent multiplication. Toxins 12:154

Nesslany F, Marzin D (1999) A micromethod for the in vitro micronucleus assay. Mutagenesis 14:403–410

Nicholson BC, Burch MD (2001) Evaluation of analytical methods for detection and quantification of cyanotoxins in relation to Australian drinking water guidelines. National Health and Medical Research Council of Australia Canberra

Nong Q, Komatsu M, Izumo K, Indo HP, Xu B, Aoyama K, Majima HJ, Horiuchi M, Morimoto K, Takeuchi T (2007) Involvement of reactive oxygen species in Microcystin-LR-induced cytogenotoxicity. Free Radic Res 41:1326–1337

Norris RLG, Seawright AA, Shaw GR, Smith MJ, Chiswell RK, Moore MR (2001) Distribution of 14C cylindrospermopsin in vivo in the mouse. Environ Toxicol 16(6):498–505

Nowruzi B, Porzani SJ (2021) Toxic compounds produced by cyanobacteria belonging to several species of the order Nostocales: a review. J Appl Toxicol 41(4):510–548

Nowruzi B, Khavari-Nejad R-A, Sivonen K, Kazemi B, Najafi F, Nejadsattari T (2012) Identification and toxigenic potential of a Nostoc sp. Algae 27:303

Nowruzi B, Khavari-Nejad RA, Sivonen K, Kazemi B, Najafi F, Nejadsattari T (2013) Identification and toxigenic potential of a cyanobacterial strain (Stigomena sp.). PBS 3:79–85

Nowruzi B, Blanco S, Nejadsattari T (2018a) Chemical and molecular evidences for the poisoning of a duck by anatoxin-a, nodularin and cryptophycin at the coast of Lake Shoormast (Mazandaran province, Iran). Int J Algae 20:359–376

Nowruzi B, Haghighat S, Fahimi H, Mohammadi E (2018b) Nostoc cyanobacteria species: a new and rich source of novel bioactive compounds with pharmaceutical potential. JPHSR 9:5–12

Nowruzi B, Wahlsten M, Jokela J (2019) A report on finding a new peptide aldehyde from cyanobacterium Nostoc sp. Bahar M by LC-MS and Marfey's analysis. IJB 17:e1853

Nowruzi B, Sarvari G, Blanco S (2020a) Applications of cyanobacteria in biomedicine. In: Handb. algal sci. microbiol. technol. med. Elsevier, Amsterdam

Nowruzi B, Sarvari G, Blanco S (2020b) The cosmetic application of cyanobacterial secondary metabolites. Algal Res 49:101959

Ohtani I, Moore RE, Runnegar MT (1992) Cylindrospermopsin: a potent hepatotoxin from the blue-green alga Cylindrospermopsis raciborskii. J Am Chem Soc 114:7941–7942

Okumura DT, Sotero-Santos RB, Takenaka RA, Rocha O (2007) Evaluation of cyanobacteria toxicity in tropical reservoirs using crude extracts bioassay with cladocerans. Ecotoxicology 16:263–270

Pace JG, Robinson NA, Miura GA, Matson CF, Geisbert TW, White JD (1991) Toxicity and kinetics of [3H] microcystin-LR in isolated perfused rat livers. Toxicol Appl Pharmacol 107(3):391–401

Paerl HW, Fulton RS, Moisander PH, Dyble J (2001) Harmful freshwater algal blooms, with an emphasis on cyanobacteria. The Scientific World JOURNAL 1:76–113

Palus J, Dziubaltowska E, Stanczyk M, Lewinska D, Mankiewicz-Boczek J, Izydorczyk K, Bonislawska A, Jurczak T, Zalewski M, Wasowicz W (2007) Biomonitoring of cyanobacterial blooms in polish water reservoir and the cytotoxicity and genotoxicity of selected cyanobacterial extracts. IJOMEH 20:48–65

Park DL, Aguirre-Flores I, Scott WF, Alterman E (1986) Evaluation of chicken embryo, brine shrimp, and bacterial bioassays for saxitoxin. J Toxicol Environ Health Part A Curr Issues 18:589–594
Parry EM, Parry JM, Corso C, Doherty A, Haddad F, Hermine TF, Johnson G, Kayani M, Quick E, Warr T, Williamson J (2002) Detection and characterization of mechanisms of action of aneugenic chemicals. Mutagenesis 17:509–521
Pegram R, Humpage A, Neilan B, Runnegar M, Nichols T, Thacker R, Pflugmacher S, Etheridge S, Love A (2008) Cyanobacterial harmful algal blooms: chapter 15: cyanotoxins workgroup report. US Environ Prot Agency Pap 34
Piyathilaka M, Pathmalal M, Tennekoon K, De Silva B, Samarakoon S, Chanthirika S (2015) Microcystin-LR-induced cytotoxicity and apoptosis in human embryonic kidney and human kidney adenocarcinoma cell lines. Microbiology 161:819–828
Porzani SJ, Lorenzi AS, Eghtedari M, Nowruzi B (2021) Interactions of dehydrogenase enzymes with nanoparticles industrial and medical applications and challenges: mini-review. Mini Rev Med Chem 21(11):1351–1366
Potterat O, Hamburger M (2006) Natural products in drug discovery-concepts and approaches for tracking bioactivity. Curr Org Chem 10:899–920
Preece E, Moore BC, Hardy FJ (2015) Transfer of microcystin from freshwater lakes to Puget Sound, WA and toxin accumulation in marine mussels (Mytilus trossulus). Ecotox Environ Safe 122:98–105
Prieto AI, Pichardo S, Jos Á, Moreno I, Cameán AM (2007) Time-dependent oxidative stress responses after acute exposure to toxic cyanobacterial cells containing microcystins in tilapia fish (Oreochromis niloticus) under laboratory conditions. Aquat Toxicol 84:337–345
Puerto M, Pichardo S, Jos Á, Cameán AM (2009) Comparison of the toxicity induced by microcystin-RR and microcystin-YR in differentiated and undifferentiated Caco-2 cells. Toxicon 54:161–169
Puerto M, Pichardo S, Jos Á, Cameán AM (2010) Microcystin-LR induces toxic effects in differentiated and undifferentiated Caco-2 cells. Arch Toxicol 84:405–410
Rajabpour N, Nowruzi B, Ghobeh M (2019) Investigation of the toxicity, antioxidant and antimicrobial activities of some cyanobacterial strains isolated from different habitats. Acta Biol Slov 62:2
Rao PL, Bhattacharya R, Parida M, Jana A, Bhaskar A (1998) Freshwater cyanobacterium Microcystis aeruginosa (UTEX 2385) induced DNA damage in vivo and in vitro. Environ Toxicol Pharmacol 5:1–6
Robinson NA, Pace JG, Matson CF, Miura GA, Lawrence WB (1990) Tissue distribution, excretion, and hepatic biotransformation of microcystin-LR in mice. Army Medical Research Inst of Infectious Diseases, Fort Detrick, MD
Rohrlack T, Christoffersen K, Dittmann E, Nogueira I, Vasconcelos V, Börner T (2005) Ingestion of microcystins by Daphnia: intestinal uptake and toxic effects. Limnol Oceanogr 50:440–448
Rosefort C, Fauth E, Zankl H (2004) Micronuclei induced by aneugens and clastogens in mononucleate and binucleate cells using the cytokinesis block assay. Mutagenesis 19:277–284
Rouhiainen L, Jokela J, Fewer DP, Urmann M, Sivonen K (2010) Two alternative starter modules for the non-ribosomal biosynthesis of specific anabaenopeptin variants in Anabaena (Cyanobacteria). Chem Biol 17:265–273
Ryan JA, Hightower LE (1994) Evaluation of heavy-metal ion toxicity in fish cells using a combined stress protein and cytotoxicity assay. Environ Toxicol Chem 13:1231–1240
Safavi M, Nowruzi B, Estalaki S, Shokri M (2019) Biological activity of methanol extract from Nostoc sp. N42 and Fischerella sp. S29 isolated from aquatic and terrestrial ecosystems. Int J Algae 21:373–391
Sanseverino I, António DC, Loos R, Lettieri T (2017) Cyanotoxins: methods and approaches for their analysis and detection. JRC Publication

Seawright AA, Nolan CC, Shaw GR, Chiswell RK, Norris RL, Moore MR, Smith MJ (1999) The oral toxicity for mice of the tropical cyanobacterium Cylindrospermopsis raciborskii (Woloszynska). Environ Toxicol 14:135–142
Sedda T, Baralla E, Varoni MV, Pasciu V, Lorenzoni G, Demontis MP (2016) Determination of microcystin-LR in clams (Tapes decussatus) of two Sardinian coastal ponds (Italy). Mar Pollut Bull 108:317–320
Singh NP, McCoy MT, Tice RR, Schneider EL (1988) A simple technique for quantitation of low levels of DNA damage in individual cells. Exp Cell Res 175:184–191
Sivonen K, Kononen K, Carmichael W, Dahlem A, Rinehart K, Kiviranta J, Niemela S (1989) Occurrence of the hepatotoxic cyanobacterium Nodularia spumigena in the Baltic Sea and structure of the toxin. Appl Environ Microbiol 55:1990–1995
Sontag E (2001) Protein phosphatase 2A: the Trojan Horse of cellular signaling. Cell Signal 13:7–16
Spoof L, Vesterkvist P, Lindholm T, Meriluoto J (2003) Screening for cyanobacterial hepatotoxins, microcystins and nodularin in environmental water samples by reversed-phase liquid chromatography–electrospray ionisation mass spectrometry. J Chromatogr A 1020:105–119
Takenaka RA (2007) Avaliação da toxicidade de Microcystis aeruginosa e de florações naturais de cianobactérias de reservatórios do rio Tietê, SP. Universidade de São Paulo
Takser L, Benachour N, Husk B, Cabana H, Gris D (2016) Cyanotoxins at low doses induce apoptosis and inflammatory effects in murine brain cells: potential implications for neurodegenerative diseases. Toxicol Rep 3:180–189
Tencalla F, Dietrich D (1997) Biochemical characterization of microcystin toxicity in rainbow trout (Oncorhynchus mykiss). Toxicon 35:583–595
Teneva I, Mladenov R, Popov N, Dzhambazov B (2005) Cytotoxicity and apoptotic effects of microcystin-LR and anatoxin-a in mouse lymphocytes. Folia Biol (Praha) 51:62
Teneva I, Stoyanov P, Mladenov R, Dzhambazov B (2013) In vitro and in vivo toxicity evaluation of the freshwater cyanobacterium Heteroleiblenia kuetzingii. Open Life Sci 8:1216–1229
Terao K, Ohmori S, Igarashi K, Ohtani I, Watanabe MF, Harada KI, Ito E, Watanabe M (1994) Electron microscopic studies on experimental poisoning in mice induced by cylindrospermopsin isolated from blue-green alga Umezakia natans. Toxicon 32:833–843
Tice RR, Agurell E, Anderson D, Burlinson B, Hartmann A, Kobayashi H, Miyamae Y, Rojas E, Ryu JC, Sasaki YF (2000) Single cell gel/comet assay: guidelines for in vitro and in vivo. Environ Mol Mutagen 35:206–221
Tillmanns AR, Pick FR, Aranda-Rodriguez R (2007) Sampling and analysis of microcystins: implications for the development of standardized methods. Environ Toxicol 22:132–143
Triantis T, Tsimeli K, Kaloudis T, Thanassoulias N, Lytras E, Hiskia A (2010) Development of an integrated laboratory system for the monitoring of cyanotoxins in surface and drinking waters. Toxicon 55(5):979–989
Valério E, Chaves S, Tenreiro R (2010) Diversity and impact of prokaryotic toxins on aquatic environments: a review. Toxins 2:2359–2410
Vermes I, Haanen C, Steffens-Nakken H, Reutellingsperger C (1995) A novel assay for apoptosis flow cytometric detection of phosphatidylserine expression on early apoptotic cells using fluorescein labelled annexin V. J Immunol Methods 184:39–51
Welker M, Von Döhren H (2006) Cyanobacterial peptides—nature's own combinatorial biosynthesis. FEMS Microbiol Rev 30:530–563
Weller MG (2013) Immunoassays and biosensors for the detection of cyanobacterial toxins in water. Sensors 13:15085–15112
Wharton RE, Cunningham BR, Schaefer AM, Guldberg SM, Hamelin EI, Johnson RC (2019) Measurement of microcystin and nodularin activity in human urine by immunocapture-protein phosphatase 2A assay. Toxins 11:729
Wiegand C, Peuthert A, Pflugmacher S, Carmeli S (2002) Effects of microcin SF608 and microcystin-LR, two cyanobacterial compounds produced by Microcystis sp., on aquatic organisms. Environ Toxicol 17:400–406

Woese CR (1987) Bacterial evolution. Microbiol Rev 51:221–271
Xu LH, Lam PK, Chen JP, Xu JM, Wong BS, Zhang YY, Wu RS, Harada KI (2000) Use of protein phosphatase inhibition assay to detect microcystins in Donghu Lake and a fish pond in China. Chemosphere 41:53–58
Yen G, Chen H, Peng H (2001) Evaluation of the cytotoxicity, mutagenicity and antimutagenicity of emerging edible plants. Food Chem Toxicol 39:1045–1053
Yoshida T, Makita Y, Tsutsumi T, Nagata S, Tashiro F, Yoshida F, Sekuima M, Tamura SI, Harada T, Maita K, Ueno Y (1998) Immunohistochemical localization of microcystin-LR in the liver of mice: a study on the pathogenesis of microcystin-LR-induced hepatotoxicity. Toxicol Pathol 26:411–418
Young FM, Micklem J, Humpage AR (2008) Effects of blue-green algal toxin cylindrospermopsin (CYN) on human granulosa cells in vitro. Reprod Toxicol 25:374–380
Zanchett G, Oliveira-Filho EC (2013) Cyanobacteria and cyanotoxins: from impacts on aquatic ecosystems and human health to anticarcinogenic effects. Toxins 5:1896–1917
Zeck A, Eikenberg A, Weller MG, Niessner R (2001) Highly sensitive immunoassay based on a monoclonal antibody specific for [4-arginine] microcystins. Anal Chim Acta 441:1–13
Žegura B, Sedmak B, Filipič M (2003) Microcystin-LR induces oxidative DNA damage in human hepatoma cell line HepG2. Toxicon 41:41–48
Žegura B, Lah TT, Filipič M (2004) The role of reactive oxygen species in microcystin-LR-induced DNA damage. Toxicology 200:59–68
Žegura B, Lah TT, Filipič M (2006) Alteration of intracellular GSH levels and its role in microcystin-LR-induced DNA damage in human hepatoma HepG2 cells. Mutat Res 611:25–33
Žegura B, Volčič M, Lah TT, Filipič M (2008) Different sensitivities of human colon adenocarcinoma (CaCo-2), astrocytoma (IPDDC-A2) and lymphoblastoid (NCNC) cell lines to microcystin-LR induced reactive oxygen species and DNA damage. Toxicon 52:518–525
Zegura B, Sever N, Kološa K, Filipic M (2009) The influence of microcystin-LR on HepG2 cell proliferation and micronuclei formation. In: Proceedings of tenth international conference on environmental mutagenesis: the renaissance of environmental mutagenesis. Firenze, Italy, pp 20–25
Žegura B, Gajski G, Štraser A, Garaj-Vrhovac V, Filipič M (2011a) Microcystin-LR induced DNA damage in human peripheral blood lymphocytes. Mutat Res 726:116–122
Žegura B, Štraser A, Filipič M (2011b) Genotoxicity and potential carcinogenicity of cyanobacterial toxins–a review. Mutat Res Rev Mutat Res 727:16–41
Zerikly M, Challis GL (2009) Strategies for the discovery of new natural products by genome mining. Chembiochem 10:625–633
Zhan L, Sakamoto H, Sakuraba M, Wu DS, Zhang LS, Suzuki T, Hayashi M, Honma M (2004) Genotoxicity of microcystin-LR in human lymphoblastoid TK6 cells. Mutat Res 557:1–6
Zurawell RW, Chen H, Burke JM, Prepas EE (2005) Hepatotoxic cyanobacteria: a review of the biological importance of microcystins in freshwater environments. J Toxicol Environ Health Part B 8:1–37

Comprehensive Review of Cadmium Toxicity Mechanisms in Male Reproduction and Therapeutic Strategies

Lijuan Xiong, Bin Zhou, Hong Liu, and Lu Cai

Contents

L. Xiong (✉)
Department of Emergency, Jiangxi Provincial Children's Hospital, Nanchang, Jiangxi, China

Pediatric Research Institute, Department of Pediatrics, University of Louisville School of Medicine, Louisville, KY, USA
e-mail: lijuanxiong87@gmail.com

B. Zhou · H. Liu
Department of Emergency, Jiangxi Provincial Children's Hospital, Nanchang, Jiangxi, China
e-mail: jxzhoubin@126.com; 13970018998@qq.com

L. Cai (✉)
Pediatric Research Institute, Department of Pediatrics, University of Louisville School of Medicine, Louisville, KY, USA

Departments of Pharmacology and Toxicology, University of Louisville School of Medicine, Louisville, KY, USA
e-mail: L0cai001@louisville.edu

P. de Voogt (ed.), *Reviews of Environmental Contamination and Toxicology Volume 258*,
Reviews of Environmental Contamination and Toxicology 258,
https://doi.org/10.1007/398_2021_75

Abstract Cadmium (Cd) has been widely studied as an environmental pollutant for many years. Numerous studies have reported that Cd exposure causes damage to the heart, liver, kidneys, and thyroid in vivo. The emerging evidence suggests that Cd exposure induces damage on male reproductive system, which is related to oxidative stress, inflammation, steroidogenesis disruption, and epigenetics. Current preclinical animal studies have confirmed a large number of proteins and intracellular signaling pathways involved in the pathological process of Cd-induced male reproductive damage and potential measures for prophylaxis and treatment, which primarily include antioxidants, anti-inflammatory agents, and essential ion supplement. However, explicit pathogenesis and effective treatments remain uncertain. This review collects data from the literatures, discusses the underlying mechanisms of Cd-induced toxicity on male reproductive function, and summarizes evidence that may provide guidance for the treatment and prevention of Cd-induced male reproductive toxicity.

Keywords Cadmium · Male reproduction · Mechanism · Offspring · Prevention · Treatment

Abbreviations

17β-HSD	17β-Hydroxysteroid dehydrogenase
3β-HSD	3β-Hydroxysteroid dehydrogenase
ABCG2	ATP-binding cassette G2
AR	Androgen receptor
BTB	Blood-testis barrier
BW	Body weight
CatSper	Sperm-specific cation channel
Cd	Cadmium
DAX-1	Dosage-sensitive sex reversal, adrenal hypoplasia critical region, on chromosome X, gene 1
DMT	Divalent metal receptor
DNMT	DNA methyltransferase
EDTA	Ethylenediaminetetraacetic acid
FSH	Follicle stimulating hormone
GPx	Glutathione peroxidase
H3K9me2	Histone H3 lysine 9 dimethylation
HPG	Hypothalamic-pituitary-gonadal
ICAM2	Intercellular adhesion molecule 2
LH	Luteinizing hormone
LHR	Luteinizing hormone receptor
lncRNA	Long non-coding RNA
LPO	Lipid peroxidation
MDA	Malondialdehyde

MMP	Mitochondrial membrane potential
MRL	Minimum risk level
MT	Metallothionein
NAC	N-acetylcysteine
NLRP3	Nucleotide-binding oligomerization domain, leucine-rich repeat, and pyrin domain-containing 3
PMCA	Plasma membrane Ca^{2+}-ATPase
QE	Quercetin
RGN	Regucalcin
ROS	Reactive oxygen species
Se	Selenium
SelP	Selenoprotein P
SOD	Superoxide dismutase
Star	Steroidogenic acute regulatory protein
STAT3	Signal transducer and activator of transcription 3
TBARS	Thiobarbituric acid reactive substances
ZIP	Zinc finger protein
Zn	Zinc

1 Introduction

Cadmium (Cd), as a heavy metal, mostly exists in nature as compounds, such as CdO, $CdCl_2$, or $CdSO_4$. Cd is mainly released from industrial emissions including electroplating, smelting, mining, manufacture of batteries and pigments, and from municipal waste incineration. The release of Cd from industrial production is a threat to the ecology because it pollutes the water, air, and soil, upon which people depend.

The absorption of Cd in humans is mainly through the consumption of contaminated dietary items and inhalation such as smoking. Smoking one pack of cigarettes per day causes an additional 1–3 μg of Cd absorption (Waalkes 2003; Wilder et al. 2001). Due to the long biological half-life (10–35 years) of Cd, prolonged exposure to Cd can lead to Cd accumulation in the body, further damaging organs and tissues. As early as in the early twentieth century, scientists discovered that chronic exposure to Cd causes Itai-itai (reported in Japan), a disease caused by long-term consumption of Cd-contaminated water, causing kidney damage that leads to osteomalacia with severe bone pain (Aoshima 2012; Nishijo et al. 2017). Subsequent studies have confirmed that Cd exposure causes damage to the liver, kidneys, and heart, as well as the reproductive system.

As infertility rates increase globally, researchers are concerned about the effects of Cd toxicity on male reproduction. Numerous experiments on animals have confirmed that Cd exposure can damage the testis by inhibiting spermatogenesis, damaging the blood-testis barrier (BTB), decreasing Sertoli and Leydig cells, affecting the synthesis of steroid hormones, and reducing sperm quality and sperm motility, thereby causing infertility (Chen et al. 2018; Elblehi et al. 2019; Ji et al. 2013). Consistent with the results of animal experiments, several epidemiological investigations have shown that the Cd concentrations in blood and seminal fluid, accumulated through either occupational or non-occupational exposure, are inversely proportional to sperm parameters and sperm enzyme activity (Sengupta et al. 2013; Wang and Tian 2004; Wijesekara et al. 2015). Other studies have found a variety of health problems in people dwelling in Cd-contaminated areas, e.g., rising incidences of abnormal menstruation, difficulty in conceiving, and increased premature and stillbirth (Wang and Tian 2004). The infertility rate increases by about 5–6 times for every ten-fold increase in the concentration of Cd in semen (Wu et al. 2008).

Thus, numerous preclinical studies have investigated treatments for Cd-induced male reproductive toxicity. Reactive oxygen species (ROS) are one of the primary causes of Cd-induced reproductive toxicity. Majority of studies suggest that a logical solution for preventing and treating male infertility is to increase the intake of antioxidants, including vitamins C, E, Co-enzyme Q10, and zinc (Zn) (Amara et al. 2008; Saha et al. 2019). Previous studies reported that antioxidants may provide cellular protection by scavenging oxygen and hydroxyl radicals. However, randomized clinical trials do not fully support this notion because antioxidants appear to also have certain adverse effects such as DNA damage in sperm, increased infertility rates, and even risks of cancer or Alzheimer's disease, if antioxidants are taken in excess or if there is a lack of synergistic compounds (Henkel et al. 2019). Therefore, there is an urgent need for an in-depth understanding of the mechanisms underlying Cd toxicity in the reproductive system.

This review summarizes data from animal experiments and clinical studies, discusses the potential mechanisms by which Cd damages the male reproductive system, and summarizes the potential prevention and therapy strategies that are currently under investigation. This review may provide guidance for the prophylaxis and treatment of reproductive toxicity caused by Cd.

2 Cd Toxicity on Male Reproduction

2.1 Animal Experiments

Numerous experimental studies both in vivo and in vitro have provided direct evidence of the Cd-induced toxic effects on reproductive function. However, these toxic effects vary with the exposure dose and exposure time to Cd. Medina et al. treated rats with Cd at a dose of 10 mg/kg body weight (BW) by gavage for 3 months and examined the specimens every month. Primary spermatocyte impairment was observed in the first month while ultrastructural changes, such as the disruption of tight junctions, appeared from the second month; for the sperm parameters in the Cd group, no significant changes were observed in the first month compared to the control group, whereas a significant decrease in the count of normal sperm was observed from the second month. In the third month, there was a significant increase in immotile sperm count and a decrease in the progressive motility of sperm. These results indicate that Cd-induced reproductive toxicity is time-dependent (Medina et al. 2017). Likewise, a study investigating the reproductive effects of Cd exposure at different doses was carried out, in which three groups of male rats were given $CdCl_2$-containing water for 100 days with the doses of 5, 50, and 100 ppm. It was found that Cd concentrations were positively related to sperm abnormality and inversely associated with sperm numbers, which presented with only mild injury in 5 ppm group and severe impairment in 100 ppm group. In other words, Cd causes reproductive toxicity in a dose-dependent manner (Rania Abdel Rahman and Mohamed Elshabrawy 2014). However, the Cd-induced toxic effects appeared only at certain critical concentrations for different target organ, suggesting the existence of dose threshold for different kinds of injures. Similar effects may be induced by either long-term low-level Cd exposure or short-term high-level Cd exposure. Hence, to mimic the Cd-induced toxic effects on human reproduction, Cd exposure in animal models is divided into three categories: acute exposure (24 h to 14 days), sub-chronic exposure (28 to 90 days), and chronic exposure ($\geq$ 6 months), treated as either high or low doses (EPA 1998).

In vivo studies have demonstrated the Cd induced toxic effects on the reproductive tract. In certain studies, adult rats that were administered with Cd (2 mg $CdCl_2$/kg.BW, intraperitoneal (IP)) for 24 h showed overt histological changes in the testes, followed by increased testicular weights owing to testicular edema and hemorrhage. Histopathological examination revealed necrosis of the spermatogenic tubules and germ cells, interstitial inflammatory cell infiltration, and disruption of the BTB in Cd-exposed rats. In addition, a decrease in the sperm parameters and a reduction in the sperm mitochondrial membrane potential (MMP), along with an impairment in the acrosomal integrity of the sperm, were observed (Chen et al. 2018; Ji et al. 2013; Liu et al. 2016b, 2020; Sen Gupta et al. 2004). In contrast, a single 0.5 mg $CdCl_2$/kg.BW/IP exposure for 24 h did not cause notable pathological changes and cell death in gonad organs, although it induced oxidative DNA base damage and altered the expression of multiple genes related to DNA repair and replication (Ates et al. 2004; Zhou et al. 2004). These results may be contributed to the different exposure doses of Cd, indicating that 0.5 mg $CdCl_2$/kg.BW/IP is low-dose exposure to Cd compared to 2 mg. Consistent with these results, toxicity caused by sub-chronic exposure has a positive correlation with the exposure doses and exposure times. Sub-chronic high-dose exposure (3 or 5 mg $CdCl_2$/kg.BW/IP) significantly decreased testicular weight, sperm motility, and density, as well as dysregulated testicular lncRNA (long non-coding RNA) or mRNA, which resulted in more severe damage compared to low-dose exposure (1 mg $CdCl_2$/kg.BW/IP) (Gao et al. 2017; Wu et al. 2017) (Table 1). Furthermore, exposure routes affect reproductive toxicity. Previous studies have reported that the oral route shows more severe damage compared to the IP route, especially when Cd is administered in a fractionated dose (Mouro et al. 2019, 2020). However, another study barely detected histopathological changes in mouse testes exposed to low dose of Cd (1.3 mg/kg.BW) via the oral route over a 3-month period (Blanco et al. 2010).

Table 1 Animal studies on Cd-induced male reproductive toxicity

Category	Model	Dose	Treatment schedule	Main findings	References
Acute exposure	Adult male Long-Evans rats	$CdCl_2$/0.3 mg/kg.BW/d/IP	Injected once and sacrificed after 2 and 3 d	Initiate copulation latency↑, the number of penile insertions↓, copulatory efficacy↓, the number of males displaying ejaculatory behavior↓	Clark et al. (1994)
	CD-1 mice	$CdCl_2$/2.0 mg/kg.BW/IP	Injected once and sacrificed after 1 d	Testicular edema, testis weights↑, germ cell apoptosis, ER stress↑, GRP78↑, sXBP-1↑, p-JNK↑, CHOP↑, p-eIF2α↑, HO-1↑,3-NT↑, GSH↓	Ji et al. (2013)
	SD rats	$CdCl_2$/2.0 mg/kg.BW/d/SC	Injected once and sacrificed after 2 d	Germinal epithelium disorganized, germ cells detached from Sertoli cells, claudin-11↓, TJ disruption, SOD↓ CAT↓, MDA↑, TGF-β↑, p38 MAPK↑	Chen et al. (2018)
	BALB/c mice	$CdCl_2$/3.0 mg/kg.BW/d/SC	Injected daily for 5 d	Spermatogenic cells arrange disordered, Leydig cells degenerated, sperm aberration rate↑, LDH↑, MDA↑, SOD↓,3β-HSD↓	Liu et al. (2020)
	Wistar rats	$CdCl_2$/50 mg/kg.BW/d/oral	Administration daily for 15 d	CAT↓, MDA↑, progressive motility of spermatozoa↓, sperm number↓, abnormal sperm↑, serum T↓, testicular Cd concentration↑	Abarikwu et al. (2017)
Sub-chronic exposure	SD rats	$CdCl_2$/1.0 mg/kg.BW/d/ intragastrically	Administration daily for 8 weeks	Reproductive organ relative weights↓, incomplete spermatogenesis, germ and Sertoli cells vacuolation, sperm count↓, sperm motility↓, alive sperm↓, seminal vesiculitis, serum T↓, GST↓, MDA↑, Star↓, CYP17α↓	Elblehi et al. (2019)
	Balb/c mice	$CdCl_2$/1.2 mg/kg.BW/IP or 24 mg/kg.BW/d/gavage	Injected once and sacrificed after 6 weeks	Zn↓(oral), SOD、MDA↓, GST↑(IP and oral), total length of seminiferous tubules↑ (oral), Leydig cells↓ (IP and oral), detachment of epithelium, germ cells degenerated	Mouro et al. (2020)

	Wistar rats	$CdCl_2$/5.0 mg/kg.BW/d/oral	Administration daily for 4 weeks	Testis and epididymis weights↓, Cd concentration↑, sperm motility↓ abnormal sperm↑, GnRH, LH, FSH and serum T↓, testicular glucose concentration↑, testicular lactate concentration↑, GPx, SOD↓, CAT↓, MDA↑, Bax/Bcl-2↑, Leydig cell↓, seminiferous epithelial height↓	Nna et al. (2017)
	C57BL/6 J mice	$CdCl_2$/1.0 or 3.0 or 5.0 mg/kg.BW/d/IP	Injected once and sacrificed after 5 weeks	Testicular weight↓(3,5 mg), sperm motility and density↓(3,5 mg), sperm abnormality (5 mg), numerous lncRNA expressed abnormally	Gao et al. (2017)
Chronic exposure	White male rats	$CdCl_2$/0.00005 or 0.0005 or 0.005 mg/kg.BW/d in drinking water	Administrated daily for 6 months	Spermatogenesis decreased (0.0005, 0.005 mg), Spermatogenic epithelium detached from the basal membrane (0.0005, 0.005 mg)	Krasovskii et al. (1976)
	Mus musculus	$CdCl_2$/0.1 mg/kg.BW/IP	Injected once a week and sacrificed after 52 weeks	No cadmium was detected in sperm, the appearance of the testis and epididymis did not change	Bench et al. (1999)
	Albino rats	$CdCl_2$/0.5 mg/kg.BW/d/SC	Administrated daily, 5 d per week, for 24 weeks	Edematous vasculitis, stromal hemorrhage, seminiferous tubules degenerated, loss of spermatogenesis	Yadav and Khandelwal (2008)
	Swiss OF-1 mice	0.015 g/L $CdCl_2$ in drinking water	Administrated daily for 6, 12, 18 months	Cd concentration↑, vascular lesions (6 months), hemorrhages, thrombus and even necrosis (12/18 months), hypertrophy and cytoplasm vacuolization in Leydig cells (6 months), apoptosis and tumors in Leydig cells (12/18 months)	Blanco et al. (2010)

Cd cadmium, *Zn* zinc, *T* testosterone, *d* day, *BW* body weight, *IP* intraperitoneal injection, *SC* subcutaneous injection, *CAT* catalase, *SOD* superoxide dismutase, *MDA* malondialdehyde, *GST* glutathione S-transferase, *HO-1* Heme Oxygenase-1, *3-NT* 3-Nitrotyrosine, *GSH* glutathione, *GPx* glutathione peroxidase, *ER* endoplasmic reticulum, *CHOP* C/EBP homologous protein, *GRP78* Glucose regulatory protein 78, *sXBP-1* X-box binding protein 1, *p-eIF2α* Phosphorylation of the eukaryotic initiation factor 2, *p-JNK* phosphorus c-Jun N-terminal kinase, *MAPK* mitogen-activated protein kinase, *CYP17α* P450-C17 alpha, *GnRH* gonadotropin-releasing hormone, *LH* luteinizing hormone, *FSH* follicle stimulating hormone, *TJ* tight junction, *lncRNA* long non-coding RNA, *3β-HSD* 3β-Hydroxysteroid dehydrogenase, ↑ increase, ↓ decrease

Chronic low-dose exposure to Cd in animals mimics the environmental Cd exposure in the general population. The reproductive toxicity of long-term exposure to Cd is primarily caused by its bioaccumulation. Testicular Cd concentrations increase with time, with a corresponding increase in reproductive toxicity. Vascular lesions and structural changes in Leydig cells were observed in mice treated orally with low-dose Cd ($CdCl_2$ 0.015 g/L in drinking water) for 6 months (chronic exposure), while the vascular damage and Leydig cells impairment were aggravated if the exposure time was extended to 12 months or 18 months, as shown by hemorrhages and necrosis in vessels, along with apoptosis and event tumors in Leydig cells (Blanco et al. 2010). Yadav et al. reported edematous vasculitis, stromal hemorrhage, and absent spermatogenesis in rats treated with Cd (subcutaneous injection of 0.5 mg $CdCl_2$/kg/day) for 18 weeks, with enhanced damage after 24 weeks (Yadav and Khandelwal 2008). Taken together, these studies clearly indicate that exposure time is also a crucial factor in inducing reproductive toxicity, expect for exposure dose. No pathomorphological changes were observed in the gonad glands when adult male rats were administered with 0.1 mg $CdCl_2$ /kg.BW/IP for 52 weeks (Bench et al. 1999). This is because the Cd exposure dose was lower than the minimum dose required for causing reproductive toxicity. Accordingly, the minimum risk level (MRL) was set for Cd exposure for the general population.

These results are in agreement with those reported in vitro. Incubated with Cd, sperm presented decreased counts and reduced motility; and interrupted tight and gap junctions between Sertoli cells were found, with significantly decreased expression of the Ocln gene, as well as overexpression of the Gja1 gene in Sertoli cells (Ramos-Trevino et al. 2017; Wang et al. 2016b). Furthermore, Cd exposure in vitro induced a reduction in Leydig cell viability, inhibited the levels of 3-β-Hydroxysteroid dehydrogenase (3β-HSD) by activating the mitogen-activated protein kinases (MAPK)-extracellular signal-regulated kinases (ERK) pathway, and increased intracellular Ca^{2+} levels to trigger oxidative stress (Khanna et al. 2016). This is inconsistent with those observed in in vivo experiments, wherein the Cd-induced effects on germ cells and stromal cells are different – germ cells are sensitive to Cd exposure while Leydig cells are resistant to Cd, despite a profound impairment of steroidogenesis (Cupertino et al. 2017).

Besides, Cd affects sexual activities, except for damaging gonads and suppressing steroidogenesis. Adult male rats showed sexual dysfunction when administered with $CdCl_2$ at a dose as low as 0.3 mg/kg.BW/IP; the number of rats displaying mounting, penile erection, and ejaculatory behavior decreased (Clark et al. 1994). In addition, Ujah et al. found that a significant increase in mount latency, intromission latency, ejaculatory latency, and post-ejaculatory interval, as well as a decrease in intromission frequency in $CdCl_2$-treated male rats (Ujah et al. 2018). These results are in agreement with Kamel et al., who suggested that Cd exposure markedly abolished aggressive behavior and increased fear and anxiety in adult male rats, in addition to suppressed sexual behavior (Kamel et al. 2011). The reduced concentration of cyclic guanosine monophosphate in the penile and suppression of testosterone contribute to the sexual dysfunction caused by Cd, which are almost irreversible even after removing Cd exposure (Ujah et al. 2018).

As mentioned above, regarding the induction of male reproductive toxicity, the exposure routes of Cd in animal experiments have primarily been IP or subcutaneous (SC) injection and oral administration, previous studies have rarely investigated inhalation route. In the IP or SC route, the lowest observed adverse effect level (LOAEL) of $CdCl_2$ for an acute exposure is 0.3 mg/kg, while 0.25 mg/kg in sub-chronic and 0.1 mg/kg in chronic exposure (Bomhard et al. 1987; Clark et al. 1994). In oral route (including gavage and drinking water), the LOAEL of $CdCl_2$ in acute exposure is 4 mg/kg, by contrast, 1 mg/kg in sub-chronic exposure and 0.0005 mg/kg in chronic exposure (Bu et al. 2011; Elblehi et al. 2019; Krasovskii et al. 1976). All these doses are the minimum doses that cause reproductive toxicity, which are low-dose exposure to Cd. In comparison, during short-term exposure, a single dose above 2 mg/kg.BW/IP or above 25 mg/kg.BW/oral can induce significant toxic effects in testes, such as acute hemorrhagic inflammation, edema, and even necrosis, but no mortality; these doses are high-dose exposure to Cd (Abarikwu et al. 2017; Bench et al. 1999; Faroon et al. 2012; Ohta et al. 2019). Thus, these low- and high-dose exposures to Cd in animal models were corresponded to environmental exposure and occupational exposure in humans. Environmental exposure includes the intake of Cd-containing water, consumption of food with a high Cd content, such as seafood, and the use of products with a high Cd content, as well as smoking, which are low-dose exposure sources for humans (Schopfer et al. 2010; Sharma et al. 1983). Occupational exposure is a high-dose exposure scenario for humans, which usually occurs in workers employed in industries involving the smelting, welding, and manufacturing of products containing Cd, such as batteries and pigments (Goyal et al. 2020; Zeng et al. 2002) The average peak blood Cd (B-Cd) concentration (2.25 μg/L) in animals exposed to low-dose Cd (3 mg/L $CdCl_2$-containing drinking water) is similar to the mean B-Cd levels (2.22 μg/L) in infertile men exposed to environmental Cd (Kiziler et al. 2007; Wang et al. 2018). In comparison, the average B-Cd level (9.51 μg/L) in animals exposed to high-dose Cd (100 mg/L $CdCl_2$-containing drinking water) for 1 week approaches to that (11.63 μg/L) in occupational-exposed workers (Alli 2015; Thijssen et al. 2007). Male workers in a smelter exposed to Cd at a dose of 3.55 mg/m^3 showed disruption of the hormone homeostasis and infertility (Zeng et al. 2002). The mean urine Cd (U-Cd) level (16.18 μg/g creatinine) in male workers was overtly higher than that in general men (0.189 μg/g creatinine) who were exposed to the environmental Cd and presented with low-quality sperm (Dhooge et al. 2001).

However, previous studies have not defined the MRL of Cd-induced reproductive toxicity in the general population and occupational workers. Considerable animal experiments will be required to confirm the LOAEL and no observed adverse effect level (NOAEL) to access the MRL of Cd-induced reproductive toxicity in humans. For example, the MRL of human health effects for acute-duration Cd inhalation was set based on the results of National Toxicology Program studies calculated using the traditional NOAEL/LOAEL approach and the regional deposited dose ratio via a complex formula, which assists in establishing relevant thresholds to reduce and prevent Cd exposure (Bellinger 2004; EPA 1998; Faroon et al. 2012).

2.2 Clinical or Epidemiological Studies

According to the recommendations of the Agency for Toxic Substances and Disease Registry (ATSDR), Cd exposure in humans is classified as acute ($\leq$14 days), intermediate (15–364 days), and chronic exposure ($\geq$1 year) based on the Cd exposure duration. The ATSDR set 0.00003 mg/m^3 and 0.00001 mg/m^3 as the inhalation MRL of human health effects for acute and chronic exposure, respectively, and 0.0005 mg/kg/day and 0.0001 mg/kg/day as the oral MRL for intermediate and chronic exposure, respectively (ATSDR 2012). Clinically, B-Cd and U-Cd as well as the seminal plasma Cd (S-Cd) levels have been commonly used as biomarkers to evaluate Cd exposure level (ATSDR 2020). The lowest Cd biological concentration reported inducing kidney damage is 1 μg/g creatinine (U-Cd), decreasing bone mineral density is 4 μg/g creatinine (U-Cd), and causing pancreatic cancer is 11.1 ng/mL serum (B-Cd). The adverse effect levels vary followed different endpoints (Marano et al. 2012). According to the results of multiple cross-sectional studies, the U-Cd, B-Cd, and S-Cd were significantly inversely associated with sperm parameters (He et al. 2020; Li et al. 2016d; Sukhn et al. 2018; Wang et al. 2016c) (Fig. 1). However, the lowest biological concentration of Cd for causing reproductive toxicity is undefined.

In general population, the minimal level of U-Cd among reported studies is 0.14 μg/L, above which was demonstrated to decrease motile sperm count (Li et al. 2017) (Table 2). And high level of U-Cd (0.84 μg/L) was found to be inversely associated with sperm motility – approximately 6.15% decrease in the sperm motility for every ten-fold increase in the U-Cd levels (Wang et al. 2016c). However, a study in China revealed that U-Cd level was not related to sperm DNA damage in infertile men who had an average U-Cd level of 0.63 μg/L (Zhou et al. 2016b). This is consistent with different dose thresholds of Cd based on the observed end points, mentioned above. Therefore, low levels of U-Cd may only affect sperm count and motility without genomic damage. The B-Cd level of $\geq$0.2 μg/L can increase serum inhibin B (Meeker et al. 2010), and a mean B-Cd level of 0.287 μg/L may impair sperm function via inducing oxidative stress, which causes higher oxidant levels in oligospermia compared to normospermia (Chinyere Nsonwu-Anyanwu et al. 2019); whereas the B-Cd level of $\geq$1.5 μg/L significantly reduces sperm count and is inversely associated with progressive motility and sperm morphology (Xu et al. 1993). These results are consistent with the study of Sukhn et al., who found an average B-Cd level of 6.8 μg/L was inversely associated with sperm viability, after adjusting for confounders (Sukhn et al. 2018). Hence, the B-Cd level of 0.2 μg/L maybe a lowest level for Cd-induced reproductive toxicity in general population. According to the study by Ikeda et al., the B-Cd level of 0.2 μg/L is equivalent to daily dietary Cd intake of 11.69 μg/day, which far exceeds the MRL for chronic-duration oral exposure to Cd (Ikeda et al. 2015). So, 11.69 μg/day may be the minimum daily Cd intake for causing reproductive toxicity. However, this minimum daily Cd intake in human is lower than the minimum oral intake in animals related to reproductive injury. This may be on account of the different bioavailability and half-

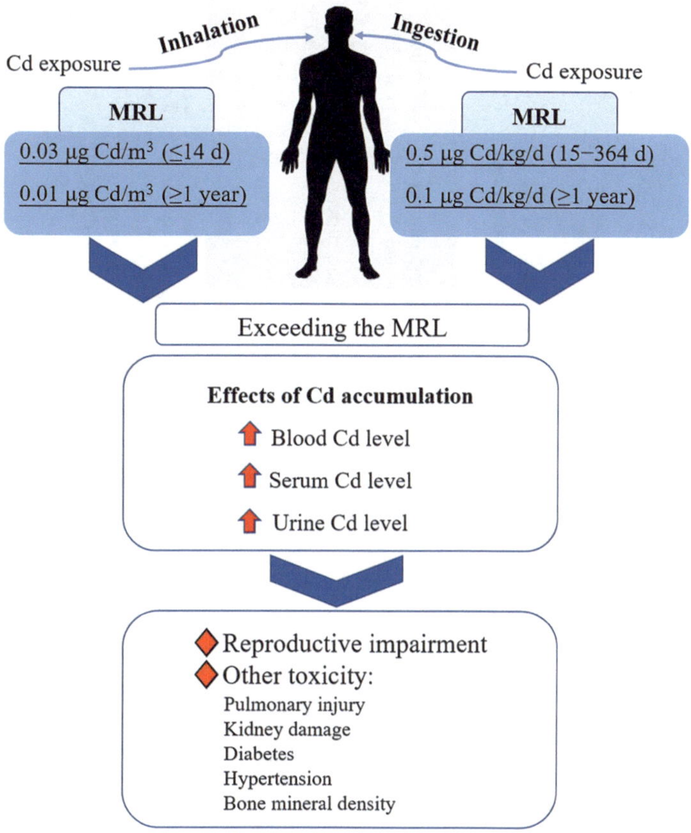

Fig. 1 Exposure dose of Cd that causes toxicity in humans. Cd accumulates in human body through inhalation and ingestion. The content of Cd in human body exceeding a certain MRL causes an increase in U-Cd, B-Cd, or S-Cd, inducing healthy damage which includes reproductive impairment, pulmonary injury, kidney damage, and other health problems (the data are based on the previous publications referred in the review)

life of Cd in animals and humans. Previous studies have reported the bioavailability of Cd is 0.5%–3% in animals and 3.0–3.5% in humans; for half-life of Cd, it is weeks to 9 years in animals contrasting to 10–35 years in humans (Bellinger 2004).

In contrast to U-Cd and B-Cd, the S-Cd level (geometric mean of 0.8 µg/L) exhibited a significant positive correlation with the percentage of immotile sperm (Mendiola et al. 2011). This suggests that there is no association exists among B-Cd, U-Cd, and S-Cd, as S-Cd in the male reproductive tract would be a better risk marker assessed than the circulating Cd concentration (Mendiola et al. 2011). The S-Cd level at as low as 0.47 µg/L damaged the sperm function by inducing oxidative stress in oligospermia (Sukhn et al. 2018); and the S-Cd level of $\geq$3.77 µg/L increased the risk of damaging sperm DNA (Taha et al. 2013). A study from Lebanon has reported that the average S-Cd level of 6.22 µg/L (in low-quality semen group) was inversely

Table 2 Clinical or epidemiological studies on Cd-induced reproductive toxicity

Test items	Mean level	Influence index	Study population	Age distribution	Adjust cofounders	References
S-Cd	3.77 μg/L	Positively related to impairment of sperm motility and vitality percentages and sperm DNA fragmentation	Idiopathic oligoasthenospermia men, 30	Mean age 34.6 years	No	Taha et al. (2013)
S-Cd	Normal-quality semen group 3.67 μg/L Low-quality semen group 6.22 μg/L	Decrease sperm viability	Low- and normal-quality semen groups, 116	18–55 years	Age, cigarette smoking, alcohol intake, and period of sexual abstinence	Sukhn et al. (2018)
S-Cd	Oligo-astheno-teratozoospermia 0.9 μg/L; normospermia 0.8 μg/L	Positively correlated with the percentage of immotile sperms	Infertile men, 30	Mean age 33.5 years	Age, BMI, and number of cigarettes per day	Mendiola et al. (2011)
B-Cd	0.2 μg/L	Increase in inhibin B	General population, 219 men	18–55 years	Age, BMI, and current smoking status	Meeker et al. (2010)
B-Cd	1.25 μg/L	B-Cd level of over 1.5 μg/L significantly reduces sperm count	Infertile men, 221	24–54 years, mean age 34.8 years	No	Xu et al. (1993)
B-Cd	1.9 μg/L	The B-Cd of $\leq$6.3 μg/L was inversely associated with semen volume progressive motility, and sperm morphology	General population, 587 men	20–59 years	Age, occupation, season of semen sample collection, abstinence intervals, smoking, alcohol drinking, and BMI	Li et al. (2016d)
B-Cd	Oligospermia 0.287 μg/L, Azoospermia 0.305 μg/L, Normospermia 0.219 μg/L	Decrease sperm count	Infertile men, 130	20–60 years	No	Chinyere Nsonwu-Anyanwu et al. (2019)

U-Cd	0.84 μg/L	Inversely correlated with progressive sperm motility and total motility	Infertile men, 1,247	Mean age 32.1 years	Age, BMI, smoking status, daily cigarette consumption, alcohol use, having ever fathered a pregnancy and urinary creatinine	Wang et al. (2016c)
U-Cd	0.63 μg/L	No relationship with sperm DNA damage	Infertile men, 207	Mean age 31.6 years	Age, BMI (continuous), smoking status, and abstinence time	Zhou et al. (2016b)
U-Cd	1.4 μg/L	The U-Cd of ≥1.4 μg/L decreases the count of motile sperm	HIV-infected population, 50 men	23–44 years, mean age 35.6 years	No	Li et al. (2017)
U-Cd		Serum testosterone levels elevated when U-Cd level at 10–20 μg/g creatinine, LH increased when the U-Cd level above 20 μg/g creatinine	Occupational exposure (from different workshops in a smelter), 166 men	21 to 78 years	Age, smoking habits, and alcohol consumption	Zeng et al. (2002)
U-Cd	14.5 μg/L	Positively related to the number of multiple morphological defects and DNA damage in the sperm and the concentration of seminal MDA	Occupational exposure (working in tea garden for more than 5 years), 100 men	Mean age 30.4 years	No	Mitra et al. (2020)

Note: All the samples are collected for 1 time

Cd cadmium, *S-Cd* seminal plasma cadmium, *B-Cd* blood cadmium, *U-Cd* urine cadmium, *BMI* body mass index associated with sperm motility, after adjusting for confounders (Chinyere Nsonwu-Anyanwu et al. 2019). Therefore, 0.47 μg/L may be the minimal S-Cd level which causes reproductive toxicity.

Concerning occupational Cd exposure, the epidemiological studies related to male reproductive function are scarce. As the above population studies, based on U-Cd level, a Chinese cross-sectional study reported that men in occupational exposure group have higher U-Cd levels (mean level 16.18 μg/g creatinine) compared to that in normal group (mean level 5.01 μg/g creatinine); U-Cd level (10–20 μg/g creatinine) was positively correlated with the concentration of serum testosterone and not associated with serum follicle stimulating hormone (FSH) level (Zeng et al. 2002). However, an Indian study measured the S-Cd concentration of 200 tea garden workers and 200 environmentally exposed donors. Compared to the donor candidates, as expected, the tea garden workers exhibited a notably high concentration of S-Cd along with a decreased sperm motility and an increased number of morphologically defective sperms. The median values of S-Cd in both garden workers (14.5 μg/L) and the normal group (0.99 μg/L) varied significantly, which is probably because of Cd accumulated in tea-estate soil and tea leaves, corresponding to a high-dose exposure in animal models (Mitra et al. 2020).

However, these results are from observational studies, which vary according to the geography, ethnicity, sample size, lifestyle, and other factors. In addition, some studies failed to find the toxic effects, even reported contradictory results. Meeker et al. studied on the semen quality of 219 participants, which include fertile and infertile men; They found that there was no correlation between the B-Cd levels and semen quality (Meeker et al. 2008). A study in German recruited 12 fertility men and 44 normozoospermia infertility men, 118 patients in an infertility clinic, as well as two workers with occupational exposure to Cd. The study did not report any correlation between the S-Cd levels and semen parameters or the fertility status (Keck et al. 1995). These contradictory results may be due to the discrepancies in the study population, the variation in the extent of Cd exposure, or the lack of adjustment for potential confounding factors. Previous studies have reported that age, body mass index, marital status, other metal exposures, and the level of FSH are independent factors affecting male reproductive function (Li et al. 2015; Zeng et al. 2002, 2004a). Meanwhile, the dietary habits and the content of Fe^{2+} in the body influence the extent of Cd intake, which lead to different effects on reproductive system (Jarup et al. 1998). Besides, these observational studies do not assess the fluctuation and duration of Cd exposure or the endpoints that induce reproductive toxicity over time. In a word, excessive Cd in the body can cause reproductive damage to different extent. Living in Cd-contaminated areas, smoking, intaking foods with high Cd content and Fe^{2+} deficiency are risk factors for high Cd exposure (Jarup et al. 1998). However, the Cd exposure threshold for reproductive system damage needs to be further studied and confirmed.

2.3 *Relevance of Studies in Animals to Human Exposure and Potential Risks*

Although the information summarized above includes both animal models and human data, there appears to be a large gap between these two groups of studies. For instance, animal models (summarized in Table 1) have the following features: (1) Of all the effect of Cd on male reproductive system, there are a few features: Cd was primarily given orally (dietary food or drinking water) or by injection via IP or SC; (2) pathogenic effects can be examined at all levels – testicular weight, pathology, spermatogenesis and cell death, seminal volume, and sperm parameters, as well as cellular and molecular variables; (3) a few studies with measurement of blood sex hormones such as luteinizing hormone (LH) and FSH; and (4) all these variables mentioned above are against the dose levels of Cd given with milligram per body weight (mg/kg), instead of testicular and seminal levels. In contrast, however, human studies (summarized in Table 3) have totally different features: (1) Cd intake is predominantly through chronic inhalation, dietary foods, water, fruits, and vegetables; (2) measurable indices are seminal volume and sperm quantity and quality; (3) blood FSH, LH, and testosterone; (4) all these variables against Cd levels in blood or urine but predominantly in semen at microgram level (μg/L, Table 3). These feature differences produce gaps between the two types of studies, which will significantly restrict the translational findings from animal studies to human situations and risk assessments.

In general, understanding how to extrapolate doses among species is a particularly important step for pharmaceutical and toxicological researchers to consider use the findings from animal studies for human health evaluations. Interspecies allometric scaling for dose conversion from animal to human studies is an empirical approach where the exchange of drug dose is based on the normalization of dose to body surface area (Nair and Jacob 2016). This approach assumes that there are some unique characteristics on anatomical, physiological, and biochemical process among species. This method is frequently used in research for experimental purpose to predict an approximate dose based on data existing in other species. The key points in scaling of dose include (a) larger animals have lower metabolic rates, (b) physiological process of larger animals is slower, (c) larger animals require smaller drug doses on a weight basis, (c) allometry accounts for the difference in physiological time among species, and (d) allometric scaling should not be applied to convert adult doses to kids (Nair et al. 2018; Nair and Jacob 2016). Therefore, dose estimation requires careful consideration of the difference in pharmacokinetics and pharmacodynamics among species; therefore, to date, there has not been a complete risk assessment for the effects that Cd has on male germ cells and sperm. This is probably due to the complication of Cd in the following aspects.

The patterns and levels of Cd exposure are different between studies on animal models and human epidemiological observations. In the former, Cd is typically administered to animals in its chemical state, such as $CdCl_2$ or $CdSO_4$, whereas in the latter, the form of Cd changed based on exposed routes of Cd. The mice were

Table 3 Main therapeutic agents for Cd-induced male reproductive toxicity

Therapeutic classification	Agents	Dose	Model	Major findings	References
Antioxidant	Curcumin	50 mg/kg.BW/d/IP for 10 d	Acute Cd exposure ($CdCl_2$ 2 mg/kg.BW/d/IP for 10 d)	Serum T↑, antioxidant capacity↑, Nrf2↑, MDA↓	Yang et al. (2019)
	Piceatannol	10 mg/kg.BW/d/oral for 28 d	Sub-chronic Cd exposure ($CdCl_2$ 5 mg/kg.BW/d/oral for 28 d)	3β-HSD and 17β-HSD↑, sperm parameters improved, serum T↑, anti-oxidant enzyme activities↑	Shi and Fu (2019)
	N-acetylcysteine	100 mg/kg.BW/d/IP for 24 h	Acute Cd exposure ($CdCl_2$ 2 mg/kg.BW/IP, once for 24 h)	Alleviates the Cd-induced testicular histopathological damage, protects against germ cell apoptosis, HO-1, and 3-NT↓, CHOP/ eIF2α/ GRP78↓	Ji et al. (2013)
	Tribulus terrestris exact	5 mg/kg.BW/d/oral for 6 w	Sub-chronic Cd exposure ($CdCl_2$ 3 mg/kg.BW/w/SC for 6 w)	LDH and ALP↓, MDA↓, antioxidant capacity↑	Rajendar et al. (2011)
	Quercetin	50 mg/kg.BW/d/oral for 4 w	Sub-chronic Cd exposure ($CdCl_2$ 5 mg/kg.BW/d/oral for 4 w)	WBC counts/Hb↑, T4/T3/fT4/fT3/TSH↑, sperm quality↑, testicular spermatogenesis scores↑	Badr et al. (2019)
	Melatonin	10 mg/kg.BW/d/IP for 7 d	Acute Cd exposure ($CdCl_2$ 2 mg/kg.BW/d/IP for 7 d)	Alleviate DNA damage, improve sperm parameters, cell apoptosis↓, mitochondrial membrane potential↑, protect acrosomal integrity, serum T↑	Li et al. (2016b)
	Rutin	25, 50 or 100 mg/kg.BW/d/oral for 15 d	Acute Cd exposure ($CdCl_2$ 50 mg/kg.BW/d/oral for 15 d)	CAT↑, GSH ↑, MDA↓, sperm quality↑, serum T↑	Abarikwu et al. (2017)
Anti-inflammation	Ganoderma triterpenoids	0.5 mL (20 mg/mL)/d/oral for 60 d	Sub-chronic Cd exposure ($CdCl_2$ 140 mg/kg.BW/d/oral for 60 d)	GPx↑, SOD↑, MDA↓, TNF-α/IL-1β/IL-6↓, Bcl-2↑, Bax↓, alleviated histo-pathological changes	Wang et al. (2019a)
	Thymoquinone	10 mg/kg.BW/d/IP for 5 d	Acute Cd exposure ($CdCl_2$ 2 mg/kg.BW/IP, injection once)	GSH↑, SOD↑, MDA↓, iNOS/TNF-α/COX-2/NF-κB↓	Fouad and Jresat 2015
	Infliximab	5 or 7 mg/kg.BW/w/IP for 3 w	Cd exposure ($CdSO_4$ 15 mg/kg.BW/d/oral for 3 w)	Body weight↑, testes weight↑, sperm count and motility↑, GSH↑, CAT↑,	Habib et al. (2019)

				MDA↓, ACP and ALP↑, 3β-HSD and 17β-HSD↑, iNOS/TNF-α/COX-2/NF-κB↓	
	Captopril	60 mg/kg.BW/d/oral for 5 d	Acute Cd exposure ($CdCl_2$ 2 mg/kg.BW/IP, injection once)	Serum T↑, MDA↓, iNOS/NF-κB↓, FasL and caspase-3↓	Fouad and Jresat (2013)
	Telmisartan	10 mg/kg.BW/d/oral for 5 d	Acute Cd exposure ($CdCl_2$ 2 mg/kg.BW/IP, injection once)	Serum T↑, MDA↓, iNOS/NF-κB↓, FasL and caspase-3↓	Fouad and Jresat (2013)
Maintain testosterone production	Shilajit	50, 100 or 200 mg/kg.BW/d/oral for 35 d	Sub-chronic Cd exposure ($CdCl_2$ 2 mg/kg.BW/d/oral for 35 d)	Testes weight↓, sperm counts↓, serum T↓, 3β-HSD/17β-HSD↓	Mishra et al. (2018)
	Moringa leaf ethanolic extract	500 mg/kg.BW/d/oral for 56 d	Sub-chronic Cd exposure ($CdCl_2$ 1 mg/kg.BW/d/oral for 56 d)	Relative testes weight↑, sperm motility↑, serum T↑, MDA↓, GSH↑, alleviate the damage of seminal vesicles, CYP17α↑, StAR↑	Elblehi et al. (2019)
Supplement trace elements	Na_2SeO_3	0.1, 0.2 or 0.4 mg/kg.BW/d/oral for 15, 25 and 35 d	Acute and sub-chronic Cd exposure ($CdCl_2$ 5 mg/kg.BW/d/oral for 15, 25 and 35 d)	Body weights↑, SOD and GPx↑, MDA↓, sperm concentration↑, spermatogenesis↑, serum T↑, StAR↑	Ren et al. (2012)
	ZnCl2	0.5 or 1.5 mg/kg.BW/d/IP for 21 d	Cd exposure ($CdCl_2$ 1 mg/kg.BW/d/IP for 21 d)	GPx↑, alleviate the testis damage, spermiogenesis↑	Babaknejad et al. (2018)
Chelation therapy	Diethyldithiocarbamate	0.1, 0.2, 0.5, or 1 mmol/kg.BW/d/IP for 7 d	Acute Cd exposure ($CdCl_2$ 3 mg/kg.BW/d/SC for 7 d)	Lipid peroxidation↓, testicular weight↑, testicular Ca^{2+} concentration↓, Fe^{2+} concentration↓	Ono et al. (1997)
	Deferasirox	0.05, 0.1, 0.25, or 0.5 mmol/kg.BW/d/IP for 7 d	Acute Cd exposure ($CdCl_2$ 3 mg/kg.BW/d/SC for 7 d)	Lipid peroxidation↓, testicular weight↑, testicular Ca^{2+} concentration↓, Fe^{2+} concentration↓	Ono et al. (1997)

Cd cadmium, *Zn* zinc, *T* testosterone, *d* day, *w* week, *BW* body weight, *IP* intraperitoneal injection, *SC* subcutaneous injection, *CAT* catalase, *SOD* superoxide dismutase, *MDA* malondialdehyde, *GST* glutathione S-transferase, *HO-1* Heme Oxygenase-1, *3-NT* 3-Nitrotyrosine, *GSH* glutathione, *GPx* glutathione peroxidase, *COX-2* cyclooxygenase-2, *ER* endoplasmic reticulum, *CHOP* C/EBP homologous protein, *GRP78* Glucose regulatory protein 78, *p-eIF2α* Phosphorylation of the eukaryotic initiation factor 2, *CYP17α* P450-C17 alpha, *StAR* steroidogenic acute regulatory protein, *Nrf2* nuclear factor erythroid 2-related factor 2, *3β-HSD* 3β-Hydroxysteroid dehydrogenase, *17β-HSD* 17β-Hydroxysteroid dehydrogenases, *iNOS* inducible nitric oxide synthase, *NF-κB* nuclear factor kappa B, *TNF-α* tumor necrosis factor alpha, *FasL* Fas ligand, *IL* interleukin, *LDH* lactate dehydrogenase, *ALP* alkaline phosphatase, *ACP* acid phosphatase, *T4* thyroxine, *T3* triiodothyronine, *fT3* free triiodothyronine, *fT4* free thyroxine, ↑ increase, ↓ decrease repeatedly administered Cd via 3 intravenous (IV) injections at 2-day intervals, and then, 2 days after the last IV injection, they were killed to observe the Cd distribution; Cd retention was the highest in the liver, followed by the kidneys (6.8%); ~0.2% of the total injected Cd was observed in the testes (Shank et al. 1977). However, for human, dietary intake of small doses of Cd is one major resource; Most of the Cd in meats, some shellfish, and even in plants and vegetables are present in the form of binding metallothionein (MT) or MT-like proteins; MT is heat resistant (over 100 °C) so that cooking procedure does not destroy the Cd-binding MT structure. Therefore, Cd intake in humans is significantly different from that in animals. Studies have confirmed that Cd in its chemical form and Cd-MT have similar absorption features, but significantly different tissue distribution features: (a) more than 73% of $CdCl_2$ was retained in the liver, whereas only 69.8% and 18% of MT-Cd were retained in the kidneys and liver, respectively; (b) similar proportions of both forms of Cd were retained in the testis (~0.1%–0.2%), according to the lowest group, in which the animals were examined at 2 weeks after five doses of Cd by gavage once a week (Cherian 1983).

Compared with that in animals, Cd deposition in human tissues has been reported as the highest concentration among those in the 50 examined species (Koizumi et al. 2008). Koizumi et al. measured Cd levels in the livers and kidneys of humans and other 50 mammalian species under normal conditions in Japan; the results revealed that the highest Cd accumulation levels in both the livers and kidneys is in humans, followed by primates, and the lowest Cd accumulation levels in mice and rats; the Cd accumulation in humans was ~1,000-fold to that in mice and rats. These cross-species discrepancies cannot be explained by differences in daily Cd intake. Therefore, we need further systemic studies on human population alongside other animal species, particularly mice or rats, under the same conditions to determine the potential mechanisms responsible for these differences in Cd accumulation.

3 Cd Toxicity on Offspring

The development of embryo relies on the placenta in mother's body. Because the placenta not only provides nutrients for the embryo, but also filter toxin from blood and prevents toxin damage to the embryo. Hence, once the structure and function of the placenta are impaired, embryonic development will be affected. During pregnancy, Cd can travel through the bloodstream to the placenta and decrease the expression of metal transporters, such as ATP-binding cassette G2 and ATP-binding cassette G4, thereby increasing Cd accumulation in the placenta (Liu et al. 2016a). Once accumulated in the placenta, Cd can decrease the proliferation of placental trophoblast cells and inhibit cell migration by downregulating the expression of several related functional genes and proteins, resulting in delayed placental formation and reduced placenta quality (Brooks and Fry 2017; Erboga and Kanter 2016; Liu et al. 2016a; Zhou et al. 2016a). Cd passes through the placenta to the embryo, further causes embryonic genome damage and abnormal embryonic development (Ling et al. 2017), leading to fetal growth restrictions and reduced birth weight and length (Romano et al. 2016; Zhang et al. 2018). However, several studies have confirmed that both the sex and gestational exposure time affect the fetal Cd toxicity (Taylor et al. 2016; Wang et al. 2016a).

After birth, a neonate whose mother was exposed to Cd during pregnancy showed a low birth weight and small head circumference, as well as damaged organs and numerous altered gene expressions. Maternal Cd exposure reduced the testis weight of fetuses, decreased the activity of 17β-Hydroxysteroid dehydrogenase (17β-HSD), and inhibited the expression of P450 cholesterol side-chain cleavage (P450scc) enzyme and steroidogenic acute regulatory (Star) protein (a key enzyme for transferring cholesterol into the mitochondria to produce steroid hormones), resulting in the decreased levels of serum and testicular testosterone in male fetuses (Ji et al. 2011). However, in another study, no statistically significant difference in testicular weight in adult male offspring was observed between control mice and mice whose mothers were exposed to Cd in gestational period, although there was a significant decrease in the testosterone levels in male fetuses whose mothers were exposed to Cd (Ji et al. 2011). These results may indicate that the reduction of testicular weight in the fetus period caused by maternal Cd exposure is reversible, whereas the damage in testicular steroidogenesis is irreversible. For female offspring in F1(the first) generation whose mother was exposed to Cd during pregnancy and lactation, decreased gonad gland weight and damaged follicles were observed. In addition, Cd can also affect the production of steroid hormones in females by decreasing the activity of 17β-HSD and 3β-HSD, as well as downregulating the expression of Star and CYP11 (Cytochrome P450, family 11), which are critical enzymes for testosterone synthesis (Pillai et al. 2010). However, the results in the F2 (the second) generation offspring were not exactly consistent with that in the F1 generation. Prenatal Cd exposure increased serum testosterone in male F2 rats and upregulated the mRNA expression of Star. Although gonad damage was observed in both F1 and F2 generations after maternal Cd exposure, the damage was less in F2 offspring. The different effects are probably because F1generation females may have a self-regulating mechanism to acclimate, thereby causing less damage in the F2 generation (Huang et al. 2020).

Maternal Cd exposure had a significantly negative effect on neurodevelopment. In vivo studies on animal models illustrate that chronic exposure to Cd during pregnancy damages the brain structure and impairs cognitive function in male offspring (Feng et al. 2019). Epidemiological research results are consistent with these results. Several prospective birth cohort studies have shown that the concentration of Cd in pregnant women has an inverse correlation with the performance intelligence quotient (IQ), development quotient (DQ), and brain-derived neurotrophic factor of offspring (Jeong et al. 2015; Kippler et al. 2016). However, there is no damage to the motor skills of school-age offspring (Taylor et al. 2018). Cd may have a selective impairment on cranial nerves; or the cognitive function damage in infants is irreversible but the motor function damage is reversible. Further research is required to verify these speculations. In addition, perinatal exposure to low-dose Cd induced inflammation and fibrosis of the prostate, increased the risk of congenital heart defects and hypertension, and impaired kidney function in offspring (Blum et al. 2015; Hudson et al. 2019; Ronco et al. 2011; Santana et al. 2016). These results clearly indicate Cd-induced toxicity in male reproduction and offspring; however, the specific mechanism is unclear. Accordingly, we reviewed current studies on the toxicity of Cd on the testis and summarized the mechanisms.

4 Mechanism of Cd Damage on Male Reproductive System

4.1 Activation of Oxidative Stress

According to a large number of studies, oxidative damage is considered as the primary reason for Cd-induced reproductive toxicity. Under normal physiological circumstances, the body remains in a redox balance, which is a state regulated by two effects: oxidation and anti-oxidation. However, when exposed to Cd, cells produce a large amount of reactive oxygen species (ROS, such as $O^{2\cdot-}$, H_2O_2, and OH·) in response to changes in the microenvironment of organisms and suppress the activities of antioxidants, which are important scavengers of ROS, resulting in ROS exceeding the scavenging capacity, eventually leading to oxidative stress (Bashir et al. 2019; Khanna et al. 2016; Oguzturk et al. 2012; Ono et al. 1997). It is known that Cd-induced excessive ROS causes lipid peroxidation (LPO) and DNA base pair damage in testes, as evidenced by the excessive production of malondialdehyde (MDA) and thiobarbituric acid reactive substances (TBARS), which are the end products of LPO (Ates et al. 2004; Bashir et al. 2019; Chen et al. 2018; Oguzturk et al. 2012).

Cd, however, does not directly produce oxygen radicals. Cd increases the production of ROS by increasing cytosolic Ca^{2+} concentrations in mouse testes and seminiferous tubules, as well as by suppressing the total antioxidant capacity (Correia et al. 2017). Both acute and chronic exposure to Cd in mouse models decreased sperm motility, reduced sperm progressive motility and viability due to decreased intracellular Ca^{2+} through transiently inhibiting the sperm-specific cation channel (CatSper) current (Wang et al. 2017). Moreover, Cd increased cytoplasmic Ca^{2+} by inhibiting the activities of dynein ATPase and plasma membrane Ca^{2+}-ATPase (PMCA), disrupting Ca^{2+} homeostasis, which decreased MMP and caused Leydig cell death (Da Costa et al. 2016; Khanna et al. 2016; Yang et al. 2003). Treatment with regucalcin (RGN) displayed a lower activity of TBARS and a higher activity of glutathione S-transferase (GST), along with an increased intracellular Ca^{2+} concentration, exhibiting an antioxidant effect (Correia et al. 2017).

In addition, Cd can activate specific signaling pathways to induce oxidative stress (Figs. 2 and 3). Nuclear factor erythroid-derived 2-like 2 (Nrf2), a transcription factor that regulates the activity of antioxidant enzymes, plays a vital role in preventing oxidative stress caused by Cd. Cd exposure downregulates nuclear Nrf2 followed by a decrease in the expression of antioxidants that are downstream of Nrf2, inducing testicular dysfunction (Shi and Fu 2019; Yang et al. 2019).

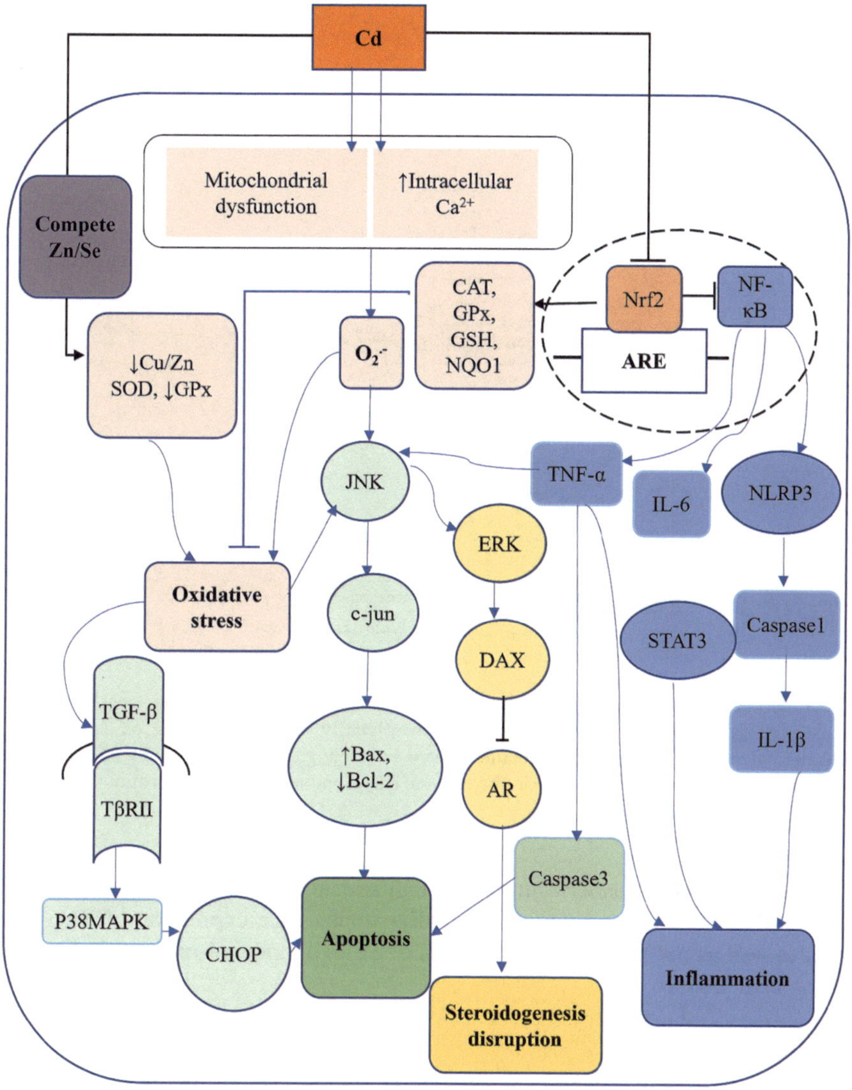

Fig. 2 Main signaling pathways that regulate the development of reproductive toxicity. Cd-induced reproductive toxicity is involved in multiple proteins and signaling pathways, as detailed in this current review (the data are based on the previous publications referred in the review)

Furthermore, the oxidative stress caused by Cd triggers the TGF-β (transforming growth factor beta)/p38 MAPK signaling pathway, leading to germ cell apoptosis and BTB disruption (Chen et al. 2018). In addition, Cd exposure causes increased apoptosis and elevated ROS production in Leydig cells by stimulating the JNK

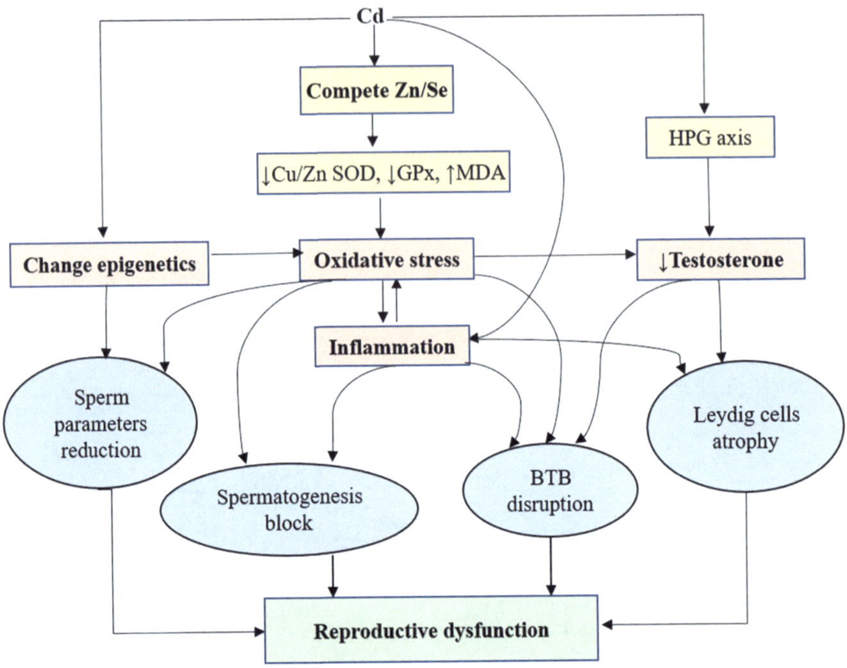

Fig. 3 Mechanisms of Cd-induced damage on male reproductive system. The mechanisms of Cd-induced damage on male reproductive system primarily include activating oxidative stress, inducing inflammation, impairing endocrine system, competing the active sites with essential trace elements, and changing the epigenetics (the data are based on the previous publications referred in the review)

(C-Jun N-terminal kinases) pathway, as antioxidant N-acetylcysteine (NAC) and JNK-IN-8 (JNK inhibitor) can substantially suppress the expression of ROS and JNK, as well as mitigate the cell apoptosis of TM3, a model of mouse Leydig cells (Wang et al. 2019b).

4.2 *Induction of Gonad Inflammation*

Inflammation is another important factor that mediates male reproductive damage caused by Cd. Cd can activate proinflammatory factors, such as tumor necrosis factor alpha (TNF-α) and nuclear factor kappa-light-chain-enhancer of activated B cells (NF-κB), and inhibit anti-inflammatory cytokines, such as interleukin-4 (IL-4), which in turn increases the expression of Fas (factor associated suicide) ligand, caspase 1, and caspase 3, resulting in germ cell apoptosis (Fouad et al. 2020; Fouad and Jresat 2013). Activated nuclear NF-κB, a transcription factor, can upregulate inflammatory cytokines, such as TNF-α and cyclooxygenase-2 (COX-2). TNF-α acts as a cell signaling proinflammatory cytokine and activates inducible nitric oxide synthase (iNOS), leading to excessive production of reactive nitrogen species (RNS), such as peroxynitrite; together with the reduced IL-4, TNF-α induces testicular tissue destruction and decreases serum testosterone in Cd-exposed males (Al-Azemi et al. 2010; Habib et al. 2019; Minutoli et al. 2015).

Previous studies have reported that Cd induced testicular inflammation by activating several signaling pathways. Cd significantly increases nucleotide-binding oligomerization domain, leucine-rich repeat, and pyrin domain-containing 3 (NLRP3) inflammasome, which was activated in response to different stressors, further upregulates inflammatory cytokines, such as interleukin-1β and interleukin-18. In addition, Cd activates the interleukin-6 (IL-6)/ signal transducer and activator of transcription 3 (STAT3) signaling pathways, wherein IL-6 binds to its receptor, and then, causes the phosphorylation of STAT3, which acts as a transcriptional factor enhancing the transcription of inflammatory genes (Chen et al. 2016; Fouad et al. 2020; Huang et al. 2016). Furthermore, in vitro studies have confirmed that the increased proinflammatory cytokine TNF-α in Cd-treated Leydig cells activates the JNK/ERK MAPK signaling pathway, further upregulating the expression of Dosage-sensitive sex reversal, Adrenal hypoplasia critical region, on chromosome X, gene 1 (DAX-1). DAX-1 is a transcription factor that can inhibit the activation of androgen receptor (AR) and regulate the transcription of genes related to androgen, leading to decreased expression of steroidogenic enzymes and reduced the production of testosterone (Minutoli et al. 2015; Sadasivam et al. 2015).

Moreover, Cd decreases sperm parameters and reduces testosterone concentration by activating undifferentiated testicular macrophages and altering their morphology, with a smooth and spherical outer surface (Chakraborty et al. 2014). Leukocyte infiltration is another important cause of Cd-induced asthenospermia, which activates intratesticular proinflammatory cytokines (Shi et al. 2012). Intercellular adhesion molecule-2 (ICAM2) plays a vital role in the migration of leukocytes across the endothelium during inflammation. Normally, ICAM2 localizes to elongating/elongated spermatids and Sertoli cells. Rats exposed to Cd showed increased ICAM2 in the restructuring process of seminiferous epithelium, with increased immunoreactivity to ICAM2 in round spermatids, leading to a disruption of spermatogenesis (Xiao et al. 2013).

4.3 Impairment of Endocrine System

The hypothalamic-pituitary-gonadal (HPG) axis plays an important role in regulating reproductive functions. The gonadotropin-releasing hormone (GnRH) produced by hypothalamus neurons activates the anterior portion of the pituitary gland to secrete LH and FSH. LH and FSH are vital hormones for testosterone production and secretion. Normally, FSH stimulates the production of androgen-binding proteins of Sertoli cells and facilitates the testicular growth. LH binds to AR on the Sertoli cells, and then activates the adenylate cyclase, to catalyze the conversion of adenosine triphosphatase (ATP) into cyclic adenosine monophosphate (cAMP), which mobilizes cholesterol through a series of intermediate steps to finally produce testosterone. Testosterone is indispensable for normal spermatogenesis including maintaining the integrity of BTB, aiding the adhesion of spermatids to Sertoli cells, supporting meiosis, and releasing sperm (Smith and Walker 2014).

Cd disrupts the secretion and regulation of reproductive hormones by influencing the HPG axis, including causing oxidative damage to the hypothalamus and pituitary (Nna et al. 2017). Low-dose Cd may affect the hypothalamus, altering the LH circadian pattern without decreasing the LH concentration and testosterone levels (Lafuente et al. 2004). In contrast, high-dose Cd exposure significantly reduced the secretion of LH and FSH, which in turn decreased the production of testosterone, leading to incomplete spermatogenesis and decreased count and motility of sperm (Badr et al. 2019). This may be because Cd exerts direct effects on the HPG axis when administered with high doses, which demonstrates Cd has a concentration-dependent biphasic effect on the reproductive system. However, Elbaghdady et al. reported that the levels of FSH and LH increased following Cd exposure, although the levels of testosterone and inhibin decreased (Elbaghdady et al. 2018). Two main causes may contribute to these discrepant results. First, under physiological conditions, inhibin suppresses the secretion of FSH; thus, the decreased inhibin expression caused by Cd induces a dysregulated increase in FSH levels. Inhibin suppresses the secretion of FSH; thus, the decreased inhibin expression caused by Cd induces a dysregulated increase in FSH levels. Second, Cd reduces the levels of FSH and LH via a negative feedback. Concurrently, Cd decreased the expression of luteinizing hormone receptor (LHR) to suppress the production of testosterone (Gunnarsson et al. 2003).

Furthermore, Cd exposure directly induced a significant decrease in the expression of 3β-HSD (3β-Hydroxysteroid dehydrogenase) and 17β-HSD (17-β-Hydroxysteroid dehydrogenases), which are rate-limiting enzymes in the conversion of cholesterol to steroid hormones, resulting in reduced testosterone and decreased sperm motility and concentration (Habib et al. 2019; Mishra et al. 2018). In addition, Cd downregulated the expression of several genes involved in testosterone production, such as the Star gene and P450-C17 alpha (CYP17α) gene. Star plays a key role in promoting the transport of cholesterol (a precursor of testosterone synthesis) from the outer mitochondrial membrane to the inner membrane in Leydig cells, and the cytochrome CYP17α codes for 17α-hydroxylase activity in the biosynthesis of androgenic hormones (Elblehi et al. 2019). In addition, Cd inhibits Sertoli cell and the germ cell protein – desert hedgehog (DHH), which in turn affects Leydig cell development and suppresses the expression of several genes involved in testosterone production, leading to decreased testosterone levels (Wu et al. 2017). In addition, Cd has androgen-like effects. For instance, Cd can mimic the androgen to occupy the binding sites of AR, inhibiting the binding of androgen to AR. With an increasing Cd concentration, a decrease in the number of binding sites and a reduction in AR proteins have been observed (Martin et al. 2002).

4.4 Competition with Essential Trace Elements

Cd binds directly to the active site of enzymes, then inactivates enzymes by replacing metal ions in the active center of the enzymes. Selenium (Se) binds to the active sites of many selenoenzymes and selenoproteins, which is crucial for maintaining normal spermatogenesis, testosterone production, and sperm quality (Boitani and Puglisi 2008). Clinical epidemiological investigations showed higher Cd concentrations along with a low Se concentrations in infertile men and smokers, demonstrating a significant inverse correlation between Se and Cd (Riaz et al. 2016; Schopfer et al. 2010). The decreased sperm parameters and testosterone in Cd-treated mice were directly positively correlated with the decreased Se concentration in testis and the downregulated expression of glutathione peroxidase (GPx), a Se-dependent antioxidant enzyme. Thus, Se supplementation could diminish the accumulation of Cd and upregulate the activity of GPx, exerting an antagonistic effect on Cd-decreased sperm parameters (Abarikwu et al. 2013; Jin et al. 2018; Said et al. 2010). These results suggest that the increased oxidative stress caused by Cd is correlated to the depletion of Se, owing to Cd occupying the Se site in antioxidant enzymes, deactivating the enzymes. Alternatively, Cd exposure decreased the concentration of Se in plasma and testicular tissues, by suppressing the Selenoprotein P (SelP) which is involved in the distribution and homeostasis of Se. Prior administration of Se before Cd treatment restored the relative percentage of spermatogenic cells and alleviated the reduction of serum testosterone levels that were affected by Cd (Messaoudi et al. 2010; Ren et al. 2012).

Evidence from experimental studies in animals has shown that Cd also mimics and antagonizes Zn (Bonda et al. 2004; Mouro et al. 2019). Zn is an active component of many enzymes and proteins that are necessary for proper physiological functions in organisms, such as DNA synthesis, cell division and gene expression, and electron transport. Zn is essential for maintaining normal spermatogenesis, male hormonal status, and sperm maturation. Several studies have reported that Zn concentration in seminal fluid is positively related to sperm amount and functions (Bonda et al. 2004; Fallah et al. 2018). As summarized in a recent review, Zn deficiency causes delayed sexual maturation, impaired spermatogenesis, and decreased testosterone levels, as well as increased cell apoptosis and poor semen quality (Beigi Harchegani et al. 2020). These outcomes may be associated with the decreased membrane-stabilization and antioxidant activity that due to the lack of Zn. Testicular Zn depletion may be a vital cause of testicular injury and lipid peroxidation induced by Cd; Zn supplementation can alleviate these effects (Bonda et al. 2004; Mouro et al. 2019). In addition, maternal Cd exposure decreased Zn transport to the fetus, thus reducing the Zn concentration in offspring testis, further damaged testicular structure and reduced testicular testosterone (Chemek et al. 2016).

The plausible mechanism of Cd-induced testicular damage may be that Cd competes with Zn for active sites in the Cu/Zn-superoxide dismutase (Cu/Zn-SOD), which is an antioxidative enzyme. Cd can inactivate this enzyme, increasing ROS and LPO, resulting in testicular oxidative damage including DNA damage in testes, and decreased sperm parameters and the testosterone concentrations (Amara et al. 2008). Similarly, Cd competes with Zn for active sites in Zn finger protein (Zip) which is a Zn transporter. Cd exerts effects on deleting Zn, then inactivates Zip 6 and Zip10 which are located on the membranes of spermatogenic cells, and finally enters cells, impairing the testicular structure (Croxford et al. 2011; He et al. 2009). Consistent with this, Cd competes with Zn for divalent metal receptor-1 sites, which are located in the duodenum and testis, for the specific transportation of Fe, Zn, and Mg. In addition, Cd increases hepatic MT, which can bind and store intracellular Zn in liver, leading to increased liver Zn concentration and decreased testicular Zn (Bonda et al. 2004; Mouro et al. 2019).

4.5 Changes in Epigenetics

Epigenetics refers to the heredity of altered gene function, without changes in the DNA code sequence, which ultimately leads to changes in the phenotype. This involves many cellular molecules and physiological processes, and slight changes in epigenetics can cause dysfunction in organisms (Nicoglou and Merlin 2017). Meanwhile, if these changes occur in germ cells or fertilized eggs, these epigenetic modifications will be passed on to the next generation, known as transgenerational epigenetic inheritance. Epigenetic mechanisms mainly include DNA modification, histone modification, non-coding RNA regulation, chromatin remodeling, and nucleosome localization. Epigenetics is widely studied in diabetes and cancer; however, it has not been adequately studied in male reproductive diseases (Basu Mallik et al. 2018; Jasperson et al. 2010).

It has been proposed that changes in epigenetics play an essential role in Cd-induced male reproductive toxicity. Inhalation exposure to Cd for 1 year in adult male rats caused global DNA hypermethylation in testes and elevated the expression of DNA methyltransferase (DNMT) 3a and DNMT 3b, enzymes that add a methyl group to unmethylated DNA (Fig. 4). However, epigenetics in the liver or kidneys are normal. These results may be attributed to the unique hypomethylated DNA domains in testis and germ cells (Nakayama et al. 2019). In contrast, acute Cd exposure during the neonatal period reduced testicular DNMT activity and suppressed the normal decrease in c-fos gene DNA methylation, which is continuously expressed throughout the spermatogenic cycle and is vital for regulating the resumption of spermatogenesis, but not altered at the global DNA methylation level. Cd-induced testicular damage may be related to decreased DNMT activity and aberrant DNA methylation of c-fos, which are expressed throughout the whole spermatogenic cycle (Zhu et al. 2011). Simultaneously, Cd disrupted BTB via hypermethylation of the multidrug resistance 1b (Mdr1b) gene promoter. It is known that Mdr1b is a membrane-localized protein expressed exclusively at the BTB. The hypermethylation of Mdr1b gene promoter leads to the failure of the recruitment of nuclear factor Ya, a critical factor for Mdr1b transcription, causing a

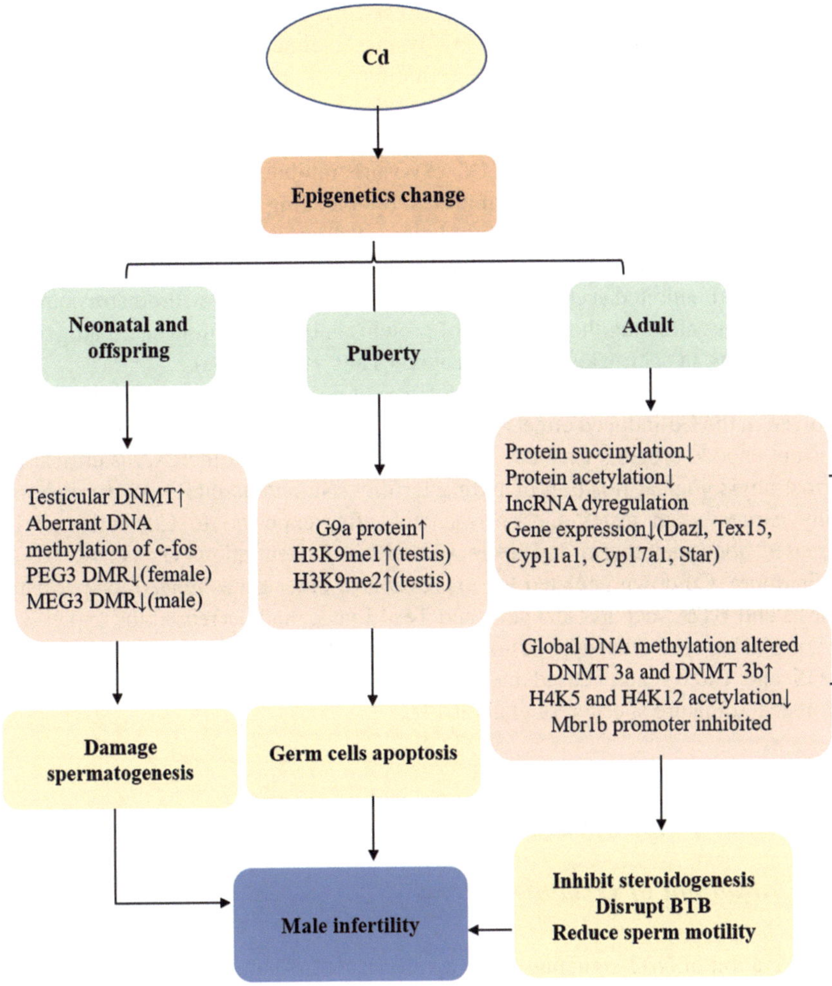

Fig. 4 Main epigenetics mechanisms of Cd damage on male reproductive system. Cd causes different changes of epigenetics in different stages of life, leading to the damage on reproductive function (the data are based on the previous publications referred in the review)

decreased expression of Mdr1b (Fang et al. 2020). In addition, several prospective cohort studies have shown that Cd can cross the placental barrier and induce sex-specific alterations at DNA methylation level in a fetus exhibiting decreased paternally expressed gene 3 (PEG3) differentially methylated regions (DMR) methylation in female offspring and reduced maternally expressed gene 3 (MEG3) DMR methylation in male offspring. The changes in DNA methylation are more pronounced in male offspring (Cowley et al. 2018; Dharmadasa et al. 2017).

Histone modification is another mechanism for Cd-induced toxicity on the reproductive system. Pubertal Cd exposure increased the level of G9a protein, which is a histone methyltransferase, followed by elevated methylation levels of several histones including histone H3 lysine 9 dimethylation (H3K9me2) and monomethylation (H3K9me1) in seminiferous epithelium. H3K9me2 and H3K9me1 in turn increased Bax levels and decreased the levels of Bcl-2, leading to induced extensive germ cell apoptosis and decreased sperm count (Li et al. 2016a). In line with the results, Cd exposure inhibited global acetylation and succinylation levels of proteins in testes by blocking tricarboxylic acid cycle, which is related to tyrosine phosphorylation of dihydrolipoamide dehydrogenase, and decreased the ATP levels, leading to decreased sperm motility (Li et al. 2016c). Moreover, Cd inhibited acetylation of several histones related to chromatin compaction, adversely affecting the assembly of proteins and chromatin in spermatogenetic cells, resulting in spermatogenesis damaged (Yang et al. 2018a).

As an epigenetic mechanism, the regulation of non-coding RNA (ncRNA) is also involved in the Cd-induced effects on the testis. lncRNA refers to a type of RNA that does not encode proteins over 200 nucleotides in length; and lncRNA is critical for normal physiological function including fertility. Numerous lncRNAs were abnormally expressed in testes and spermatozoa after exposure to Cd, followed by increased abnormal sperm and decreased sperm concentration (Gao et al. 2017). Furthermore, Cd downregulated the expression of other genes related to spermatogenesis and BTB, such as Dazl gene and Tex15 gene, and decreased the expression of hormone genes related to testosterone production. Moreover, the deletion of Tex15 and Dazl genes caused meiotic arrest and decreased spermatogenic cells and reduced mature sperm (Hu et al. 2014).

5 Potential Therapeutic Strategies

5.1 *Inhibition of Oxidative Stress*

The activation of Nrf2 signaling pathway alleviates Cd-induced testicular injury by enhancing the expression of antioxidants. Several plant extracts, such as curcumin, grapeseed proanthocyanidins, and sulforaphane, act as antioxidants by activating the Nrf2 signaling pathway (Shi and Fu 2019; Yang et al. 2018b, 2019). Curcumin is a chemical component extracted from Zingiberaceae and Araceae, with protective benefits that attenuate Cd-induced histopathological damage, abnormal sperm morphology, decreased serum testosterone and germ cell apoptosis, which have been demonstrated in Cd-induced testicular damage mouse models (Aktas et al. 2012; Momeni and Eskandari 2020; Oguzturk et al. 2012). Administration with sulforaphane, which is extracted from cruciferous vegetables, significantly upregulated Nrf2 transcriptional activity, along with the activated Nrf2/ARE (antioxidant responsive element) signaling pathway, thus preventing Cd-induced testicular oxidative damage and Sertoli cell apoptosis (Yang et al. 2019).

Increased intracellular Ca^{2+}, followed by oxidative damage, has been observed in Cd-treated rats and Leydig cells incubated with Cd. However, the administration of

RGN or NAC can abrogate oxidative damage. Overexpression of RGN in rats showed high antioxidant capacity normalizing the levels of oxidative stress and subsequently preventing incomplete spermatogenesis and germ cell apoptosis. Pretreatment with the NAC in vitro rescued cell viability which is reduced by Cd and increased the expression of 3-βHSD. These results suggest that RGN or NAC effectively protect testis from Cd-induced oxidative damage by normalizing the intracellular Ca^{2+} (Bashir et al. 2019; Correia et al. 2017; Ji et al. 2013).

In addition, natural antioxidant compounds, such as flavonoids, have direct antioxidant effects and protective effects on male reproductive function. Administration with Tribulus terrestris Linn reversed Cd-induced increase in MDA, as well as Cd-induced decrease in SOD and glutathione (GSH), preventing testicular oxidative damage (Rajendar et al. 2011). Quercetin (QE) and Rutin abrogated testicular damage caused by Cd through upregulating the expression of the enzymatic and non-enzymatic antioxidants, such as catalase and GSH, thus restoring normal tissue structure, increasing the concentration of testosterone, and improving sperm concentration and sperm quality (Nna et al. 2017). However, several randomized controlled trials have failed to demonstrate positive correlation between semen parameters and antioxidant treatment, and even reported negative outcomes (Hawkes et al. 2009; Henkel et al. 2019). It is reported that a healthy couple with recurrent pregnancy failures showed high levels of sperm DNA damage in male partner after taking high doses of antioxidants. When no antioxidant was consumed for 3 months, the wife became pregnant spontaneously (Henkel et al. 2019). This case potentially implies that high antioxidant levels may result in male infertility or negatively affect early embryo development.

5.2 *Anti-Gonad Inflammation*

Suppressing proinflammatory cytokines may protect the male reproductive system from Cd damage. In animal experiments, many plants and plant extracts mitigate Cd-induced testicular damage (Table 3). Ganoderma is a medicinal mushroom with several beneficial properties, such as antitumor, antioxidant, and anti-inflammatory (Tung et al. 2013). Pretreating Cd-exposed animals with 20 mg/mL Ganoderma Triterpenoids decreased the expression of proinflammatory cytokines such as TNF-α and prevented testicular damage (Wang et al. 2019a). Similarly, thymoquinone, which is extracted from Nigella sativa seed oil, significantly alleviated testicular damage by reducing the levels of several proinflammatory cytokines such as iNOS and COX-2 (Fouad and Jresat 2015). Furthermore, several chemical medicines, such as captopril, telmisartan, and infliximab reduce Cd-induced testicular toxicity by inhibiting inflammation. The protective characteristics of infliximab, such as antioxidative, anti-inflammation, and anti-apoptosis effects have been demonstrated in Cd-induced testicular damage models. Infliximab decreased proinflammatory cytokines to create an anti-inflammatory environment for the testis, further inhibiting the production of ROS to minimize oxidative damage, and abrogated Cd-induced testicular damage (Habib et al. 2019).

Previous studies have reported that angiotensin II, which binds to its receptor AT1, causes testicular inflammation by activating the NF-κB signaling pathway. Captopril decreases the production of the proinflammatory cytokines and inhibits the NF-κB signaling pathway by reducing angiotensin II, further alleviating testicular damage. Telmisartan specifically binds to the AT1-receptor and partially agitates the activity of the nuclear peroxisome proliferator-activated receptor gamma (PPAR-γ), thereby downregulating proinflammatory responses (Fouad and Jresat 2013). In addition, pretreatment with melatonin significantly reduced TNF-α and interleukin-1 and suppressed oxidative stress induced by Cd, resulting in an increased sperm count, sperm motility and serum testosterone (Li et al. 2016b). All these results illustrate that anti-inflammatory agents are beneficial to the reproductive function as they reduce Cd toxicity.

5.3 *Recovery of Endocrine System*

The HPG axis is vital for testosterone production. Stimulation of LH synthesis and secretion alleviated Cd-induced reduced testosterone and reproductive toxicity. Pretreatment with melatonin in Cd-exposed mice showed a significant amelioration in decreased LH, inhibin B, and testosterone via antioxidative and anti-inflammatory actions (Li et al. 2016b). Consistent with the effect of melatonin, QE-treated mice showed a significant increment in the levels of LH and FSH, as well as increased serum testosterone and sperm quality (Nna et al. 2017). These results may suggest that melatonin and QE can maintain the function of the HPG axis and recover testicular endocrine which is disrupted by Cd via anti-inflammatory and antioxidant effects.

The increased activities of steroidogenic enzymes play a crucial role in ameliorating Cd toxicity. Infliximab significantly upregulated the expression of 3β-HSD and 17β-HSD gene, thus increasing testosterone, preserving the normal spermatogenesis, and minimizing oxidative damage and apoptosis (Habib et al. 2019). Shilajit has also been scientifically validated as a potent medicine for Cd-induced male reproductive dysfunction. It can increase sperm concentration and motility and restore the functional status of male reproductive organs by eliminating the inhibition of steroidogenesis (Babaknejad et al. 2018; Gunnarsson et al. 2003). In addition, the stimulatory effect on steroidogenic genes such as Star and CYP17α and LHR gene increased testosterone in Moringa leaf ethanolic extract treated rats, which showed recovered gonad weight, complete spermatogenesis, and increased sperm parameters (Lafuente et al. 2004).

5.4 Supplementation with Essential Trace Elements

Many studies have confirmed the efficacy of essential elements, such as Se and Zn, in resisting the toxic effects of Cd; these elements are essential for maintaining physiological function in the body. Se concentration is positively related to sperm parameters and total antioxidant status (Riaz et al. 2016; Schopfer et al. 2010). Se acts as an active site for many proteins and is essential for normal spermatogenesis and male fertility. There are two selenoproteins that mediate the crucial role of Se – GPx4 and SelP. SelP delivers Se to the testis by binding to its receptor apolipoprotein E receptor 2, which is located in testis. SelP knockout mice and mice with Se deficiency showed abnormal spermatozoa with flagellar structural defects and sperm defects in sequential development (Boitani and Puglisi 2008; Olson et al. 2004). Se, as a cofactor of GPx, is vital for maintaining the activity of GPx, thus protecting sperm from oxidative stress. Numerous studies have confirmed that Se supplements can upregulate the expression of GPx and increase sperm motility, which were adverse in Cd exposure mice. Furthermore, supplementation with Se can prevent Cd-induced inhibition of spermatogenesis by upregulating Star and increasing testosterone synthetic enzyme activity to activate testosterone synthesis (Messaoudi et al. 2010). Several randomized clinical trials have demonstrated that supplemental Se administered daily increased sperm concentration and improved sperm motility in infertile men, thereby leading to increased pregnancy rate in their partners (Ardestani Zadeh et al. 2019; Moslemi and Tavanbakhsh 2011).

Adequate dietary Zn is important for normal fertility. Zn is assembled in the late type B spermatogonia, the spermatids and the Leydig cells, to maintain compete spermatogenesis and aid in the production and secretion of testosterone (Ren et al. 2012). Marginal to severe Zn deficiency can render the testes more susceptible to Cd toxicity and cause damage to the testes. Biologic Zn supplementation can reverse testicular pathological changes and impairments in the production of testosterone as well as improve sperm concentration and motility at the testicular, epididymis, and ejaculate levels (Babaknejad et al. 2018; Gunnarsson et al. 2003). The therapeutic effect of Zn may be due to the following factors. Zn primarily exerts a direct antioxidant effect. Zn is known as a structural and catalytic component of the Cu/Zn-SOD enzyme that protects against Cd toxicity by maintaining or increasing the concentration of sulfhydryl (SH) and Cu/Zn-SOD, which scavenge ROS (Chemek et al. 2016; Sankako et al. 2012). Furthermore, Zn has anti-inflammatory effect. Cd-treated rats showed enhanced expressions of inflammation markers, such as NF-κB, iNOS, in the testicular tissues, which in turn induced oxidative stress. However, Zn supplementation decreased the levels of proinflammatory cytokines and inhibited the activity of NF-κB and iNOS (Habib et al. 2019; Prasad 2014). In addition, Zn can upregulate the expression of testicular MT protein, which is a protein that participates in metal detoxification, binding Cd tightly via sulfhydryl groups of MT to mitigate Cd toxicity (McCarty 2012; Mikowska et al. 2018).

Higher PEG3 and MEG3 methylation levels were observed in offspring whose mother exposed to a high Zn concentration than those belonging to the Cd exposure group (Vidal et al. 2015). Creatinine-adjusted urinary Se and Zn levels were inversely related to serum testosterone and sperm parameters, whereas urinary Cd had a positive correlation with serum testosterone (Zeng et al. 2015; Zeng et al. 2013). Zn supplements increased sperm concentration and motility by upregulating the expression of several enzymes related to the urate pathway in the seminal plasma of asthenozoospermia (Alsalman et al. 2020). These results are consistent with several cross-sectional studies, which reported that Zn supplements significantly increased sperm motility as well as the activities of 5′-nucleotidase (a metalloprotein with a Zn active site in seminal plasma) and restored the adenosine deaminase and xanthine oxidase, which significantly elevated in asthenozoospermia patients (Alsalman et al. 2020; Ardestani Zadeh et al. 2019).

6 Conclusion and Recommendations

With increasing infertility in humans, Cd-induced reproductive toxicity has attracted increasing attention. As mentioned above, chronic environmental exposure is the most common pattern of Cd exposure in general population, although no consensus has been reached on how to classify Cd exposure. Current knowledge on Cd-induced reproductive toxicity is primarily derived from studies on Cd-induced testicular effects. However, Cd also affects other gonad organs in the male reproductive system. Long-term environmental Cd exposure may impair prostate function by interfering with the secretion of the prostate antigen, and even cause prostate cancer. The rats treated with 1 mg/kg CdC12 exhibited decreased intraluminal secretions and atrophic epithelial papillae in the seminal vesicles, as well as atrophic epididymis with shrinking lumen; however, the underlying mechanisms remain obscure (Andreucci et al. 2015; Jahan et al. 2014; Moslemi and Tavanbakhsh 2011; Zeng et al. 2004b). Hence, additional studies on other gonad glands for Cd exposure are required.

Current treatment regimens for Cd toxicity primarily address anti-inflammation, antioxidant agents, and trace element supplementation. As previously reported, chelators are effective in treating Cd-induced organ toxicity. Ethylenediaminetetraacetic acid, diethyldithiocarbamate, deferasirox, and deferiprone are chelators that can combine with Cd and remove Cd ions from the body. However, chelating agents are effective only during early administration after Cd exposure (approximately 1 week) and have no effect on chronic Cd exposure cases (Deng et al. 2019; Jamilaldin Fatemi et al. 2011; Ono et al. 1997). This may be due to the different cellular locations of Cd accumulation and chelator distribution. All these chelating agents are mainly distributed in the extracellular space, whereas Cd is mostly bound intracellularly to MT (Smith 2013). Therefore, agents that chelate Cd and bind to MT should be urgently investigated. In addition, the pharmacological inhibitor of G9a activity, such as BIX-01294, and inhibitors of the dihydrolipoamide dehydrogenase tyrosine phosphorylation, such as bovine serum albumin, have been shown to decrease H3K9me1and H3K9me2 and prevent the inhibition of energy metabolism, thus abrogating reproductive toxicity and infertility caused by Cd (Fang et al. 2020; Zhu et al. 2011). This means that maintaining normal epigenetics may be a new treatment for Cd-induced reproductive toxicity.

In conclusion, Cd induces the obstruction of spermatogenesis, disruption of BTB, decreased sperm quality, and infertility. Several potential mechanisms associated with Cd-induced reproductive toxicity include oxidative stress, inflammation, steroidogenesis impairment, essential trace element competition, and epigenetic changes related to numerous proteins and signaling pathways. However, effective prevention and treatment methods remain controversial. Therefore, a focus on sensitive proteins and intracellular signaling pathways that participate in the pathological processes of Cd-induced reproductive toxicity is required, and optimal strategies for prevention and treatment of Cd-induced reproductive toxicity are expected.

Acknowledgments We would like to thank Editage (www.editage.cn) for English language editing.

Authors' Contributions L.J.X., H.L., and B.Z. researched data for the article and wrote the manuscript. L.C. contributed substantially to the discussion of its contents. All authors reviewed and edited the manuscript before submission.

Conflict of Interest The authors declare that they have no conflict of interest.

Availability of Data and Materials All data were available by PubMed search since all data used in this review were obtained from PubMed with proper citations.

Funding The authors are grateful to the U.S-China Pediatric Research Exchange Training Program. The personnel expenses for LX and BZ when they worked in the University of Louisville (2018–2021) were in part provided by Jiangxi Provincial Children's Hospital, Nanchang, Jiangxi, China, according to this Exchange Program Agreement.

References

Abarikwu SO, Iserhienrhien BO, Badejo TA (2013) Rutin- and selenium-attenuated cadmium-induced testicular pathophysiology in rats. Hum Exp Toxicol 32:395–406. https://doi.org/10.1177/0960327112472995

Abarikwu SO, Olufemi PD, Lawrence CJ et al (2017) Rutin, an antioxidant flavonoid, induces glutathione and glutathione peroxidase activities to protect against ethanol effects in cadmium-induced oxidative stress in the testis of adult rats. Andrologia 49. https://doi.org/10.1111/and.12696

Aktas C, Kanter M, Erboga M et al (2012) Anti-apoptotic effects of curcumin on cadmium-induced apoptosis in rat testes. Toxicol Ind Health 28:122–130. https://doi.org/10.1177/0748233711407242

Al-Azemi M, Omu FE, Kehinde EO et al (2010) Lithium protects against toxic effects of cadmium in the rat testes. J Assist Reprod Genet 27:469–476. https://doi.org/10.1007/s10815-010-9426-3

Alli LA (2015) Blood level of cadmium and lead in occupationally exposed persons in Gwagwalada, Abuja, Nigeria. Interdiscip Toxicol 8:146–150. https://doi.org/10.1515/intox-2015-0022

Alsalman ARS, Almashhedy LA, Alta'ee AH et al (2020) Effect of zinc supplementation on urate pathway enzymes in spermatozoa and seminal plasma of Iraqi asthenozoospermic patients: a randomized controlled trial. Int J Fertil Steril 13:315–323. https://doi.org/10.22074/ijfs.2020.5760

Amara S, Abdelmelek H, Garrel C et al (2008) Preventive effect of zinc against cadmium-induced oxidative stress in the rat testis. J Reprod Dev 54:129–134. https://doi.org/10.1262/jrd.18110

Andreucci A, Mocevic E, Jonsson BA et al (2015) Cadmium may impair prostate function as measured by prostate specific antigen in semen: a cross-sectional study among European and Inuit men. Reprod Toxicol 53:33–38. https://doi.org/10.1016/j.reprotox.2015.01.010

Aoshima K (2012) Itai-itai disease: cadmium-induced renal tubular osteomalacia. Nihon Eiseigaku Zasshi 67:455–463. https://doi.org/10.1265/jjh.67.455

Ardestani Zadeh A, Arab D, Kia NS et al (2019) The role of vitamin E - selenium - folic acid supplementation in improving sperm parameters after varicocelectomy: a randomized clinical trial. Urol J 16:495–500. https://doi.org/10.22037/uj.v0i0.4653

Ates I, Suzen HS, Aydin A et al (2004) The oxidative DNA base damage in testes of rats after intraperitoneal cadmium injection. Biometals 17:371–377. https://doi.org/10.1023/b:biom.0000029416.95488.5f

ATSDR (2012) ToxGuide™ for cadmium. https://www.atsdr.cdc.gov/toxguides/toxguide-5.pdf. Accessed 11 Nov 2020

ATSDR (2020) Minimal risk levels (MRLs) for hazardous substances. https://www.atsdr.cdc.gov/toxguides/toxguide-5.pdf. Accessed 11 Nov 2020

Babaknejad N, Bahrami S, Moshtaghie AA et al (2018) Cadmium testicular toxicity in male Wistar rats: protective roles of zinc and magnesium. Biol Trace Elem Res 185:106–115. https://doi.org/10.1007/s12011-017-1218-5

Badr GM, Elsawy H, Sedky A et al (2019) Protective effects of quercetin supplementation against short-term toxicity of cadmium-induced hematological impairment, hypothyroidism, and testicular disturbances in albino rats. Environ Sci Pollut Res Int 26:8202–8211. https://doi.org/10.1007/s11356-019-04276-1

Bashir N, Shagirtha K, Manoharan V et al (2019) The molecular and biochemical insight view of grape seed proanthocyanidins in ameliorating cadmium-induced testes-toxicity in rat model: implication of PI3K/Akt/Nrf-2 signaling. Biosci Rep 39. https://doi.org/10.1042/BSR20180515

Basu Mallik S, Jayashree BS, Shenoy RR (2018) Epigenetic modulation of macrophage polarization- perspectives in diabetic wounds. J Diabetes Complicat 32:524–530. https://doi.org/10.1016/j.jdiacomp.2018.01.015

Beigi Harchegani A, Dahan H, Tahmasbpour E et al (2020) Effects of zinc deficiency on impaired spermatogenesis and male infertility: the role of oxidative stress, inflammation and apoptosis. Hum Fertil (Camb) 23:5–16. https://doi.org/10.1080/14647273.2018.1494390

Bellinger (2004) WHO food additives series: 52-cadmium. http://www.inchem.org/documents/jecfa/jecmono/v52je22.htm. Accessed 11 Nov 2020

Bench G, Corzett MH, Martinelli R et al (1999) Cadmium concentrations in the testes, sperm, and spermatids of mice subjected to long-term cadmium chloride exposure. Cytometry 35:30–36. https://doi.org/10.1002/(sici)1097-0320(19990101)35:1<30::aid-cyto5>3.3.co;2-d

Blanco A, Moyano R, Molina Lopez AM et al (2010) Preneoplastic and neoplastic changes in the Leydig cells population in mice exposed to low doses of cadmium. Toxicol Ind Health 26:451–457. https://doi.org/10.1177/0748233710371111

Blum JL, Edwards JR, Prozialeck WC et al (2015) Effects of maternal exposure to cadmium oxide nanoparticles during pregnancy on maternal and offspring kidney injury markers using a murine model. J Toxicol Environ Health A 78:711–724. https://doi.org/10.1080/15287394.2015.1026622

Boitani C, Puglisi R (2008) Selenium, a key element in spermatogenesis and male fertility. Adv Exp Med Biol 636:65–73. https://doi.org/10.1007/978-0-387-09597-4_4
Bomhard E, Vogel O, Loser E (1987) Chronic effects on single and multiple oral and subcutaneous cadmium administrations on the testes of Wistar rats. Cancer Lett 36:307–315. https://doi.org/10.1016/0304-3835(87)90024-3
Bonda E, Wlostowski T, Krasowska A (2004) Testicular toxicity induced by dietary cadmium is associated with decreased testicular zinc and increased hepatic and renal metallothionein and zinc in the bank vole (Clethrionomys glareolus). Biometals 17:615–624. https://doi.org/10.1007/s10534-004-1226-8
Brooks SA, Fry RC (2017) Cadmium inhibits placental trophoblast cell migration via miRNA regulation of the transforming growth factor beta (TGF-beta) pathway. Food Chem Toxicol 109:721–726. https://doi.org/10.1016/j.fct.2017.07.059
Bu T, Mi Y, Zeng W et al (2011) Protective effect of quercetin on cadmium-induced oxidative toxicity on germ cells in male mice. Anat Rec (Hoboken) 294:520–526. https://doi.org/10.1002/ar.21317
Chakraborty S, Gang S, Sengupta M (2014) Functional status of testicular macrophages in an immunopriviledged niche in cadmium intoxicated murine testes. Am J Reprod Immunol 72:14–21. https://doi.org/10.1111/aji.12224
Chemek M, Mimouna SB, Boughammoura S et al (2016) Protective role of zinc against the toxicity induced by exposure to cadmium during gestation and lactation on testis development. Reprod Toxicol 63:151–160. https://doi.org/10.1016/j.reprotox.2016.06.005
Chen H, Lu Y, Cao Z et al (2016) Cadmium induces NLRP3 inflammasome-dependent pyroptosis in vascular endothelial cells. Toxicol Lett 246:7–16. https://doi.org/10.1016/j.toxlet.2016.01.014
Chen N, Su P, Wang M et al (2018) Ascorbic acid inhibits cadmium-induced disruption of the blood-testis barrier by regulating oxidative stress-mediated p38 MAPK pathways. Environ Sci Pollut Res Int 25:21713–21720. https://doi.org/10.1007/s11356-018-2138-4
Cherian MG (1983) Absorption and tissue distribution of cadmium in mice after chronic feeding with cadmium chloride and cadmium-metallothionein. Bull Environ Contam Toxicol 30:33–36. https://doi.org/10.1007/BF01610095
Chinyere Nsonwu-Anyanwu A, Raymond Ekong E, Jeremiah Offor S et al (2019) Heavy metals, biomarkers of oxidative stress and changes in sperm function: a case-control study. Int J Reprod Biomed 17. https://doi.org/10.18502/ijrm.v17i3.4515
Clark JT, Jimenez B, Evans SL et al (1994) Cadmium-induced sexual dysfunction does not involve increased hepatic metabolism of testosterone nor increased circulating levels of corticosterone. Physiol Behav 56:975–981. https://doi.org/10.1016/0031-9384(94)90332-8
Correia S, Vaz CV, Silva AM et al (2017) Regucalcin counteracts tert-butyl hydroperoxide and cadmium-induced oxidative stress in rat testis. J Appl Toxicol 37:159–166. https://doi.org/10.1002/jat.3333
Cowley M, Skaar DA, Jima DD et al (2018) Effects of cadmium exposure on DNA methylation at imprinting control regions and genome-wide in mothers and newborn children. Environ Health Perspect 126:037003. https://doi.org/10.1289/EHP2085
Croxford TP, McCormick NH, Kelleher SL (2011) Moderate zinc deficiency reduces testicular Zip6 and Zip10 abundance and impairs spermatogenesis in mice. J Nutr 141:359–365. https://doi.org/10.3945/jn.110.131318
Cupertino MC, Novaes RD, Santos EC et al (2017) Differential susceptibility of germ and leydig cells to cadmium-mediated toxicity: impact on testis structure, adiponectin levels, and steroidogenesis. Oxidative Med Cell Longev 2017:3405089. https://doi.org/10.1155/2017/3405089
Da Costa R, Botana D, Pinero S et al (2016) Cadmium inhibits motility, activities of plasma membrane Ca(2+)-ATPase and axonemal dynein-ATPase of human spermatozoa. Andrologia 48:464–469. https://doi.org/10.1111/and.12466
Deng YF, Weng ZW, Liang AJ et al (2019) Shenjing Guben pills for repairing testicular spermatogenic injury induced by oxidative stress in rats. Zhonghua Nan Ke Xue 25:1118–1125

Dharmadasa P, Kim N, Thunders M (2017) Maternal cadmium exposure and impact on foetal gene expression through methylation changes. Food Chem Toxicol 109:714–720. https://doi.org/10.1016/j.fct.2017.09.002

Dhooge W, Stuyvaert S, Kaufman JM et al (2001) Observations as to male fertility in the Flemish environment and health studies. Folia Histochem Cytobiol 39(Suppl 2):38–39

Elbaghdady HAM, Alwaili MA, El-Demerdash RS (2018) Regenerative potential of bone marrow mesenchymal stem cells on cadmium chloride-induced hepato-renal injury and testicular dysfunction in Sprague dawley rats. Ecotoxicol Environ Saf 164:41–49. https://doi.org/10.1016/j.ecoenv.2018.07.019

Elblehi SS, El Euony OI, El-Nahas AF (2019) Partial ameliorative effect of Moringa leaf ethanolic extract on the reproductive toxicity and the expression of steroidogenic genes induced by subchronic cadmium in male rats. Environ Sci Pollut Res Int 26:23306–23318. https://doi.org/10.1007/s11356-019-05607-y

EPA (1998) Series 870 - health effects test guidelines. https://www.epa.gov/test-guidelines-pesticides-and-toxic-substances/series-870-health-effects-test-guidelines. Accessed 1 Dec 2020

Erboga M, Kanter M (2016) Effect of cadmium on trophoblast cell proliferation and apoptosis in different gestation periods of rat placenta. Biol Trace Elem Res 169:285–293. https://doi.org/10.1007/s12011-015-0439-8

Fallah A, Mohammad-Hasani A, Colagar AH (2018) Zinc is an essential element for male fertility: a review of Zn roles in men's health, germination, sperm quality, and fertilization. J Reprod Infertil 19:69–81

Fang Y, Xiang Y, Lu X et al (2020) Epigenetic dysregulation of Mdr1b in the blood-testis barrier contributes to dyszoospermia in mice exposed to cadmium. Ecotoxicol Environ Saf 190:110142. https://doi.org/10.1016/j.ecoenv.2019.110142

Faroon O, Ashizawa A, Wright S et al (2012). Toxicological profile for cadmium. Agency for Toxic Substances and Disease Registry (ATSDR) Toxicological Profiles, Atlanta, GA. https://www.ncbi.nlm.nih.gov/pubmed/24049863. Accessed 11 Nov 2020

Feng J, Chen S, Wang Y et al (2019) Maternal exposure to cadmium impairs cognitive development of male offspring by targeting the Coronin-1a signaling pathway. Chemosphere 225:765–774. https://doi.org/10.1016/j.chemosphere.2019.03.094

Fouad AA, Jresat I (2013) Captopril and telmisartan treatments attenuate cadmium-induced testicular toxicity in rats. Fundam Clin Pharmacol 27:152–160. https://doi.org/10.1111/j.1472-8206.2011.00974.x

Fouad AA, Jresat I (2015) Thymoquinone therapy abrogates toxic effect of cadmium on rat testes. Andrologia 47:417–426. https://doi.org/10.1111/and.12281

Fouad AA, Abdel-Aziz AM, Hamouda AAH (2020) Diacerein downregulates NLRP3/Caspase-1/IL-1beta and IL-6/STAT3 pathways of inflammation and apoptosis in a rat model of cadmium testicular toxicity. Biol Trace Elem Res 195:499–505. https://doi.org/10.1007/s12011-019-01865-6

Gao F, Zhang P, Zhang H et al (2017) Dysregulation of long noncoding RNAs in mouse testes and spermatozoa after exposure to cadmium. Biochem Biophys Res Commun 484:8–14. https://doi.org/10.1016/j.bbrc.2017.01.091

Goyal T, Mitra P, Singh P et al (2020) Assessement of blood lead and cadmium levels in occupationally exposed workers of Jodhpur, Rajasthan. Indian J Clin Biochem. https://doi.org/10.1007/s12291-020-00878-6

Gunnarsson D, Nordberg G, Lundgren P et al (2003) Cadmium-induced decrement of the LH receptor expression and cAMP levels in the testis of rats. Toxicology 183:57–63. https://doi.org/10.1016/s0300-483x(02)00440-7

Habib R, Wahdan SA, Gad AM et al (2019) Infliximab abrogates cadmium-induced testicular damage and spermiotoxicity via enhancement of steroidogenesis and suppression of inflammation and apoptosis mediators. Ecotoxicol Environ Saf 182:109398. https://doi.org/10.1016/j.ecoenv.2019.109398

Hawkes WC, Alkan Z, Wong K (2009) Selenium supplementation does not affect testicular selenium status or semen quality in North American men. J Androl 30:525–533. https://doi.org/10.2164/jandrol.108.006940

He L, Wang B, Hay EB et al (2009) Discovery of ZIP transporters that participate in cadmium damage to testis and kidney. Toxicol Appl Pharmacol 238:250–257. https://doi.org/10.1016/j.taap.2009.02.017

He Y, Zou L, Luo W et al (2020) Heavy metal exposure, oxidative stress and semen quality: exploring associations and mediation effects in reproductive-aged men. Chemosphere 244:125498. https://doi.org/10.1016/j.chemosphere.2019.125498

Henkel R, Sandhu IS, Agarwal A (2019) The excessive use of antioxidant therapy: a possible cause of male infertility? Andrologia 51:e13162. https://doi.org/10.1111/and.13162

Hu H, Lu X, Cen X et al (2014) RNA-Seq identifies key reproductive gene expression alterations in response to cadmium exposure. Biomed Res Int 2014:529271. https://doi.org/10.1155/2014/529271

Huang G, Yuan M, Zhang J et al (2016) IL-6 mediates differentiation disorder during spermatogenesis in obesity-associated inflammation by affecting the expression of Zfp637 through the SOCS3/STAT3 pathway. Sci Rep 6:28012. https://doi.org/10.1038/srep28012

Huang Y, Zhu J, Li H et al (2020) Cadmium exposure during prenatal development causes testosterone disruption in multigeneration via SF-1 signaling in rats. Food Chem Toxicol 135:110897. https://doi.org/10.1016/j.fct.2019.110897

Hudson KM, Belcher SM, Cowley M (2019) Maternal cadmium exposure in the mouse leads to increased heart weight at birth and programs susceptibility to hypertension in adulthood. Sci Rep 9:13553. https://doi.org/10.1038/s41598-019-49807-5

Ikeda M, Nakatsuka H, Watanabe T et al (2015) Estimation of daily cadmium intake from cadmium in blood or cadmium in urine. Environ Health Prev Med 20:455–459. https://doi.org/10.1007/s12199-015-0479-x

Jahan S, Zahra A, Irum U et al (2014) Protective effects of different antioxidants against cadmium induced oxidative damage in rat testis and prostate tissues. Syst Biol Reprod Med 60:199–205. https://doi.org/10.3109/19396368.2014.912363

Jamilaldin Fatemi S, Saljooghi AS, Balooch FD et al (2011) Chelation of cadmium by combining deferasirox and deferiprone in rats. Toxicol Ind Health 27:371–377. https://doi.org/10.1177/0748233710388451

Jarup L, Berglund M, Elinder CG et al (1998) Health effects of cadmium exposure--a review of the literature and a risk estimate. Scand J Work Environ Health 24(Suppl 1):1–51

Jasperson KW, Tuohy TM, Neklason DW et al (2010) Hereditary and familial colon cancer. Gastroenterology 138:2044–2058. https://doi.org/10.1053/j.gastro.2010.01.054

Jeong KS, Park H, Ha E et al (2015) Performance IQ in children is associated with blood cadmium concentration in early pregnancy. J Trace Elem Med Biol 30:107–111. https://doi.org/10.1016/j.jtemb.2014.11.007

Ji YL, Wang H, Liu P et al (2011) Effects of maternal cadmium exposure during late pregnant period on testicular steroidogenesis in male offspring. Toxicol Lett 205:69–78. https://doi.org/10.1016/j.toxlet.2011.05.233

Ji YL, Wang H, Zhang C et al (2013) N-acetylcysteine protects against cadmium-induced germ cell apoptosis by inhibiting endoplasmic reticulum stress in testes. Asian J Androl 15:290–296. https://doi.org/10.1038/aja.2012.129

Jin X, Jia T, Liu R et al (2018) The antagonistic effect of selenium on cadmium-induced apoptosis via PPAR-gamma/PI3K/Akt pathway in chicken pancreas. J Hazard Mater 357:355–362. https://doi.org/10.1016/j.jhazmat.2018.06.003

Kamel M, Razek A, Ahmed K et al (2011) Exposure of adult male rats to cadmium: assessment of sexual behaviour, fertility, aggression as well as anxiety like behaviour with special reference to biochemical and pathological alterations. Life Science Journal 8:106–119

Keck C, Bramkamp G, Behre HM et al (1995) Lack of correlation between cadmium in seminal plasma and fertility status of nonexposed individuals and two cadmium-exposed patients. Reprod Toxicol 9:35–40. https://doi.org/10.1016/0890-6238(94)00053-y

Khanna S, Mitra S, Lakhera PC et al (2016) N-acetylcysteine effectively mitigates cadmium-induced oxidative damage and cell death in Leydig cells in vitro. Drug Chem Toxicol 39:74–80. https://doi.org/10.3109/01480545.2015.1028068

Kippler M, Bottai M, Georgiou V et al (2016) Impact of prenatal exposure to cadmium on cognitive development at preschool age and the importance of selenium and iodine. Eur J Epidemiol 31:1123–1134. https://doi.org/10.1007/s10654-016-0151-9

Kiziler AR, Aydemir B, Onaran I et al (2007) High levels of cadmium and lead in seminal fluid and blood of smoking men are associated with high oxidative stress and damage in infertile subjects. Biol Trace Elem Res 120:82–91. https://doi.org/10.1007/s12011-007-8020-8

Koizumi N, Murata K, Hayashi C et al (2008) High cadmium accumulation among humans and primates: comparison across various mammalian species--a study from Japan. Biol Trace Elem Res 121:205–214. https://doi.org/10.1007/s12011-007-8048-9

Krasovskii GN, Varshavskaya SP, Borisov AI (1976) Toxic and gonadotropic effects of cadmium and boron relative to standards for these substances in drinking water. Environ Health Perspect 13:69–75. https://doi.org/10.1289/ehp.761369

Lafuente A, Gonzalez-Carracedo A, Romero A et al (2004) Cadmium exposure differentially modifies the circadian patterns of norepinephrine at the median eminence and plasma LH, FSH and testosterone levels. Toxicol Lett 146:175–182. https://doi.org/10.1016/j.toxlet.2003.10.004

Li CJ, Yeh CY, Chen RY et al (2015) Biomonitoring of blood heavy metals and reproductive hormone level related to low semen quality. J Hazard Mater 300:815–822. https://doi.org/10.1016/j.jhazmat.2015.08.027

Li M, Liu C, Yang L et al (2016a) G9a-mediated histone methylation regulates cadmium-induced male fertility damage in pubertal mice. Toxicol Lett 252:11–21. https://doi.org/10.1016/j.toxlet.2016.04.004

Li R, Luo X, Li L et al (2016b) The protective effects of melatonin against oxidative stress and inflammation induced by acute cadmium exposure in mice testis. Biol Trace Elem Res 170:152–164. https://doi.org/10.1007/s12011-015-0449-6

Li X, Wang L, Li Y et al (2016c) Tyrosine phosphorylation of dihydrolipoamide dehydrogenase as a potential cadmium target and its inhibitory role in regulating mouse sperm motility. Toxicology 357–358:52–64. https://doi.org/10.1016/j.tox.2016.06.003

Li Y, Wu J, Zhou W et al (2016d) Association between environmental exposure to cadmium and human semen quality. Int J Environ Health Res 26:175–186. https://doi.org/10.1080/09603123.2015.1061115

Li R, Zhao L, Li L et al (2017) A preliminary study about the potential effects of heavy metals on the human male reproductive parameters in HIV-infected population in China. Biol Trace Elem Res 180:39–47. https://doi.org/10.1007/s12011-017-0998-y

Ling LB, Chang Y, Liu CW et al (2017) Oxidative stress intensity-related effects of cadmium (Cd) and paraquat (PQ) on UV-damaged-DNA binding and excision repair activities in zebrafish (Danio rerio) embryos. Chemosphere 167:10–18. https://doi.org/10.1016/j.chemosphere.2016.09.068

Liu L, Zhou L, Hu S et al (2016a) Down-regulation of ABCG2 and ABCB4 transporters in the placenta of rats exposed to cadmium. Oncotarget 7:38154–38163. https://doi.org/10.18632/oncotarget.9415

Liu XR, Wang YY, Fan HR et al (2016b) Preventive effects of beta-cryptoxanthin against cadmium-induced oxidative stress in the rat testis. Asian J Androl 18:920–924. https://doi.org/10.4103/1008-682X.173449

Liu D, Wan J, Liu Z et al (2020) Determination of cadmium induced acute and chronic reproductive toxicity with Raman spectroscopy. Lasers Med Sci. https://doi.org/10.1007/s10103-020-02976-6

Marano KM, Naufal ZS, Kathman SJ et al (2012) Cadmium exposure and tobacco consumption: biomarkers and risk assessment. Regul Toxicol Pharmacol 64:243–252. https://doi.org/10.1016/j.yrtph.2012.07.008
Martin MB, Voeller HJ, Gelmann EP et al (2002) Role of cadmium in the regulation of AR gene expression and activity. Endocrinology 143:263–275. https://doi.org/10.1210/endo.143.1.8581
McCarty MF (2012) Zinc and multi-mineral supplementation should mitigate the pathogenic impact of cadmium exposure. Med Hypotheses 79:642–648. https://doi.org/10.1016/j.mehy.2012.07.043
Medina MF, Arrieta MC, Villafane MN et al (2017) Early signs of toxicity in testes and sperm of rats exposed to low cadmium doses. Toxicol Ind Health 33:576–587. https://doi.org/10.1177/0748233716689524
Meeker JD, Rossano MG, Protas B et al (2008) Cadmium, lead, and other metals in relation to semen quality: human evidence for molybdenum as a male reproductive toxicant. Environ Health Perspect 116:1473–1479. https://doi.org/10.1289/ehp.11490
Meeker JD, Rossano MG, Protas B et al (2010) Environmental exposure to metals and male reproductive hormones: circulating testosterone is inversely associated with blood molybdenum. Fertil Steril 93:130–140. https://doi.org/10.1016/j.fertnstert.2008.09.044
Mendiola J, Moreno JM, Roca M et al (2011) Relationships between heavy metal concentrations in three different body fluids and male reproductive parameters: a pilot study. Environ Health 10:6. https://doi.org/10.1186/1476-069X-10-6
Messaoudi I, Banni M, Said L et al (2010) Involvement of selenoprotein P and GPx4 gene expression in cadmium-induced testicular pathophysiology in rat. Chem Biol Interact 188:94–101. https://doi.org/10.1016/j.cbi.2010.07.012
Mikowska M, Dziublinska B, Swiergosz-Kowalewska R (2018) Variation of metallothionein I and II gene expression in the bank vole (Clethrionomys glareolus) under environmental zinc and cadmium exposure. Arch Environ Contam Toxicol 75:66–74. https://doi.org/10.1007/s00244-017-0485-7
Minutoli L, Micali A, Pisani A et al (2015) Flavocoxid protects against cadmium-induced disruption of the blood-testis barrier and improves testicular damage and germ cell impairment in mice [corrected]. Toxicol Sci 148:311–329. https://doi.org/10.1093/toxsci/kfv185
Mishra RK, Jain A, Singh SK (2018) Profertility effects of Shilajit on cadmium-induced infertility in male mice. Andrologia 50:e13064. https://doi.org/10.1111/and.13064
Mitra S, Varghese AC, Mandal S et al (2020) Lead and cadmium exposure induces male reproductive dysfunction by modulating the expression profiles of apoptotic and survival signal proteins in tea-garden workers. Reprod Toxicol. https://doi.org/10.1016/j.reprotox.2020.09.006
Momeni HR, Eskandari N (2020) Curcumin protects the testis against cadmium-induced histopathological damages and oxidative stress in mice. Hum Exp Toxicol 39:653–661. https://doi.org/10.1177/0960327119895564
Moslemi MK, Tavanbakhsh S (2011) Selenium-vitamin E supplementation in infertile men: effects on semen parameters and pregnancy rate. Int J Gen Med 4:99–104. https://doi.org/10.2147/IJGM.S16275
Mouro VGS, Martins ALP, Silva J et al (2019) Subacute testicular toxicity to cadmium exposure intraperitoneally and orally. Oxidative Med Cell Longev 2019:3429635. https://doi.org/10.1155/2019/3429635
Mouro VGS, Siman VA, da Silva J et al (2020) Cadmium-induced testicular toxicity in mice: subacute and subchronic route-dependent effects. Biol Trace Elem Res 193:466–482. https://doi.org/10.1007/s12011-019-01731-5
Nair AB, Jacob S (2016) A simple practice guide for dose conversion between animals and human. J Basic Clin Pharm 7:27–31. https://doi.org/10.4103/0976-0105.177703
Nair A, Morsy MA, Jacob S (2018) Dose translation between laboratory animals and human in preclinical and clinical phases of drug development. Drug Dev Res 79:373–382. https://doi.org/10.1002/ddr.21461

Nakayama SMM, Nakata H, Ikenaka Y et al (2019) One year exposure to Cd- and Pb-contaminated soil causes metal accumulation and alteration of global DNA methylation in rats. Environ Pollut 252:1267–1276. https://doi.org/10.1016/j.envpol.2019.05.038

Nicoglou A, Merlin F (2017) Epigenetics: a way to bridge the gap between biological fields. Stud Hist Phil Biol Biomed Sci 66:73–82. https://doi.org/10.1016/j.shpsc.2017.10.002

Nishijo M, Nakagawa H, Suwazono Y et al (2017) Causes of death in patients with Itai-itai disease suffering from severe chronic cadmium poisoning: a nested case-control analysis of a follow-up study in Japan. BMJ Open 7:e015694. https://doi.org/10.1136/bmjopen-2016-015694

Nna VU, Ujah GA, Mohamed M et al (2017) Cadmium chloride-induced testicular toxicity in male wistar rats; prophylactic effect of quercetin, and assessment of testicular recovery following cadmium chloride withdrawal. Biomed Pharmacother 94:109–123. https://doi.org/10.1016/j.biopha.2017.07.087

Oguzturk H, Ciftci O, Aydin M et al (2012) Ameliorative effects of curcumin against acute cadmium toxicity on male reproductive system in rats. Andrologia 44:243–249. https://doi.org/10.1111/j.1439-0272.2012.01273.x

Ohta H, Qi Y, Ohba K et al (2019) Role of metallothionein-like cadmium-binding protein (MTLCdBP) in the protective mechanism against cadmium toxicity in the testis. Ind Health 57:570–579. https://doi.org/10.2486/indhealth.2018-0177

Olson GE, Winfrey VP, Hill KE et al (2004) Sequential development of flagellar defects in spermatids and epididymal spermatozoa of selenium-deficient rats. Reproduction 127:335–342. https://doi.org/10.1530/rep.1.00103

Ono H, Funakoshi T, Shimada H et al (1997) Comparative effects of disulfiram and diethyldithiocarbamate against testicular toxicity in rats caused by acute exposure to cadmium. J Toxicol Environ Health 50:389–399. https://doi.org/10.1080/009841097160429

Pillai P, Pandya C, Gupta S et al (2010) Biochemical and molecular effects of gestational and lactational coexposure to lead and cadmium on ovarian steroidogenesis are associated with oxidative stress in F1 generation rats. J Biochem Mol Toxicol 24:384–394. https://doi.org/10.1002/jbt.20351

Prasad AS (2014) Zinc is an antioxidant and anti-inflammatory agent: its role in human health. Front Nutr 1:14. https://doi.org/10.3389/fnut.2014.00014

Rajendar B, Bharavi K, Rao GS et al (2011) Protective effect of an aphrodisiac herb Tribulus terrestris Linn on cadmium-induced testicular damage. Indian J Pharmacol 43:568–573. https://doi.org/10.4103/0253-7613.84974

Ramos-Trevino J, Bassol-Mayagoitia S, Ruiz-Flores P et al (2017) In vitro evaluation of damage by heavy metals in tight and gap junctions of sertoli cells. DNA Cell Biol 36:829–836. https://doi.org/10.1089/dna.2017.3839

Rania Abdel Rahman E, Mohamed Elshabrawy G (2014) Effect of long term cadmium chloride exposure on testicular functions in male albino rats. Am J Anim Vet Sci 9. https://doi.org/10.3844/ajavsp.2014.182.188

Ren XM, Wang GG, Xu DQ et al (2012) The protection of selenium on cadmium-induced inhibition of spermatogenesis via activating testosterone synthesis in mice. Food Chem Toxicol 50:3521–3529. https://doi.org/10.1016/j.fct.2012.07.021

Riaz M, Mahmood Z, Shahid M et al (2016) Impact of reactive oxygen species on antioxidant capacity of male reproductive system. Int J Immunopathol Pharmacol 29:421–425. https://doi.org/10.1177/0394632015608994

Romano ME, Enquobahrie DA, Simpson C et al (2016) Maternal body burden of cadmium and offspring size at birth. Environ Res 147:461–468. https://doi.org/10.1016/j.envres.2016.02.029

Ronco AM, Montenegro M, Castillo P et al (2011) Maternal exposure to cadmium during gestation perturbs the vascular system of the adult rat offspring. Toxicol Appl Pharmacol 251:137–145. https://doi.org/10.1016/j.taap.2011.01.001

Sadasivam M, Ramatchandirin B, Balakrishnan S et al (2015) TNF-alpha-mediated suppression of Leydig cell steroidogenesis involves DAX-1. Inflamm Res 64:549–556. https://doi.org/10.1007/s00011-015-0835-8

Saha R, Roychoudhury S, Kar K et al (2019) Coenzyme Q10 ameliorates cadmium induced reproductive toxicity in male rats. Physiol Res 68:141–145. https://doi.org/10.33549/physiolres.934000

Said L, Banni M, Kerkeni A et al (2010) Influence of combined treatment with zinc and selenium on cadmium induced testicular pathophysiology in rat. Food Chem Toxicol 48:2759–2765. https://doi.org/10.1016/j.fct.2010.07.003

Sankako MK, Garcia PC, Piffer RC et al (2012) Possible mechanism by which zinc protects the testicular function of rats exposed to cigarette smoke. Pharmacol Rep 64:1537–1546. https://doi.org/10.1016/s1734-1140(12)70951-9

Santana VP, Salles ES, Correa DE et al (2016) Long-term effects of perinatal exposure to low doses of cadmium on the prostate of adult male rats. Int J Exp Pathol 97:310–316. https://doi.org/10.1111/iep.12193

Schopfer J, Drasch G, Schrauzer GN (2010) Selenium and cadmium levels and ratios in prostates, livers, and kidneys of nonsmokers and smokers. Biol Trace Elem Res 134:180–187. https://doi.org/10.1007/s12011-010-8636-y

Sen Gupta R, Kim J, Gomes C et al (2004) Effect of ascorbic acid supplementation on testicular steroidogenesis and germ cell death in cadmium-treated male rats. Mol Cell Endocrinol 221:57–66. https://doi.org/10.1016/j.mce.2004.03.012

Sengupta M, Deb I, Sharma GD et al (2013) Human sperm and other seminal constituents in male infertile patients from arsenic and cadmium rich areas of southern Assam. Syst Biol Reprod Med 59:199–209. https://doi.org/10.3109/19396368.2013.783143

Shank KE, Vetter RJ, Ziemer PL (1977) Uptake and distribution of cadmium in mice following repeated administrations. Arch Environ Contam Toxicol 6:63–68. https://doi.org/10.1007/BF02097750

Sharma RP, Kjellstrom T, McKenzie JM (1983) Cadmium in blood and urine among smokers and non-smokers with high cadmium intake via food. Toxicology 29:163–171. https://doi.org/10.1016/0300-483x(83)90048-3

Shi X, Fu L (2019) Piceatannol inhibits oxidative stress through modification of Nrf2-signaling pathway in testes and attenuates spermatogenesis and steroidogenesis in rats exposed to cadmium during adulthood. Drug Des Devel Ther 13:2811–2824. https://doi.org/10.2147/DDDT.S198444

Shi TY, Chen G, Huang X et al (2012) Effects of reactive oxygen species from activated leucocytes on human sperm motility, viability and morphology. Andrologia 44(Suppl 1):696–703. https://doi.org/10.1111/j.1439-0272.2011.01252.x

Smith SW (2013) The role of chelation in the treatment of other metal poisonings. J Med Toxicol 9:355–369. https://doi.org/10.1007/s13181-013-0343-6

Smith LB, Walker WH (2014) The regulation of spermatogenesis by androgens. Semin Cell Dev Biol 30:2–13. https://doi.org/10.1016/j.semcdb.2014.02.012

Sukhn C, Awwad J, Ghantous A et al (2018) Associations of semen quality with non-essential heavy metals in blood and seminal fluid: data from the environment and male infertility (EMI) study in Lebanon. J Assist Reprod Genet 35:1691–1701. https://doi.org/10.1007/s10815-018-1236-z

Taha EA, Sayed SK, Ghandour NM et al (2013) Correlation between seminal lead and cadmium and seminal parameters in idiopathic oligoasthenozoospermic males. Cent European J Urol 66:84–92. https://doi.org/10.5173/ceju.2013.01.art28

Taylor CM, Golding J, Emond AM (2016) Moderate prenatal cadmium exposure and adverse birth outcomes: a role for sex-specific differences? Paediatr Perinat Epidemiol 30:603–611. https://doi.org/10.1111/ppe.12318

Taylor CM, Emond AM, Lingam R et al (2018) Prenatal lead, cadmium and mercury exposure and associations with motor skills at age 7years in a UK observational birth cohort. Environ Int 117:40–47. https://doi.org/10.1016/j.envint.2018.04.032

Thijssen S, Maringwa J, Faes C et al (2007) Chronic exposure of mice to environmentally relevant, low doses of cadmium leads to early renal damage, not predicted by blood or urine cadmium levels. Toxicology 229:145–156. https://doi.org/10.1016/j.tox.2006.10.011

Tung NT, Cuong TD, Hung TM et al (2013) Inhibitory effect on NO production of triterpenes from the fruiting bodies of Ganoderma lucidum. Bioorg Med Chem Lett 23:1428–1432. https://doi.org/10.1016/j.bmcl.2012.12.066

Ujah GA, Nna VU, Agah MI et al (2018) Effect of quercetin on cadmium chloride-induced impairments in sexual behaviour and steroidogenesis in male Wistar rats. Andrologia 50. https://doi.org/10.1111/and.12866

Vidal AC, Semenova V, Darrah T et al (2015) Maternal cadmium, iron and zinc levels, DNA methylation and birth weight. BMC Pharmacol Toxicol 16:20. https://doi.org/10.1186/s40360-015-0020-2

Waalkes MP (2003) Cadmium carcinogenesis. Mutat Res 533:107–120. https://doi.org/10.1016/j.mrfmmm.2003.07.011

Wang X, Tian J (2004) Health risks related to residential exposure to cadmium in Zhenhe County, China. Arch Environ Health 59:324–330. https://doi.org/10.3200/AEOH.59.6.324-330

Wang H, Liu L, Hu YF et al (2016a) Maternal serum cadmium level during pregnancy and its association with small for gestational age infants: a population-based birth cohort study. Sci Rep 6:22631. https://doi.org/10.1038/srep22631

Wang L, Li Y, Fu J et al (2016b) Cadmium inhibits mouse sperm motility through inducing tyrosine phosphorylation in a specific subset of proteins. Reprod Toxicol 63:96–106. https://doi.org/10.1016/j.reprotox.2016.05.018

Wang YX, Sun Y, Feng W et al (2016c) Association of urinary metal levels with human semen quality: a cross-sectional study in China. Environ Int 91:51–59. https://doi.org/10.1016/j.envint.2016.02.019

Wang HF, Chang M, Peng TT et al (2017) Exposure to cadmium impairs sperm functions by reducing CatSper in mice. Cell Physiol Biochem 42:44–54. https://doi.org/10.1159/000477113

Wang H, Zhang L, Abel GM et al (2018) Cadmium exposure impairs cognition and olfactory memory in male C57BL/6 mice. Toxicol Sci 161:87–102. https://doi.org/10.1093/toxsci/kfx202

Wang H, Zhang R, Song Y et al (2019a) Protective effect of ganoderma triterpenoids on cadmium-induced testicular toxicity in chickens. Biol Trace Elem Res 187:281–290. https://doi.org/10.1007/s12011-018-1364-4

Wang S, Ren X, Hu X et al (2019b) Cadmium-induced apoptosis through reactive oxygen species-mediated mitochondrial oxidative stress and the JNK signaling pathway in TM3 cells, a model of mouse Leydig cells. Toxicol Appl Pharmacol 368:37–48. https://doi.org/10.1016/j.taap.2019.02.012

Wijesekara GU, Fernando DM, Wijerathna S et al (2015) Environmental and occupational exposures as a cause of male infertility. Ceylon Med J 60:52–56. https://doi.org/10.4038/cmj.v60i2.7090

Wilder LC, Metcalf SW, Orloff KG et al (2001) ASTDR and the trinity industry case study. Agency for Toxic Substances and Disease Registry. AIHAJ 62:557

Wu HM, Lin-Tan DT, Wang ML et al (2008) Cadmium level in seminal plasma may affect the pregnancy rate for patients undergoing infertility evaluation and treatment. Reprod Toxicol 25:481–484. https://doi.org/10.1016/j.reprotox.2008.04.005

Wu X, Guo X, Wang H et al (2017) A brief exposure to cadmium impairs Leydig cell regeneration in the adult rat testis. Sci Rep 7:6337. https://doi.org/10.1038/s41598-017-06870-0

Xiao X, Cheng CY, Mruk DD (2013) Intercellular adhesion molecule-2 is involved in apical ectoplasmic specialization dynamics during spermatogenesis in the rat. J Endocrinol 216:73–86. https://doi.org/10.1530/JOE-12-0434

Xu B, Chia SE, Tsakok M et al (1993) Trace elements in blood and seminal plasma and their relationship to sperm quality. Reprod Toxicol 7:613–618. https://doi.org/10.1016/0890-6238(93)90038-9

Yadav N, Khandelwal S (2008) Effect of Picroliv on cadmium induced testicular damage in rat. Food Chem Toxicol 46:494–501. https://doi.org/10.1016/j.fct.2007.08.035
Yang JM, Arnush M, Chen QY et al (2003) Cadmium-induced damage to primary cultures of rat Leydig cells. Reprod Toxicol 17:553–560. https://doi.org/10.1016/s0890-6238(03)00100-x
Yang Q, Li P, Wen Y et al (2018a) Cadmium inhibits lysine acetylation and succinylation inducing testicular injury of mouse during development. Toxicol Lett 291:112–120. https://doi.org/10.1016/j.toxlet.2018.04.005
Yang SH, Yu LH, Li L et al (2018b) Protective mechanism of sulforaphane on cadmium-induced Sertoli cell injury in mice testis via Nrf2/ARE signaling pathway. Molecules 23. https://doi.org/10.3390/molecules23071774
Yang SH, He JB, Yu LH et al (2019) Protective role of curcumin in cadmium-induced testicular injury in mice by attenuating oxidative stress via Nrf2/ARE pathway. Environ Sci Pollut Res Int 26:34575–34583. https://doi.org/10.1007/s11356-019-06587-9
Zeng X, Lin T, Zhou Y et al (2002) Alterations of serum hormone levels in male workers occupationally exposed to cadmium. J Toxicol Environ Health A 65:513–521. https://doi.org/10.1080/15287390252807975
Zeng X, Jin T, Buchet JP et al (2004a) Impact of cadmium exposure on male sex hormones: a population-based study in China. Environ Res 96:338–344. https://doi.org/10.1016/j.envres.2004.02.004
Zeng X, Jin T, Jiang X et al (2004b) Effects on the prostate of environmental cadmium exposure--a cross-sectional population study in China. Biometals 17:559–565. https://doi.org/10.1023/b:biom.0000045739.89653.67
Zeng Q, Zhou B, Feng W et al (2013) Associations of urinary metal concentrations and circulating testosterone in Chinese men. Reprod Toxicol 41:109–114. https://doi.org/10.1016/j.reprotox.2013.06.062
Zeng Q, Feng W, Zhou B et al (2015) Urinary metal concentrations in relation to semen quality: a cross-sectional study in China. Environ Sci Technol 49:5052–5059. https://doi.org/10.1021/es5053478
Zhang Y, Xu X, Chen A et al (2018) Maternal urinary cadmium levels during pregnancy associated with risk of sex-dependent birth outcomes from an e-waste pollution site in China. Reprod Toxicol 75:49–55. https://doi.org/10.1016/j.reprotox.2017.11.003
Zhou T, Jia X, Chapin RE et al (2004) Cadmium at a non-toxic dose alters gene expression in mouse testes. Toxicol Lett 154:191–200. https://doi.org/10.1016/j.toxlet.2004.07.015
Zhou T, Wang H, Zhang S et al (2016a) S100P is a potential molecular target of cadmium-induced inhibition of human placental trophoblast cell proliferation. Exp Toxicol Pathol 68:565–570. https://doi.org/10.1016/j.etp.2016.09.002
Zhou Y, Fu XM, He DL et al (2016b) Evaluation of urinary metal concentrations and sperm DNA damage in infertile men from an infertility clinic. Environ Toxicol Pharmacol 45:68–73. https://doi.org/10.1016/j.etap.2016.05.020
Zhu H, Li K, Liang J et al (2011) Changes in the levels of DNA methylation in testis and liver of SD rats neonatally exposed to 5-aza-2′-deoxycytidine and cadmium. J Appl Toxicol 31:484–495. https://doi.org/10.1002/jat.1673

Micronuclei in Fish Erythrocytes as Genotoxic Biomarkers of Water Pollution: An Overview

Francesco D'Agostini and Sebastiano La Maestra

Contents

Abstract Freshwater and marine water bodies receive chemical contaminants from industrial, agricultural, urban, and domestic wastes. Eco-genotoxicity assays are useful tools to assess the cumulative genotoxicity of these pollutants. Fish are suitable indicators for biomonitoring of mutagenic and carcinogenic pollution.

In this review, we present a complete overview of the studies performed so far using the micronucleus test in peripheral erythrocytes of fish exposed to polluted water. We have listed all the species of fish used and the geographical distribution of the investigations. We have analyzed and discussed all technical aspects of using this test in fish, as well as the advantages and disadvantages of the different experimental protocols. We have reported the results of all studies. This assay has become, for years, one of the simplest, fastest, and most cost-effective for assessing genotoxic risk in aquatic environments. However, there are still several factors influencing the variability of the results. Therefore, we have given indications and suggestions to achieve a standardization of experimental procedures and ensure uniformity of future investigations.

F. D'Agostini (✉) · S. La Maestra
Department of Health Sciences (DISSAL), University of Genoa, Genoa, Italy
e-mail: fda@unige.it; Sebastiano.Lamaestra@edu.unige.it

P. de Voogt (ed.), *Reviews of Environmental Contamination and Toxicology Volume 258*, Reviews of Environmental Contamination and Toxicology 258,
https://doi.org/10.1007/398_2021_76

Keywords Biomarker · Biomonitoring · Erythrocytes · Fish · Genotoxicity · Micronuclei · Water pollution

1 Introduction

Many freshwater and marine water bodies receive, directly or indirectly, enormous amounts of chemical pollutants from industrial, agricultural, urban, and domestic wastes.

Water pollution may pose relevant public health problems. Surface waters and groundwater are the primary sources of drinking water supply and are extensively used for irrigation, food preparation and treatment, as well as for cultural and recreational purposes. Moreover, many aquatic organisms are a relevant food resource for humans. In addition to human health issues, we also need to account for damages to ecosystems and the survival of vegetal and animal species (Bickham et al. 2000).

Liquid effluents, solid wastes, and airborne emissions release chemical contaminants in the environment. Standard chemical analyses supply insufficient information on the risk of these pollutants. Moreover, chemical interactions may generate unknown mixtures and cause unpredictable effects. Therefore, it is necessary to evaluate the overall toxic impact even without knowing the chemical composition of these mixtures, using suitable biomarkers (van der Oost et al. 2003; Hook et al. 2014; Kroon et al. 2017), to quantify and manage the risk (Beyer et al. 2014; Kienzler et al. 2016).

Studies evaluating environmental quality are increasingly using biomarkers. However, a biological response does not always translate into a toxicological effect. Therefore, several researchers proposed a novel approach: using effect-based methods (mainly in vitro cell-based assays and small-scale in vivo bioassays) to derive effect-based trigger values, which allow discrimination between acceptable and unacceptable effect levels. This strategy, which is becoming increasingly popular for water quality assessment, is easily adaptable to any water type and guideline or regulatory framework. Furthermore, it is possible to expand it from the protection of human health to environmental protection. (Escher et al. 2015, 2018; van der Oost et al. 2017a, b; Brack et al. 2019).

Long since, attention has been paid to genotoxic substances and compounds, due to their potential mutagenic and carcinogenic role, even below their detection limits (De Flora et al. 1991a; Stahl 1991; Houk 1992; Würgler and Kramers 1992; Claxton et al. 1998; White and Rasmussen 1998). Genotoxins are chemical and physical agents capable of inducing mutations and related genetic changes in germinal and somatic cells. Many short-term tests predictive of mutagenic and DNA-damaging effects have been developed in recent decades. These include bacterial tests (*Salmonella* mutagenicity test, SOS Chromotest/*umu*-test, Mutatox test, p53 CALUX®) and in vitro or in vivo assays in cell lines or living organisms (DNA adducts, cytogenetic alterations, DNA strand break detection, oncogene expression, and more). Various reviews report a detailed description of these assays (Montesano et al. 1987; Feron et al. 1999; McGregor et al. 1999; Müller 2009; van der Linden et al. 2014).

It is almost impossible to show the genotoxic activity of each agent using bacterial and in vitro tests. Our knowledge of the long-term implications and effects of mutagens in the environment is still incomplete. Therefore, it is necessary to find the global genotoxic risk in the ecological context.

Eco-genotoxicity assays have increasingly proved to be useful tools to assess the cumulative genotoxicity of pollutants to which aquatic organisms are exposed. Several biomonitoring studies have been performed both in vitro and in vivo, using different biomarkers (Lyne et al. 1992; Dixon et al. 2002; Jha 2004; Ohe et al. 2004; Tabrez et al. 2011; Han et al. 2016). These studies provide early warning signs of hazardous long-term effects and became part of the strategy of environmental risk assessment and management for human health, as indicated in different official documents (European Union Water Framework Directive 2000; European Community Marine Strategy Framework Directive 2008).

Fish are aquatic organisms virtually inhabiting every freshwater or saltwater environment. They play an ecological key role in all food chains, and they also are a relevant food resource for humans. Exposure of fish to water chemical pollutants and their consequent absorption and bioaccumulation may occur directly from water, by the respiratory route through the gills, or indirectly by ingestion of contaminated food or organisms. Fish are susceptible to genetic effects induced by xenobiotics (Belfiore and Anderson 2001; Cajaraville et al. 2003), and their metabolic systems and defense mechanisms respond similarly to other vertebrates (Stegeman and Lech 1991). Therefore, despite some limitation, fish are considered very proper experimental models in several different research areas (Powers 1989). Fish are suitable indicators for biomonitoring of mutagenic and carcinogenic pollution in freshwater and marine water bodies (Hawkins et al. 1988; Dashwood and Bailey 1998; Ohe et al. 2004; Monserrat et al. 2007; Colin et al. 2016; Kroon et al. 2017).

Different assays (cytogenetic tests, Comet assay, DNA adducts, quantitative PCR, unscheduled DNA synthesis) aimed at evaluating the genotoxic potential of pollutants have been widely used in fish (Kligerman and Bloom 1975; Kligerman et al. 1975; Kligerman 1982; Al-Sabti 1991; De Flora et al. 1991b; Hose 1994; Jha 2004; Ohe et al. 2004; Frenzilli et al. 2009; Meyer 2010; Lapuente et al. 2015; Pampanin et al. 2017; Gajski et al. 2019).

Several reviews have been published about the use of micronucleus test in fish (Al-Sabti and Metcalfe 1995; Ulupinar and Okumus 2002; Cajaraville et al. 2003; Ohe et al. 2004; Mijat 2005; Lehtonen et al. 2006; Udroiu 2006; Monserrat et al. 2007; Bolognesi and Hayashi 2011; Obiakor et al. 2012; Arslan and Parlak 2017; Kroon et al. 2017). However, all these reviews reported only a limited number of studies dealing with biomonitoring of water pollution.

We searched different databases (MEDLINE/PubMed, Scopus, Google Scholar) using different keywords (fish, micronuclei, genotoxicity, clastogenicity, biomonitoring). We then analyzed the bibliography of each article to find further references not retrieved with our database search. We have obtained a total of 262 experimental studies and 63 reviews/methodological papers. We have included all the articles reporting experimental studies in this review. We did not mention studies aimed solely at evaluating the genotoxicity of arbitrarily chosen chemical or physical agents under laboratory conditions.

We have reported, for the first time in the literature, a complete list of studies published so far on micronuclei in fish blood erythrocytes and water pollution. We believe that this extensive review will help researchers to find all studies of their interest easily and quickly. Furthermore, we give recommendations and suggestions on methods and experimental protocols to achieve better standardization of future investigations using this assay.

2 Micronucleus Test

Micronuclei (MN) are chromosome fragments or whole chromosomes excluded from the daughter nuclei after cell division. They are visible as single or multiple rounded bodies located in the cell cytoplasm with similar staining features but smaller size than the main nucleus. The induction of MN is related either to chromosomal damage, causing acentric fragments (clastogenicity), or to dysfunctions of the spindle apparatus resulting in lagging chromosomes at anaphase (aneugenicity). The MN test was originally developed in rodent bone marrow erythrocytes (Schmid 1975) and peripheral blood erythrocytes (MacGregor et al. 1980; Salamone et al. 1980). Subsequently, it was used in a variety of cell types, both in vivo and in vitro, becoming one of the most popular methods to detect genotoxicity of different chemical or physical agents, as well as for in vivo biomonitoring (Fenech 2000; Luzhna et al. 2013; Araldi et al. 2015; Hayashi 2016; Knasmüller and Fenech 2019; Sommer et al. 2020).

Hooftman and De Raat (1982) modified the original MN test for application in fish. It was then validated by exposing different fish species to many genotoxic agents under laboratory conditions (Al-Sabti and Metcalfe 1995; Bolognesi and Hayashi 2011; Farag and Alagawany 2018).

MN can be detected in cells from different fish organs and tissues (gill, fin, liver), as well as in erythrocytes from different hemopoietic organs (spleen, kidney). However, such techniques are more complex and time-consuming. Moreover, they involve fish sacrifice (Al-Sabti and Metcalfe 1995; Cajaraville et al. 2003; Udroiu 2006; Bolognesi and Hayashi 2011). Peripheral blood sampling is technically simple and allows analysis of cells, also subsequently, without killing the animals. MN scoring is easy to learn and perform. Therefore, MN test in fish blood erythrocytes has become one of the simplest, fastest, and most cost-effective of in vivo assays for assessing genotoxic risk in aquatic environments, as showed in previous reviews (Al-Sabti and Metcalfe 1995; Ulupinar and Okumus 2002; Udroiu 2006; Bolognesi and Hayashi 2011; Obiakor et al. 2012; Hayashi 2016; Arslan and Parlak 2017) and confirmed by the high number of studies reported in this review. Moreover, the MN test can effectively complement other in vitro genotoxicity assays (*Salmonella* mutagenicity assay, SOS Chromotest, Mutatox® test, *umu*-C test, p53 CALUX®).

MN in peripheral circulating erythrocytes reflects events that occurred during their long lifespan time (Schlegel and MacGregor 1982, 1983). As the mean lifespan of these cells is even longer in fish (Soldatov 1995; Witeska 2013), this assay is particularly suitable for conditions of chronic exposure, such as ecotoxicological studies in aquatic environments.

A remarkable improvement of the rodent bone marrow MN test was the introduction of the acridine orange fluorescent staining (Hayashi et al. 1983). This method offers advantages over the original Giemsa staining. Acridine orange is a DNA-specific ultraviolet fluorochrome which does not stain non-nucleic acid artifacts, such as protein granules, which are difficult to discriminate from MN, allowing faster and error-free scorings. This staining protocol has been proposed as a routine procedure for the analysis of MN in fish erythrocytes (Polard et al. 2011b; Costa and Costa 2007), but it was used only in a limited number of studies (Acolas et al. 2020; Beaton et al. 2019; Bolognesi et al. 2006a; Bresler et al. 1999; Costa et al. 2008, 2011, 2012; Deguchi et al. 2007; Deutschmann et al. 2016; Floehr et al. 2015; Flores-Galván et al. 2020; Fuzinatto et al. 2013; Gusso-Choueri et al. 2016; Hayashi et al. 1998; Jovanović et al. 2018; Kanev et al. 2016; Kolarević et al. 2016; Kračun-Kolarević et al. 2016; Pinheiro-Sousa et al. 2019; Polard et al. 2011a, b; Torres-Bugarìn et al. 2007). That may be due to the need for a fluorescence microscope. Moreover, acridine orange staining is not permanent, and slides must be examined within 24 h.

An evolution of the original MN assay was the detection and recording of nuclear abnormalities (NA), induced by genotoxic damage, and appearing as modifications in the nucleus shape. These lesions are readily identifiable without any added sample preparation. Various authors (Hooftman and De Raat 1982; Manna et al. 1985; Al-Sabti 1986a, b; Das and Nanda 1986; Metcalfe 1988; Carrasco et al. 1990; Serrano-Garcìa and Montero-Montoya 2001; Fenech et al. 2003, 2011; Lindberg et al. 2007; Cheong et al. 2013) discussed and standardized their meaning, origin, and identification criteria (i.e., binucleated cells, nuclear buds, and blebbed, lobed, notched, or vacuolated nuclei). The analysis of NA in peripheral blood erythrocytes is a variant of the standard MN test and may complement it, as shown by several studies here reported (Tables 1, 2, and 3).

3 Fish Species

A total of 185 different freshwater (128) and saltwater (57) fish species have been used in the 262 experimental studies here reported (Tables 1, 2, and 3).

The most investigated fish species were *Oreochromis niloticus* (34 studies), *Platichthys flesus* (12 studies), *Clarias gariepinus* and *Cyprinus carpio* (11 studies), *Geophagus brasiliensis* and *Anguilla anguilla* (ten studies).

Table 1 Studies performed on wild fish collected from polluted waters

Site of fish collection	Fish species	Results and notes	Reference
Europe			
Gironde estuary (France)	*Acipenser sturio*	MN induction (seasonal variations); Comet assay	Acolas et al. (2020)
Sitnica river (Kosovo)	*Cyprinus carpio*	NA/MN induction	Aliko et al. (2015)
Swedish lakes (Sweden)	*Esox lucius*	Variable MN induction	Al-Sabti (1992a, b)
Baltic Sea (Sweden)	*Perca fluviatilis*	MN induction in all waste sites	Al-Sabti and Hardig (1990)
Drenica/Sitnica rivers (Kosovo)	*Rutilus rutilus*	MN induction; Comet assay	Asllani et al. (2019)
Trubia/Gafo rivers (Spain)	*Salmo trutta*	MN induction	Ayllón et al. (2000)
Nemunas river/Ginkunai lake/Svede and Bubliai ponds (Lithuania)	*Perca fluviatilis; Stiziostedion lucioperca; Rutilus rutilus; Abramis brama; Blicca bjoerkna*	MN induction	Bagdonas et al. (2003)
North Sea/Baltic Sea (Norway)	*Platichthys flesus; Symphodus melops*	MN induction in all polluted sites	Baršienė et al. (2004)
Baltic Sea (Lithuania)	*Platichthys flesus*	MN induction (seasonal variations)	Baršienė et al. (2006)
Baltic Sea (Denmark, Sweden, Finland, Russia, Baltic Republics, Poland)	*Platichthys flesus; Clupea harengus; Zoarces viviparus*	MN induction in all polluted sites and all fish species	Baršienė et al. (2012)
North Sea/Atlantic Ocean (UK, Norway, Denmark, Germany, Iceland)	*Limanda limanda; Melanogrammus aeglefinus*	NA/MN induction (in different polluted sites)	Baršienė et al. (2013)
Baltic Sea (Denmark, Sweden, Poland, Germany)	*Platichthys flesus; Clupea harengus; Gadus morhua*	NA/MN induction in all polluted sites	Baršienė et al. (2014)
Baltic Sea (Denmark, Sweden, Finland, Russia, Baltic Republics, Poland)	*Platichthys flesus; Clupea harengus; Zoarces. viviparus*	NA/MN induction	Baršienė et al. (2015)
Baltic Sea (Denmark, Sweden, Poland and Germany)	*Clupea harengus*	NA/MN induction in all polluted sites	Baršienė et al. (2016)
Danube river (Germany)	*Barbus barbus*	MN induction; Comet assay	Boettcher et al. (2010)
Ligurian Sea (Italy)	*Boops boop; Mullus barbatus; Uranoscopus scaber*	MN induction in a single fish species	Bolognesi et al. (2006b)
Firth of Forth (Scotland)	*Pholis gunnellus*	NA/MN induction; Comet assay	Bombail et al. (2001)
North Sea (Germany)	*Platichthys flesus; Limanda limanda*	MN induction	Bresler et al. (1999)

(continued)

Table 1 (continued)

Site of fish collection	Fish species	Results and notes	Reference
Baltic Sea (Latvia, Estonia)	*Platichthys flesus; Clupea harengus; Zoarces viviparus*	NA/MN induction in all fish species	Butrimavičienė et al. (2018)
Mondego/Douro/Ave estuaries (Portugal)	*Mugil cephalus*	NA/MN induction in all polluted sites	Carrola et al. (2014)
Bay of Biscay (Spain)	*Solea solea*	MN/NA baseline (seasonal variations)	Cuevas and Zorita (2018)
Danube river (Germany, Austria, Croatia, Serbia, Romania)	*Alburnus alburnus*	MN induction in a single polluted site; Comet assay	Deutschmann et al. (2016)
Kasperivtsi reservoir (Ukraine)	*Carassius auratus gibelio*	MN induction	Falfushynska et al. (2019)
Mediterranean Sea (Italy)	*Coris julis*	NA/MN induction; Comet assay	Fasulo et al. (2010)
Cecina river (Italy)	*Leuciscus cephalus*	MN induction; Comet assay	Frenzilli et al. (2008)
Ria de Aveiro lagoon (Portugal)	*Liza aurata*	NA induction (seasonal variations)	Guilherme et al. (2008)
Baltic Sea (Sweden)	*Perca fluviatilis*	MN induction	Hansson et al. (2014)
Bregalnica river (Macedonia)	*Barbus peloponnesius*	NA/MN induction	Ivanova et al. (2016)
Velika Morava river basin (Serbia)	*Alburnus alburnus*	No significant MN induction; Comet assay	Jovanović et al. (2018)
Rivne oblast rivers (Ukraine)	*Perca fluviatilis*	MN induction in all polluted sites	Klymenko and Biedunkova (2016)
Rivne oblast rivers (Ukraine)	*Alburnus alburnus; Rutilus rutilus; Scardinius erythrophthalmus; Carassius auratus; Abramis brama; Perca fluviatilis*	MN induction in all polluted sites and all fish species	Klymenko et al. (2018)
Elbe estuary (Germany)	*Platichthys flesus*	MN induction	Köhler and Ellesat (2008)
Sava river (Croatia)	*Leuciscus cephalus*	MN induction	Kolak et al. (1999)
Sava river (Slovenia, Croatia, Serbia, Bosnia)	*Alburnus alburnus; Alburnoides bipunctatus*	No significant MN induction; Comet assay	Kolarević et al. (2016)
Gulf of Gdansk (Poland)	*Platichthys flesus*	MN induction (seasonal variations)	Kopecka et al. (2006)
Adige river basin (Italy)	*Salmo cenerinus; Salmo marmoratus*	MN induction; no NA induction; Comet assay	Kračun-Kolarević et al. (2016)

(continued)

Table 1 (continued)

Site of fish collection	Fish species	Results and notes	Reference
Tajo river (Spain)	*Cyprinus carpio*	MN induction	Llorente et al. (2002)
Ripoll River (Spain)	*Barbus meridionalis*	NA induction; no MN induction	Maceda-Veiga et al. (2010)
Ripoll River (Spain)	*Squalius laietanus*	No MN/NA induction	Maceda-Veiga et al. (2013)
Mignone/Tevere rivers (Italy)	*Barbus plebejus*	MN induction in polluted river	Minissi et al. (1996)
Ria de Aveiro lagoon (Portugal)	*Dicentrarchus labrax*	NA induction (seasonal variations)	Mohmood et al. (2008)
Sitnica river (Kosovo)	*Carassius carassius; Gobio obtusirostris*	MN induction in both fish species	Morina et al. (2012)
Gulf of Gdansk (Poland)	*Platichthys flesus*	NA/MN induction	Napierska et al. (2009)
Ria de Aveiro lagoon (Portugal)	*Liza aurata*	MN/NA induction	Oliveira et al. (2010a)
Ria de Aveiro lagoon (Portugal)	*Liza aurata; Dicentrarchus labrax*	MN/NA induction	Oliveira et al. (2010b)
Ria de Aveiro lagoon (Portugal)	*Liza aurata; Liza ramada; Liza saliens*	NA induction	Pacheco et al. (2005)
Narva fish farm/Peipsi lake (Estonia)	*Oncorhynchus mykiss; Perca fluviatilis; Rutilus rutilus*	MN induction	Palm and Krause (1995)
Sava river (Croatia)	*Squalius cephalus*	MN induction; Comet assay	Pavlica et al. (2011)
Bregalnica river (Macedonia)	*Barbus peloponnesius*	NA/MN induction	Rebok et al. (2017)
Nalon, Trubia, Cancienes and Sella rivers (Spain)	*Salmo trutta; Anguilla anguilla; Phoxinus phoxinus*	MN induction in a single fish species	Rodriguez-Cea et al. (2003)
Sarno river (Italy)	*Gambusia holbrooki*	MN induction; Comet assay	Russo et al. (2004)
Baltic Sea/North Sea (UK, Germany, Netherlands, Denmark, Sweden, Lithuania, Poland, Russia)	*Platichthys flesus; Limanda limanda; Gadus morhua*	NA/MN induction in all polluted sites and all fish species	Rybakovas et al. (2009)
Piloña/Güeña/Sella rivers (Spain)	*Salmo trutta*	MN induction in all polluted sites	Sánchez-Galán et al. (1998)
Raices/Ferrerias/Sella rivers (Spain)	*Anguilla anguilla*	MN induction in all polluted sites	Sánchez-Galán et al. (2001)
Baltic Sea (Germany)	*Zoarces viviparus*	MN induction (seasonal variations)	Schiedek et al. (2006)
Adriatic Sea (Croatia)	*Sparus aurata*	NA/MN induction (seasonal variations)	Strunjak-Perovic et al. (2009a)

(continued)

Table 1 (continued)

Site of fish collection	Fish species	Results and notes	Reference
Adriatic Sea (Croatia)	*Dicentrarchus labrax; Mugil spp.*	NA/MN induction (seasonal variations)	Strunjak-Perovic et al. (2009b)
Seferani Lake (Albania)	*Carassius carassius*	NA/MN induction	Sula et al. (2019)
Sicily Ionic coast (Italy)	*Coris julis*	NA/MN induction; Comet assay	Tomasello et al. (2012)
Saronikos Gulf (Greece)	*Mugil cephalus*	NA/MN induction	Tsangaris et al. (2011b)
Narva river (Estonia)	*Oncorhynchus mykiss; Perca fluviatilis; Rutilus rutilus*	No NA/MN induction	Tuvikene et al. (1999)
Baltic Sea (Poland, Lithuania)	*Platichthys flesus; Clupea harengus; Gadus morhua*	NA/MN induction in all polluted sites and all fish species	Valskienė et al. (2019)
Po river (Italy)	*Chondrostoma söetta; Leuciscus cephalus; Barbus plebejus*	No statistically significant MN induction	Viganò et al. (2002)
Ableitergraben/Kraaker/Haubach creeks (Germany)	*Gasterosteus aculeatus*	MN induction (seasonal variations); Comet assay	Wirzinger et al. (2007)
Africa			
Nile river (Egypt)	*Oreochromis niloticus*	NA/MN induction	Abdel-Khalek (2015)
Nile river (Egypt)	*Oreochromis niloticus*	NA/MN induction; Comet assay	Abdel-Khalek et al. (2020)
Nile river/Mariout lake/Abou Homos and Kafr Eldawar drainages (Egypt)	*Oreochromis niloticus; Oreochromis aureus; Tilapia zilli; Clarias gariepinus*	MN induction in all polluted sites and all fish species	Ali et al. (2008)
Apodu reservoir (Nigeria)	*Clarias gariepinus; Synodontis gambiensis; Sarotherodon galilaeus; Protopterus annectens*	MN/NA induction in all fish species	Anifowoshe et al. (2018)
Abu-Qir Bay (Egypt)	*Mugil cephalus*	NA/MN induction	Hafez (2009)
Cairo wastewater treatment plant effluent (Egypt)	*Clarias gariepinus; Oreochromis niloticus*	MN induction	Lasheen et al. (2012)
Anambra river (Nigeria)	*Synodontis clarias; Tilapia nilotica*	MN induction (seasonal variations)	Obiakor et al. (2014)
Qaroun lake (Egypt)	*Oreochromis niloticus; Mugil cephalus*	MN/NA induction in both fish species	Omar et al. (2012)
Nile river (Egypt)	*Clarias gariepinus*	MN/NA induction	Osman and Kloas (2010)
Nile river (Egypt)	*Oreochromis niloticus; Clarias gariepinus*	MN/NA induction	Osman et al. (2011)

(continued)

Table 1 (continued)

Site of fish collection	Fish species	Results and notes	Reference
Asia			
Izmir Bay (Turkey)	*Gobius niger*	MN/NA induction	Arslan et al. (2010)
Aliaga Bay (Turkey)	*Pagellus erythrinus; Diplodus vulgaris; Serranus cabrilla; Boops boop; Solea solea*	MN/NA induction in all fish species	Arslan et al. (2015)
Çine stream (Turkey)	*Lepomis gibbosus*	MN induction	Başımoğlu Koca et al. (2005)
Mediterranean coast (Turkey)	*Mugil cephalus*	NA/MN induction (seasonal variations)	Çavas and Ergene-Gözükara (2005a)
Goa coast (India)	*Arius arius*	MN induction; Comet assay	D'Costa et al. (2017)
Goksu river (Turkey)	*Clarias gariepinus; Alburnus orontis; Mugil cephalus*	NA/MN induction in all fish species	Ergene et al. (2007b)
Yangtze river (China)	*Pelteobagrus vacchellii*	MN induction	Floehr et al. (2015)
Karaj river (Iran)	*Salmo trutta*	No MN induction; Comet assay	Hariri et al. (2018)
Tomio river/ Mochimune Harbor (Japan)	*Carassius sp.; Zacco platypus; Leiognathus nuchalis; Ditrema temmincki*	MN induction (seasonal variations)	Hayashi et al. (1998)
Chenab river (Pakistan)	*Labeo rohita*	NA/MN induction; Comet assay	Hussain et al. (2018)
Chenab river (Pakistan)	*Wallago attu; Cirrhinus mrigala*	NA/MN induction	Hussain et al. (2020)
Tom river (Russia)	*Esox lucius*	MN induction	Ilyinskikh et al. (1998)
Buyuk Menderes river (Turkey)	*Chondrostoma nasus; Barbus capito pectoralis*	No MN induction	Koca et al. (2008)
Melet river (Turkey)	*Alburnus chalcoides*	MN induction; Comet assay	Kontaş and Bostancı (2020)
Shitovskoe lake (Russia)	*Abramis brama; Perca fluviatilis; Rutilus rutilus; Esox lucius*	No statistically significant MN induction	Lugas'kova (2003)
Hoogly/Matlah rivers (India)	*Lates calcarifer; Liza parsia; Liza tade; Mugil cephalus; Rhinomugil corsula; Terapon jarbua; Scatophagus argus*	NA/MN induction	Mallick and Khuda-Bukhsh (2003)

(continued)

Table 1 (continued)

Site of fish collection	Fish species	Results and notes	Reference
Ganges river (India)	*Channa punctatus*	MN induction (seasonal variations); Comet assay	Nagpure et al. (2015)
Polluted sewage waters (India)	*Lepidocephalus spp.*	MN induction	Poongothai et al. (1996)
Brantas river delta (Indonesia)	*Oreochromis niloticus*	MN induction	Risjani et al. (2020)
Uluabat lake (Turkey)	*Cyprinus carpio*	MN induction	Saleh (2010)
Uluabat lake (Turkey)	*Cyprinus carpio*	MN induction	Saleh and Alshehri (2011)
Sincheon stream (Korea)	*Carassius auratus*	NA/MN induction	Samanta et al. (2018)
Techa river (Russia)	*Rutilus rutilus; Perca fluviatilis; Esox lucius*	MN induction in all fish species	Smagin et al. (2005)
Calcutta sewage-fed fish-ponds (India)	*Labeo bata*	NA/MN induction	Talapatra and Banerjee (2007)
Norilo-Pyasinsk water system (Russia)		MN induction (seasonal variations)	Zakhidov et al. (1996)
North America			
Ashtabula/Detroit/Genesee/Milwakee/St. Louis rivers (USA)	*Ameiurus nebulosus; Catostomus commersonii; Micropterus salmoides; Micropterus dolomieu*	NA/MN induction in all polluted sites and all fish species	Braham et al. (2017)
California coast (USA)	*Genyonemus lineatus*	MN/NA induction	Carrasco et al. (1990)
Androscoggin river (USA)	*Micropterus dolomieu*	MN induction; Comet assay	Chamberland et al. (2002)
Southern California coast (USA)	*Genyonemus lineatus; Paralabrax clathratus*	MN induction in both fish species	Hose et al. (1987)
North Atlantic coast (USA)	*Pseudopleuronectes americanus*	MN induction	Hughes and Hebert (1991)
San Francisco Bay (USA)	*Platichthys stellatus*	MN/NA induction	Long and Buchman (1989)
Mid-Atlantic coast (USA)	*Scomber scombrus; Scophthalmus aquosus; Pseudopleuronectes americanus*	MN induction in all polluted sites and all fish species	Longwell et al. (1983, 1992)
Superior/Ontario lakes (Canada)	*Ictalurus nebulosus; Catostomus commersoni*	No MN/NA induction	Smith (1990)
San Francisco Bay (USA)	*Platichthys stellatus*	MN induction	Spies et al. (1990)

(continued)

Table 1 (continued)

Site of fish collection	Fish species	Results and notes	Reference
Middle America			
Colorado/Hardy rivers, Reforma channel, Xochimilco lagoon (Mexico)	*Lepomis microlophus; Dorosoma petenense; Micropterus salmoides; Cyprinus carpio; Ameiurus natalis; Poecilia latipinna; Cyprinella lutrensis; Mugil cephalus; Lepomis gulosus; Carassius auratus; Tilapia sp.; Lepomis macrochirus*	MN induction	Flores-Galván et al. (2020)
South America			
Santos/Itanhaém estuaries (Brazil)	*Micropogonias furnieri*	No NA/MN induction	Abujamara et al. (2011)
Belem river (Brazil)	*Poecilia vivipara*	MN induction	Adam et al. (2010)
Patos lagoon estuary (Brazil)	*Paralichthys orbignyanus*	MN induction; Comet assay	Amado et al. (2006a)
Patos lagoon estuary (Brazil)	*Micropogonias furnieri*	MN induction (seasonal variations); Comet assay	Amado et al. (2006b)
Tramandaì/Mampituba estuaries (Brazil)	*Mugil sp.; Netuma sp.*	MN induction; Comet assay	Andrade et al. (2004)
Iguaçu river (Brazil)	*Rhamdia quelen*	MN induction in polluted sites	Arantes et al. (2016)
Santos-São Vicente/Cananéia estuaries (Brazil)	*Cathorops spixii*	MN/NA induction (seasonal variations)	Azevedo et al. (2012)
Extremoz lake (Brazil)	*Oreochromis niloticus*	No MN induction; Comet assay	Barbosa et al. (2010)
Corrente river (Brazil)	*Tilapia rendalli; Hoplias malabaricus*	MN induction	Batista et al. (2016)
Camacho/Santa Marta lakes (Brazil)	*Geophagus brasiliensis*	MN induction; Comet assay	Benincá et al. (2012)
Engano river (Brazil)	*Astyanax bimaculatus*	MN induction	Bogoni et al. (2014)
Canela National Forest reservoirs (Brazil)	*Hyphessobrycon luetkenii*	MN induction	Bühler et al. (2014)
Uberabinha river (Brazil)	*Rhamdia quelen*	MN induction	Campos Júnior et al. (2014)
Mumbuca stream (Brazil)	*Rhamdia quelen; Geophagus brasiliensis*	NA/MN induction in both fish species	Campos Júnior et al. (2015)
Paranaìba river (Brazil)	*Rhamdia quelen; Geophagus brasiliensis*	MN induction in both fish species	Campos Júnior et al. (2016)
Ambude river (Brazil)	*Colossoma macropomum*	MN/NA induction	Carvalho-Neta et al. (2015)

(continued)

Table 1 (continued)

Site of fish collection	Fish species	Results and notes	Reference
Tapajòs river (Brazil)	*Cichla sp.*	No MN induction	Castilhos et al. (2004)
Imboassica lagoon (Brazil)	*Hoplias malabaricus*	MN induction	Coimbra et al. (2018)
Ocoa river (Colombia)	*Astyanax gr. bimaculatus; A. metae*	NA/MN induction in both fish species	Corredor-Santamaria et al. (2016)
Ilha river (Brazil)	*Bryconamericus iheringii*	NA/MN induction (seasonal variations)	Dalzochio et al. (2017)
Sinos river (Brazil)	*Bryconamericus iheringii*	NA/MN induction	Dalzochio et al. (2018)
Flamengo Bay (Brazil)	*Atherinella brasiliensis*	No MN/NA induction	Fernandez et al. (2011)
Rio Santa Catarina river (Brazil)	*Boulengerella spp.*	MN induction	Ferreira and Nepomuceno (2008)
Iguaçu river (Brazil)	*Astyanax sp.*	No MN/NA induction; Comet assay	Freire et al. (2015)
Guanabara/Sepetiba/Ilha Grande coast (Brazil)	*Genidens genidens*	NA/MN induction	Freire et al. (2020)
Todos os Santos Bay (Brazil)	*Bathygobius soporator*	MN induction; no NA induction (seasonal variations)	Galindo and Moreira (2009)
Campo river (Brazil)	*Astyanax altiparanae*	NA/MN induction	Ghisi and Oliveira (2013)
Campo river (Brazil)	*Astyanax paranae*	NA/MN induction (seasonal variations); Comet assay	Ghisi et al. (2014)
Pirapó river (Brazil)	*Hypostomus ancistroides*	NA/MN induction (seasonal variations); Comet assay	Ghisi et al. (2016)
Uberabinha river (Brazil)	*Phalloceros caudimaculatus; Poecilia reticulata*	NA/MN induction	Gomes-Silva et al. (2020)
Paranoà lake (Brazil)	*Tilapia rendalli; Oreochromis niloticus; Cyprinus carpio*	No MN induction	Grisolia and Starling (2001)
Paranoà lake (Brazil)	*Geophagus brasiliensis; Cichla temensis; Hoplias malabaricus; Astyanax bimaculatus; Oreochromis niloticus; Cyprinus carpio; Steindachnerina insculpita*	NA/MN induction in three fish species; Comet assay	Grisolia et al. (2009)
Cananéia-Iguape-Peruíbe lagoon (Brazil)	*Cathorops spixii*	NA/MN induction (seasonal variations); Comet assay	Gusso-Choueri et al. (2016)

(continued)

Table 1 (continued)

Site of fish collection	Fish species	Results and notes	Reference
Rìo de la Plata subestuaries (Uruguay)	*Odontesthes argentinensis; Paralichthys orbignyanus; Mugil platanus; Micropogonias furnieri*	MN induction in all polluted sites and all fish species	Gutiérrez et al. (2015)
Porce river (Colombia)	*Brycon henni*	MN induction	Hurtado-Alarcón et al. (2007)
Contas river (Brazil)	*Serrasalmus brandtii; Hoplias malabaricus; Oreochromis niloticus; Geophagus brasiliensis*	NA/MN induction in all fish species	Jesus et al. (2016)
Paranaguà Bay (Brazil)	*Cathorops spixii*	NA/MN induction; Comet assay	Katsumiti et al. (2009)
Cananéia/São Vicente estuaries (Brazil)	*Centropomus parallelus*	NA/MN induction (seasonal variations)	Kirschbaum et al. (2009)
Itapemirim river (Brazil)	*Geophagus brasiliensis; Harttia sp.; Leporinus copellandii*	NA/MN induction (seasonal variations)	Lacerda et al. (2020)
Bom Jardin brook (Brazil)	*Astyanax jacuhiensis*	NA/MN induction	Lemos et al. (2008)
Javaes/Formoso rivers (Brazil)	*Cichla sp.*	No MN induction	Lima et al. (2018)
Goiana/Jaguaribe/Capibaribe/Sirinhaèm/Formoso estuaries (Brazil)	*Mugil curema*	MN induction; Comet assay	Lima et al. (2019)
Açu lake (Brazil)	*Prochilodus lacustris*	NA/MN induction (seasonal variations)	Lima Cardoso et al. (2018)
Paraiba do Sul river (Brazil)	*Oreochromis niloticus*	No statistically significant MN induction	Linde-Arias et al. (2008)
Uruguai river (Brazil)	*Astyanax jacuhiensis*	NA/MN induction (seasonal variations)	Loro et al. (2015)
Eastern Amazon rivers (Brazil)	*Rhamphichthys marmoratus; Steatogenys elegans; Sternopygus macrurus; Parapteronotus hasemani; Gymnotus mamiraua; Gymnotus arapaima; Brachyhypopomus beebei; Brachyhypopomus sp*	NA/MN induction in a single polluted site	Melo et al. (2013)
Mumbuca river (Brazil)	*Geophagus brasiliensis*	MN induction	Morais et al. (2016)

(continued)

Table 1 (continued)

Site of fish collection	Fish species	Results and notes	Reference
Cachimbero/Ayapel lakes (Colombia)	*Cyphocharax magdalenae; Prochilodus magdalenae; Hoplias malabaricus; Pimelodus blochii; Caquetaia krausii; Astyanax caucanus; Triportheus magdalenae; Ctenolucius hujeta; Trachelyopterus insignis; Eigenmania virescens*	MN induction in some fish species	Palacio-Betancur et al. (2009)
Japaratuba river (Brazil)	*Astyanax bimaculatus; Hoplias malabaricus*	MN induction in a single fish species	Pantaleão et al. (2006)
Paraguay river (Brazil)	*Pimelodus maculatus; Leporinus friderici*	MN induction in both fish species	Pimenta et al. (2013)
Ambude river (Brazil)	*Colossoma macropomum*	MN/NA induction	Pinheiro-Sousa et al. (2019)
Madeira river (Brazil)	*Prochilodus nigricans; Mylossoma duriventris; Hoplias malabaricus*	MN induction in all fish species	Porto et al. (2005)
Cangüiri river (Brazil)	*Astyanax sp.*	MN/NA induction; Comet assay	Ramsdorf et al. (2012)
São Francisco river (Brazil)	*Astyanax paranae*	MN induction	Ribeiro et al. (2014)
Tietê river (Brazil)	*Oreochromis niloticus*	MN induction Comet assay	Rocha et al. (2009)
Marajò archipelago (Brazil)	*Plagioscion squamosissimus*	MN/NA induction	Rocha et al. (2016)
Lambedor river (Brazil)	*Cyprinus carpio; Hypostomus punctatus; Rhamdia quelen; Oreochromis niloticus*	MN induction	Salvagni et al. (2011)
São Francisco river (Brazil)	*Prochilodus argenteus; Pimelodus maculatus; Myleus micans*	NA/MN induction in two fish species	Seriani et al. (2011)
Girassol pond (Brazil)	*Oreochromis niloticus*	No statistically significant MN/NA induction	Seriani et al. (2012)
Tietê river (Brazil)	*Oreochromis niloticus*	NA/MN induction	Seriani et al. (2015)
Barbacena farm reserve (Brazil)	*Astyanax sp.*	NA/MN induction (seasonal variations)	Silva et al. (2014)
Santo André pond (Brazil)	*Oreochromis niloticus*	NA induction, no MN induction	Silva et al. (2018)
Vitoria Bay/Santa Cruz estuary (Brazil)	*Centropomus parallelus*	NA induction, no MN induction (seasonal variations)	Souza et al. (2013)

(continued)

Table 1 (continued)

Site of fish collection	Fish species	Results and notes	Reference
Dourados/Brilhante rivers (Brazil)	*Astyanax lacustris*	No MN induction; NA induction (seasonal variations); Comet assay	Sposito et al. (2019)
Amambai river (Brazil)	*Pterygoplichthys ambrosetti; Prochilodus lineatus*	MN induction, no NA induction	Viana et al. (2017)
Paranà river (Brazil)	*Astyanax lacustris; Hypostomus ancitroides; Rhamdia quelen*	MN induction, no NA induction	Viana et al. (2018a)
Jirau Alto river (Brazil)	*Astyanax bifasciatus*	NA/MN induction; Comet assay	Wachtel et al. (2019)
Rio Grande (Colombia)	*Brycon henni*	MN induction (seasonal variations)	Zapata-Restrepo et al. (2017)
Oceania			
Spencer Gulf/Gulf St. Vincent (Australia)	*Aldrichetta forsteri; Sillago schomburgkii*	MN induction in both fish species	Edwards et al. (2001)
Spencer Gulf/Gulf St. Vincent (Australia)	*Aldrichetta forsteri; Sillago schomburgkii*	MN induction in both fish species	Hamann et al. (1999)

MN Micronuclei, *NA* Nuclear abnormalities

Selection of the right fish species is a crucial point. Remarkable differences exist about fish basic physiological features and behavior (van der Oost et al. 2003).

In various studies herein reported, authors based their choice on different criteria: wide distribution in water environments (Abdel-Khalek et al. 2020; Cuevas and Zorita 2018; Hansson et al. 2014; Kopecka et al. 2006; Llorente et al. 2002; Pantaleão et al. 2006; Pavlica et al. 2011; Sánchez-Galán et al. 1998; Souza et al. 2013; Sposito et al. 2019; Tomasello et al. 2012; Walia et al. 2015), limited mobility and migration (Cuevas and Zorita 2018; Fasulo et al. 2010; Rebok et al. 2017; Schiedek et al. 2006), convenient size (Ghisi et al. 2014; Viana et al. 2018b), ease of catching and handling (Benincá et al. 2012; Ghisi et al. 2014; Hansson et al. 2014; Klobučar et al. 2010; Viana et al. 2018b), amenability of maintenance and rearing under laboratory conditions (Ghisi et al. 2014; Lemos et al. 2007; SanJuan-Reyes et al. 2015; Talapatra and Banerjee 2007; Viana et al. 2018b), high sensitivity to water contaminants (Morais et al. 2016; Risjani et al. 2020). Fish species being a common food resource or of great economic and commercial relevance for human consumption have been chosen for both field and laboratory studies (Abdel-Khalek et al. 2020; Alimba et al. 2017; Andrade et al. 2004; Arslan et al. 2015; Başımoğlu Koca et al. 2005; Coimbra et al. 2018; D'Costa et al. 2017; Edwards et al. 2001; Gallão et al. 2019; Kirschbaum et al. 2009; Porto et al. 2005). In field biomonitoring studies, bottom-dwelling species are often preferred, due to their prolonged contact with polluted sediments (Azevedo et al. 2012; Boettcher et al. 2010; Bombail et al.

Table 2 Studies on fish exposed *in situ* to polluted waters

Site of exposure	Fish species	Results	Reference
Europe			
Ljubljanica river (Slovenia)	*Carassius gibelio*	MN induction	Al-Sabti et al. (1994)
Cunha Baixa uranium mine ponds (Portugal)	*Carassius auratus*	MN induction	Bessa et al. (2016)
Trasimeno lake (Italy)	*Cyprinus carpio*	MN induction; Comet assay	Buschini et al. (2004)
Ria de Aveiro lagoon (Portugal)	*Anguilla anguilla*	No MN induction; NA induction	Castro et al. (2018)
Sado estuary (Portugal)	*Solea senegalensis*	MN induction; Comet assay	Costa et al. (2011, 2012)
Po river (Italy)	*Oncorhynchus mykiss*	MN induction	De Flora et al. (1993)
Ria de Aveiro lagoon (Portugal)	*Liza aurata*	No MN induction; NA induction	Guilherme et al. (2008)
Drava river (Croatia)	*Cyprinus carpio*	MN induction; Comet assay	Klobučar et al. (2010)
Aveiro lagoon (Portugal)	*Anguilla anguilla*	No MN induction	Maria et al. (2003b)
Vouga river (Portugal)	*Anguilla anguilla*	NA induction	Maria et al. (2004)
Pulp mill effluents/Vouga river (Portugal)	*Anguilla anguilla*	NA induction	Pacheco and Santos (1999)
Aveiro lagoon (Portugal)	*Anguilla anguilla*	No NA induction	Pacheco and Santos (2001)
Oil-shale mine drainage (Estonia)	*Oncorhynchus mykiss*	MN induction	Palm et al. (1992)
Kaštela Bay (Croatia)	*Dicentrarchus labrax*	No MN induction; Comet assay	Šrut et al. (2010)
Pateira de Fermentelos wetland (Portugal)	*Anguilla anguilla*	No NA induction	Teles et al. (2007)
Narva river (Estonia)	*Oncorhynchus mykiss; Perca fluviatilis; Rutilus rutilus*	No MN induction	Tuvikene et al. (1999)
Asia			
Gomti river (India)	*Channa punctatus; Mystus vittatus*	MN induction; Comet assay	Kushwaha et al. (2012)
North America			
Pacific Northwest lake (USA)	*Oncorhynchus kisutch*	No MN induction	Barbee et al. (2008)
Maskinonge/Perch lakes, Duke swamp, Perch creek (Canada)	*Pimephales promelas*	MN induction; Comet assay	Beaton et al. (2019)

(continued)

Table 2 (continued)

Site of exposure	Fish species	Results	Reference
South America			
Água das Araras stream (Brazil)	*Astyanax altiparanae*	MN/NA induction	Vieira et al. (2014)
Godoy stream (Brazil)	*Prochilodus lineatus*	No MN induction; NA induction	Vieira et al. (2016)
Antarctica			
Admiralty Bay (King George Island)	*Trematomus newnesi*	MN/NA induction	Van Ngan et al. (2007)

MN Micronuclei, *NA* Nuclear abnormalities

Table 3 Studies on fish exposed in the laboratory to polluted waters

Site of water sampling	Fish species	Results	Reference
Europe			
Sado estuary (Portugal)	*Solea senegalensis*	MN induction; Comet assay	Costa et al. (2008, 2011, 2012)
Aveiro lagoon (Portugal)	*Dicentrarchus labrax*	NA induction	Gravato et al. (2000)
Aveiro lagoon (Portugal)	*Dicentrarchus labrax*	MN/NA induction	Gravato and Santos (2002, 2003)
Offshore platform effluent, Adriatic Sea (Italy)	*Dicentrarchus labrax*	No MN induction; Comet assay	Gorbi et al. (2009)
Wastewater treatment work effluent (UK)	*Rutilus rutilus*	MN induction	Liney et al. (2006)
Vouga river (Portugal)	*Anguilla anguilla*	MN induction	Maria et al. (2003a)
Wool shrinkproofing effluent (Spain)	*Oncorhynchus mykiss*	MN induction	Marlasca et al. (1998)
Pulp mill effluents (Portugal)	*Anguilla anguilla*	NA induction	Pacheco and Santos (1996)
Pulp mill effluents/Vouga river (Portugal)	*Anguilla anguilla*	NA induction	Pacheco and Santos (1999)
Save river (France)	*Carassius carassius*	MN induction (seasonal variations); Comet assay	Polard et al. (2011a, 2011b)
Aveiro lagoon (Portugal)	*Anguilla anguilla*	NA induction	Pacheco and Santos (2001)
Dnieper and Desna rivers (Ukraine)	*Carassius gibelio*	NA induction	Tsangaris et al. (2011a)
Vilnius wastewater treatment plant (Lithuania)	*Oncorhynchus mykiss*	MN induction, no NA induction	Vosyliene et al. (2010)
Africa			
Olosun/Aba Ku landfills (Nigeria)	*Clarias gariepinus*	MN/NA induction	Alimba and Bakare (2016)
Oyo State hospital effluent (Nigeria)	*Clarias gariepinus*	MN/NA induction	Alimba et al. (2017)

(continued)

Table 3 (continued)

Site of water sampling	Fish species	Results	Reference
Pharmaceutical industry effluent (Nigeria)	*Clarias gariepinus*	MN/NA induction	Alimba et al. (2019)
Textile industry effluent (Nigeria)	*Clarias gariepinus*	MN/NA induction	Ayoola et al. (2012)
Oyi river (Nigeria)	*Clarias gariepinus*	MN induction	Obiakor et al. (2010)
Brewery/textile mills effluents (Nigeria)	*Clarias lazera*	MN induction	Odeigah and Osanyinpeju (1995)
Asia			
Textile dyeing mill effluent (Turkey)	*Oreochromis niloticus*	MN/NA induction	Çavas and Ergene-Gözükara (2003)
Petroleum refinery/chromium plant (Turkey)	*Oreochromis niloticus*	MN/NA induction	Çavas and Ergene-Gözükara (2005b)
Sambalpur paper mill effluent (India)	*Heteropneustes fossilis*	MN/NA induction	Das and Nanda (1986)
Landfill leachates (Japan)	*Carassius auratus*	MN induction; Comet assay	Deguchi et al. (2007)
Berdan river (Turkey)	*Oreochromis niloticus*	MN/NA induction	Ergene et al. (2007a)
Textile industry effluent (Sri Lanka)	*Oreochromis niloticus*	MN/NA induction; Comet assay	Hemachandra and Pathiratne (2016)
Kelani river (Sri Lanka)	*Oreochromis niloticus*	MN/NA induction; Comet assay	Hemachandra and Pathiratne (2017)
Ergene river (Turkey)	*Gambusia affinis*	MN/NA induction	Kanev et al. (2016)
Tannery effluent (India)	*Labeo rohita*	MN induction	Praveena et al. (2014)
Nilufer stream (Turkey)	*Oreochromis niloticus*	MN/NA induction	Summak et al. (2010)
Tannery industry effluent (India)	*Labeo rohita*	MN induction	Walia et al. (2015)
Textile factory/wastewater treatment plant effluent (China)	*Danio rerio*	MN induction; Comet assay	Zhang et al. (2012, 2014)
Middle America			
Toluca NSAID-manufacturing plant effluent (Mexico)	*Cyprinus carpio*	MN induction; Comet assay	SanJuan-Reyes et al. (2015)
South America			
Espìrito Santo stone processing waste (Brazil)	*Oreochromis niloticus*	MN/NA induction	Aguiar et al. (2016)
Reconquista river (Argentina)	*Cnesterodon decemmaculatus*	No MN induction; Comet assay	Baudou et al. (2019)
Sinos river (Brazil)	*Leporinus obtusidens*	No MN/NA induction	Bergamaschi et al. (2015)
Sinos river (Brazil)	*Astyanax jacuhiensis*	No MN induction; Comet assay	Bianchi et al. (2015)

(continued)

Table 3 (continued)

Site of water sampling	Fish species	Results	Reference
Itajaì-Açù river (Brazil)	*Oreochromis niloticus*	MN induction	Bücker and Conceição (2012)
Los Padres pond (Argentina)	*Cheirodon interruptus*	MN induction	Campana et al. (2001)
Agua Boa stream (Brazil)	*Astyanax altiparanae*	MN induction; Comet assay	Dourado et al. (2017)
Jacuném pond (Brazil)	*Oreochromis niloticus*	MN/NA induction; Comet assay	Duarte et al. (2012)
Curitiba hospital waste landfill (Brazil)	*Astyanax sp.*	No MN/NA induction; Comet assay	Erbe et al. (2011)
Jordão river (Brazil)	*Astyanax altiparanae*	MN induction	Francisco et al. (2019)
Cubatão do Sul river (Brazil)	*Oreochromis niloticus*	MN induction (seasonal variations)	Fuzinatto et al. (2013)
Vitoria sanitary landfill leachate (Brazil)	*Oreochromis niloticus*	MN induction; Comet assay	Gallão et al. (2019)
Doce river (Brazil)	*Geophagus brasiliensis*	MN induction; Comet assay	Gomes et al. (2019)
Brasilia domestic sewage wastewater (Brazil)	*Oreochromis niloticus; Tilapia rendalli*	No MN induction	Grisolia et al. (2005)
Atibaia river (Brazil)	*Oreochromis niloticus*	MN/NA induction	Hoshina et al. (2008)
Cai river (Brazil)	*Pimephales promelas*	MN induction (seasonal variations)	Lemos et al. (2007)
Còrrego dos Bagres stream (Brazil)	*Oreochromis niloticus*	MN/NA induction; Comet assay	Matsumoto et al. (2006)
Gasoline stations wastewater (Brazil)	*Oreochromis niloticus*	NA induction, no MN induction	Oliveira-Martins and Grisolia (2009)
Reconquista river (Argentina)	*Cnesterodon decemmaculatus*	NA induction, no MN induction; Comet assay	Ossana et al. (2016, 2019)
Dourados river (Brazil)	*Astyanax lacustris*	MN induction (seasonal variations); Comet assay	Rocha et al. (2018)
Paraìba do Sul river (Brazil)	*Oreochromis niloticus*	MN/NA induction (seasonal variations)	Silva Souza and Fontanetti (2006)
Bacanga river (Brazil)	*Danio rerio*	MN/NA induction	Sousa et al. (2020)
Ornamental stone processing waste (Brazil)	*Geophagus brasiliensis*	MN induction; Comet assay	Venturoti et al. (2019a, b)
Paranà river (Brazil)	*Astyanax lacustris*	MN/NA induction	Viana et al. (2018b)
Antarctica			
Admiralty Bay (King George Island)	*Trematomus newnesi*	MN/NA induction	Van Ngan et al. (2007)

MN Micronuclei, *NA* Nuclear abnormalities

2001; Costa et al. 2008; Cuevas and Zorita 2018; Guilherme et al. 2008; Gusso-Choueri et al. 2016; Kopecka et al. 2006; Maria et al. 2003b).

Various authors investigated spontaneous MN frequencies in wild unexposed specimens. The analysis of the "baseline" MN frequencies shows significant variations amongst distinct species, as well as within the same species. These differences could be due to several factors such as age, sex, genetic profile, reproductive status, mobility, location, diet, trophic level, season, water temperature, oxygen content, and habitat features (Jha 2004; Torres-Bugarìn et al. 2007; Bolognesi and Hayashi 2011; Melo et al. 2013; Furnus et al. 2014; Cuevas and Zorita 2018).

Karyotype and ploidy are other relevant choice criteria. Many fish species have a high number of small chromosomes (Ojima et al. 1976; Hooftman and Vink 1981) and MN arising after clastogenic damage could be very small, escaping direct microscopical detection. Therefore, species with fewer and larger chromosomes are recommended (Kligerman 1982; Carrasco et al. 1990; Udroiu 2006). MN induction is also influenced by erythropoiesis (Udroiu 2006) and directly associated with polyploidy (Strunjak-Perovic et al. 2003).

Fish species living in the same environment may also have different sensitivity to aquatic genotoxins (Hayashi et al. 1998; Bolognesi et al. 2006a; Grisolia et al. 2009). Some species are less sensitive than others to MN induction: this could mainly be due to differences in metabolism, DNA repair efficiency, defense mechanisms, and responses to genotoxic damage (Al-Sabti and Metcalfe 1995; van der Oost et al. 2003; Bolognesi and Hayashi 2011).

4 Experimental Conditions

Genotoxicity of polluted waters has been evaluated both in freshwater (187 studies) and brackish or saltwater environments (75 studies) of all continents (Tables 1, 2, and 3).

All studies here listed were performed under three different experimental conditions: (a) collection of wild fish in polluted waters, (b) in situ exposure of caged fish in polluted waters, or (c) laboratory exposure of fish to sampled polluted waters.

Most of the studies (181 out of 262) used fish caught in polluted waters (Table 1). Twenty-two studies were carried out in situ, placing caged fish in polluted and unpolluted waters (Table 2). Fifty-nine studies were carried out by exposing fish to polluted water samples under controlled laboratory conditions (Table 3).

As mentioned earlier, the present review does not report laboratory studies testing single or combined chemical agents at arbitrarily chosen concentrations. Other authors extensively reviewed laboratory exposure investigations evaluating the genotoxic activity of different chemical or physical agents on fish erythrocytes (Al-Sabti and Metcalfe 1995; Bolognesi and Hayashi 2011; Farag and Alagawany 2018).

4.1 Catching of Wild Fish in Polluted Waters

These studies compared one or more polluted sites with unpolluted reference areas. In several studies evaluating river pollution, fish were caught upstream and downstream a specific pollution source.

Studies involving the catching of wild fish offer a more realistic view of the ecological context (Suter 2006). However, animal migration could be a confounding factor, and it is not always possible to precisely link the observed effect with the site of capture and duration of exposure. Also, habitat, food availability, feeding, breeding, trophic level in the food web, and predation may stand for relevant variability factors (Kroon et al. 2017). Another problem is the need to transfer the samples quickly to the laboratory for processing. Furthermore, it is impossible to have a real negative control: in fact, negative controls are specimens from pristine reference waters, often different and far from the polluted water bodies object of the study. In general, these studies are less standardized, repeatable, and predictable. Nevertheless, they are much more numerous than other types, as evidenced in the present review and confirmed by Colin et al. (2016).

4.2 In Situ Exposure of Caged Fish in Polluted Waters

In situ tests using caged fish are particularly useful, as they consider all biotic, physical, and chemical variables. These studies offer many advantages (Oikari 2006). It is possible to standardize nearly all experimental parameters while preserving ecological features. Feral fish are usually caught and then acclimatized in the laboratory; alternatively, commercially available species can be used. Animals are housed in a plastic or soft net cage, placed in established sites of the water body. Negative controls are matching fish specimens kept into unpolluted waters. This experimental method of exposure allows deciding the exact place and duration of exposure, reducing inter-individual variability, and minimizing the impact of adaptive mechanisms. However, some disadvantages are reported: unsuitability of some fish species for caging, animal stress, constant feeding necessity, logistical difficulties and practical problems, excessive costs, unfavorable environmental conditions (storms, floods), and the possibility of vandalism (Oikari 2006). Also, for these studies, there is a need for rapid sample transfer to the laboratory.

4.3 Laboratory Exposure of Fish to Sampled Polluted Waters

Laboratory exposure studies are simpler to set up, cheaper, and less susceptible to environmental variables. Usually, wild fish specimens are caught in unpolluted waters and transferred alive to the laboratory. However, some fish species are too hard to capture or to keep in the laboratory. Therefore, the choice often falls on species commercially available or supplied by research fish farms, which are not resident in the polluted water under evaluation and are hardly comparable with wild aquatic populations. Fish are kept into large glass/plastic tanks filled with water sampled from polluted areas. It is possible to have negative control (pure groundwater or filtered dechlorinated water) and positive control (a known genotoxic agent). Animal dimensions, weight, and age can be precisely standardized. Feeding, aeration, pH, chemical composition, temperature, and lighting conditions are highly standardized. However, laboratory exposure is less ecologically realistic than the field in situ exposure.

Moreover, all studies (field or laboratory) may involve further organizational, bureaucratic, and logistic problems (obtaining of authorizations, the involvement of extraneous people and use of suitable tools for fish catching) (Oikari 2006; Suter 2006).

5 Results of the Studies

Most studies (88%) showed induction of MN and/or NA in fish exposed to polluted water (Tables 1, 2, and 3). Eighteen studies (10%) performed on collected wild fish (Table 1), six studies (3%) performed on fish exposed in situ (Table 2), and six studies (10%) performed on laboratory-exposed fish (Table 3) reported negative results.

Negative results could be due to several reasons. Some fish species are more sensitive than others to genotoxic damage, due to physiological response to chemical pollutants and less efficient DNA repair mechanisms (Bolognesi and Hayashi 2011). The extent of cytogenetic damage may vary with sex, hormonal status, age, or diet (Al-Sabti and Metcalfe 1995; Udroiu 2006).

Inhibition of erythropoiesis by elevated levels of toxic chemicals and faster removal of apoptotic cells could also play a role, reducing the actual number of micronucleated erythrocytes in peripheral blood (Witeska 2013; Farag and Alagawany 2018). Moreover, mechanisms of action of pollutants and their concentrations could not be sufficient to induce a significant clastogenic or aneugenic damage.

MN formation is a cell cycle-dependent event, and short-term exposures may not induce it. As previously discussed, this assay is particularly suitable for chronic exposures (Udroiu 2006).

Several studies showed seasonal variations of MN and NA frequencies (Tables 1, 2, and 3). In general, genotoxic effects in peripheral blood erythrocytes are higher in spring and even more in summer, whilst lower values are reported in winter. That may depend on environmental temperature (Strunjak-Perovic et al. 2009a). In particular, the water temperature may influence fish metabolism, erythropoiesis, cell replication, mitotic index, and DNA repair mechanisms (van der Oost et al. 2003; Udroiu 2006). Higher water temperature could also increase oxygen consumption and induce reactive oxygen species (ROS) generation (Ross et al. 2001; Cheng et al. 2015; Okoye et al. 2019).

Increases in MN/NA frequency values are reported in the dry season (when water levels are lower), while decreases are seen during the rainy season (with higher water levels). Likely, genotoxic pollutant concentrations may vary according to water levels. Moreover, rainfalls and river runoffs may "wash" soil contaminants, transferring them into water bodies and therefore raising their concentrations. This aspect could be particularly relevant for coastal marine environments and estuaries (Andrade et al. 2004; Carrola et al. 2014; Gutiérrez et al. 2015).

6 Conclusions

MN test in peripheral blood fish erythrocytes has gained popularity and diffusion. This assay offers several advantages (simplicity, rapidity, low costs, biological and genotoxicological significance). However, some limitations and drawbacks remain. Therefore, a complete procedure standardization is necessary to compare the results adequately. All researchers should use similar sample preparation and staining protocols, as well as analysis techniques.

In our opinion, acridine orange staining is the "gold standard," and all researchers should adopt it. As proved by different authors (Polard et al. 2011b; Costa and Costa 2007), this method reduces the detection of false-positive artifacts. Moreover, it ensures a better and faster scoring process.

NA analysis and recording give fuller information than MN recording alone. Therefore, we believe that all researchers should adopt NA analysis, using the same identification and classification criteria proposed by different authors (Carrasco et al. 1990; Fenech et al. 2003; Carrola et al. 2014).

Most researchers analyze a thousand cells per animal. We suggest the scoring and analysis of a higher number of cells (many thousands or even tens of thousands), which ensures greater statistical significance. This goal can be achieved using automated scoring techniques or flow cytometry (Schultz et al. 1993; Ahmad et al. 2002; Dertinger et al. 2011). However, these procedures are more expensive, not yet fully standardized and still present some technical problems (Sommer et al. 2020).

The most suitable and sensitive fish species should be used. However, it is not easy to find a single species representative of all geographical areas. Therefore, for studies based on the catching of wild fish, it is almost impossible to recommend a preferential species. We suggest choosing the most common fish species in that area, preferably with known sensitivity for MN assay. Various authors give further useful suggestions about species choice in biomonitoring surveys (van der Oost et al. 2003; Suter 2006).

For in situ caging or laboratory exposure studies we suggest choosing among fish species indicated by Oikari (2006). According to this author, commercially available species should be chosen from hatcheries logistically close enough to the research site. Species tolerating handling by an experimenter are preferable. Various local fish species may be caught from clean natural habitat by non-damaging fishing techniques for later caging. Furthermore, for better standardization, reproducibility, and comparability of the results, it would be advisable to always use the same species, chosen from those recognized as most sensitive to the MN test.

As previously discussed, intraspecific variability factors (age, sex, metabolism, diet), as well as environmental variables (water temperature and levels, salinity, UV radiation), should also be considered and standardized. Therefore, we suggest always using fish of the same sex and size. Various studies show seasonal variability of results. However, there are no indications of the best season. We suggest performing fish catches, in situ caging, and water sampling always in the same season of the year to guarantee uniformity and comparability of the results.

Considering the minimum time necessary to induce the formation of MN in fish (Udroiu 2006; Polard et al. 2011b), we recommend an exposure time of at least 4–6 days.

Another relevant point is the choice of experimental conditions. As previously highlighted, each method (feral fish catching, in situ caging, laboratory exposure) has specific advantages and disadvantages (Oikari 2006; Suter 2006). The ideal strategy would be a combination of different protocols, as performed in some studies here reviewed (Pacheco and Santos 2001; Van Ngan et al. 2007; Guilherme et al. 2008; Costa et al. 2011, 2012). However, this is not always possible, due to logistic and organizational problems, as well as higher costs. In situ or laboratory studies should always be preferred, because they allow a better standardization of parameters and reproducibility of the results. We suggest choosing the field catching studies whenever ecologically realistic information represents the main target. In fact, these studies take into account interactions among biotic, physical, and chemical variables (Bonnineau et al. 2012). In addition, all studies can be useful for better drafting of laws or regulations concerning water and environmental quality (Castaño et al. 2003; Suter 2006).

Like all other biomarkers, MN test alone gives partial information about the genotoxic effect of pollutants. As previously discussed, this assay reveals only clastogenic (i.e., chromosome breakage) or aneugenic (i.e., mitotic spindle dysfunctions) lesions, which are not repairable damages. Furthermore, as already mentioned, it is best suited for long-term chronic exposures, rather than for detection of early effects. Therefore, it would be advisable to associate at least another test to assess DNA damage and repair. The single-cell gel electrophoresis assay, also known as Comet test, is an excellent experimental method for assessing earlier repairable lesions, such as DNA strand breaks (Dhawan et al. 2009). This assay has long been used in water eco-genotoxicology studies (Mitchelmore and Chipman 1998; Cotelle and Ferard 1999; Lee and Steinert 2003; Lapuente et al. 2015). MN and Comet tests used together may give complementary information and offer a fuller picture of the genotoxic potential of xenobiotics (Araldi et al. 2015). Sixty-five studies have adopted this combined approach, as reported in Tables 1, 2, and 3. Furthermore, combining MN and Comet tests with simpler and cheaper in vitro genotoxicity assays (*Salmonella* mutagenicity assay, SOS Chromotest, Mutatox® test, *umu*-C test, p53 CALUX®) can give additional useful information. In fact, the use of these systems in ecotoxicological studies allows extrapolation from in vitro to in vivo and gives precious information on biological responses (Houk 1992; Claxton et al. 1998; Castaño et al. 2003; Jha 2004; Tabrez et al. 2011; van der Linden et al. 2014).

As previously said, effect-based monitoring tools are increasingly being used and can be successfully applied to complement chemical analysis in routine water quality monitoring. Furthermore, these methods may supply useful guidance in a regulatory context (Escher and Leusch 2012; van der Burg et al. 2013; Escher et al. 2015, 2018; Leusch and Snyder 2015; Prasse et al. 2015; Wernersson et al. 2015; van der Oost et al. 2017a, b; Brack et al. 2019). Effect-based methods have also long been used in fish, and their added value for water risk assessment appears from numerous studies (reviewed by van der Oost et al. 2003). Several studies reported an evident dose-response increase in MN frequencies in fish erythrocytes after laboratory exposure to selected physical or chemical genotoxic agents, as well as polluted water samples (reviewed in Al-Sabti and Metcalfe 1995, Bolognesi and Hayashi 2011, Farag and Alagawany 2018). However, no study has proposed to derive effect-based trigger values for the MN test in fish erythrocytes in our literature analysis. We believe it appropriate to include this assay among those routinely used in effect-based method strategies. That would allow a better application of this test in assessing human and environmental risks from genotoxic water pollution.

At present, MN analysis is one of the most useful biomonitoring methods for the evaluation of water genotoxic pollution. Knowledge of all studies published so far on this subject will surely ease the improvement and standardization of technical procedures, as well as the uniformity and reproducibility of results.

Conflict of Interest The authors declare that they have no conflict of interest.

References

Abdel-Khalek AA (2015) Antioxidant responses and nuclear deformations in freshwater fish, *Oreochromis niloticus*, facing degraded environmental conditions. Bull Environ Contam Toxicol 94:701–708

Abdel-Khalek AA, Dajem SB, Morsy K (2020) The long-term exposure to discharges of Sabal drain induces genotoxic effects on *Oreochromis niloticus*. Bull Environ Contam Toxicol 104:858–863

Abujamara LD, Seriani R, Ranzani-Paiva MJ, Moreira LB, Abessa DMS, Pereira CDS (2011) Análise de parâmetros citogenotóxicos da corvina, *Micropogonias furnieri*, provenientes de dois estuários da Baixada Santista, SP. Rev Ceciliana Dez 3:17–20 (in Portuguese)

Acolas M-L, Davail B, Gonzalez P, Jean S, Clérandeau C, Morin B, Gourves P-Y, Daffe G, Labadie P, Perrault A, Lauzent M, Pierre M, Le Barh R, Baudrimont M, Peluhet L, Le Menach K, Budzinski H, Rochard E, Cachot J (2020) Health indicators and contaminant levels of a critically endangered species in the Gironde estuary, the European sturgeon. Environ Sci Pollut Res 27:3726–3745

Adam ML, Torres RA, Sponchiado G, Motta TS, Oliveira CMR, Carvalho-Filho MA, Correia MTS (2010) Environmental degradation at a public park in Southern Brazil as revealed through a genotoxicity test (MN) on peripheral blood cells from *Poecilia vivipara* (Teleostei). Water Air Soil Pollut 211:61–68

Aguiar LL, Tonon CB, Nunes ET, Braga ACA, Neves MA, David JAO (2016) Mutagenic potential of fine wastes from dimension stone industry. Ecotoxicol Environ Saf 125:116–120

Ahmad W, Ali NM, Farah AM, Ateeq B (2002) Computerized automated morphometric assay including frequency estimation of pentachlorophenol induced nuclear anomalies (micronucleus) in catfish *Heteropneustes fossilis*. Chromosoma 110:570–574

Ali FK, El-Shehawi AM, Seehy MA (2008) Micronucleus test in fish genome: a sensitive monitor for aquatic pollution. Afr J Biotechnol 7:606–612

Aliko V, Morina V, Sula E (2015) Cellular and nuclear alterations of carp fish (*Cyprinus carpio*) erythrocytes as sensitive biomarkers of citotoxicity and genotoxicity of Sitnica River waters (Kosovo). Aktet 8:163–168

Alimba CG, Bakare AA (2016) In vivo micronucleus test in the assessment of cytogenotoxicity of landfill leachates in three animal models from various ecological habitats. Ecotoxicology 25:310–319

Alimba CG, Ajiboye RD, Fagbenro OS (2017) Dietary ascorbic acid reduced micronucleus and nuclear abnormalities in *Clarias gariepinus* (Burchell 1822) exposed to hospital effluent. Fish Physiol Biochem 43:1325–1335

Alimba CG, Adekoya KO, Soyinka OO (2019) Exposure to effluent from pharmaceutical industry induced cytogenotoxicity, hematological and histopathological alterations in *Clarias gariepinus*. Excli J 18:63–78

Al-Sabti K (1986a) Clastogenic effects of five carcinogenic mutagenic chemicals on the cells of the common carp, *Cyprinus carpio* L. Comp Biochem Physiol C 85:5–9

Al-Sabti K (1986b) Comparative micronucleated erythrocyte cell induction in three cyprinids by five carcinogenic-mutagenic chemicals. Cytobios 47:147–154

Al-Sabti K (1991) Handbook of genotoxic effects and fish chromosomes. J Stefan Institute Press, Yugoslavia

Al-Sabti K (1992a) Micronuclei induction in pike (*Esox lucius*) in Swedish lakes contaminated with radiocesium. Cytobios 70:27–32

Al-Sabti K (1992b) Monitoring the genotoxicity of radiocontaminants in Swedish lakes by fish micronuclei. Cytobios 70:101–106

Al-Sabti K, Hardig J (1990) Micronucleus test in fish for monitoring the genotoxic effects of industrial waste products in the Baltic Sea, Sweden. Comp Biochem Physiol C 97:179–182

Al-Sabti K, Metcalfe CD (1995) Fish micronuclei for assessing genotoxicity in water. Mutat Res 343:121–135

Al-Sabti K, Franko M, Andrijanic B, Knez S, Stegnar P (1994) Chromium-induced micronuclei in fish. J Appl Toxicol 14:333–336

Amado LL, Robaldo RB, Geracitano LA, Monserrat JM, Bianchini A (2006a) Biomarkers of exposure and effect in the Brazilian flounder *Paralichthys orbignyanus* (Teleostei: Paralichthyidae) from the Patos Lagoon estuary (Southern Brazil). Mar Pollut Bull 52:207–213

Amado LL, Rosa CE, Leite AM, Moraes L, Pires WV, Pinho GLL, Martins CMG, Robaldo RB, Nery LEM, Monserrat JM, Bianchini A, Martinez PE, Geracitano LA (2006b) Biomarkers in croakers *Micropogonias furnieri* (Teleostei: Sciaenidae) from polluted and non-polluted areas from the Patos Lagoon estuary (Southern Brazil): evidences of genotoxic and immunological effects. Mar Pollut Bull 52:199–206

Andrade VM, Silva J, Silva FR, Heuser VD, Dias JF, Yoneama ML, Freitas TRO (2004) Fish as bioindicators to assess the effects of pollution in two southern brazilian rivers using the comet assay and micronucleus test. Environ Mol Mutagen 44:459–468

Anifowoshe AT, Oladipo SO, Owolodun OA, Akinseye MK, Olafimihan TF, Sidiq AG, Muhammed AA, Asa TA, Mustapha MK (2018) Ecogenotoxicogical assessments of some selected fish species from Apodu Reservoir, Malete, North Central, Nigeria. Manila J Sci 11:1–14

Araldi RP, Melo TC, Mendes TB, Sà Jùnior PL, Nozima BHN, Ito ET, Carvalho RF, Souza EB, Stocco RC (2015) Using the comet and micronucleus assay for genotoxicity studies: a review. Biomed Pharmacother 72:74–82

Arantes ACR, Adam ML, Souza JRB, Bastos LP, Jacobina UP, Torres RA (2016) Frequency of fish micronuclei to diagnose aquatic environmental conditions from Brazilian megacities: a case study of Iguaçu river, Southern Brazil. R Bras Bioci 14:111–116

Arslan ÖC, Parlak H (2017) Micronucleus test good biomarker for determination of genetic changes in aquatic organisms. J Aquat Pollut Toxicol 1:18

Arslan ÖC, Parlak H, Katalay S, Boyacioglu M, Karaaslan MA, Guner H (2010) Detecting micronuclei frequency in some aquatic organisms for monitoring pollution of Izmir Bay (Western Turkey). Environ Monit Assess 165:55–56

Arslan ÖC, Boyacioglu M, Parlak H, Katalay S, Karaaslan MA (2015) Assessment of micronuclei induction in peripheral blood and gill cells of some fish species from Aliaga Bay Turkey. Mar Pollut Bull 94:48–54

Asllani FH, Schürz M, Bresgen N, Eckl PM, Alija AJ (2019) Genotoxicity risk assessment in fish (*Rutilus rutilus*) from two contaminated rivers in the Kosovo. Sci Total Environ 676:429–435

Ayllón F, Suciu R, Gephard S, Juanes F, Garcia-Vazquez E (2000) Conventional armament wastes induce micronuclei in wild brown trout *Salmo trutta*. Mutat Res 470:169–176

Ayoola SO, Bassey BO, Alimba CG, Ajani EK (2012) Textile effluent induced genotoxic effect and oxidative stress in *Clarias gariepinus*. Pak J Biol Sci 15:804–812

Azevedo JS, Braga ES, Ribeiro CAO (2012) Nuclear abnormalities in erythrocytes and morphometric indexes in the catfish *Cathorops spixii* from different sites on the Southeastern Brazilian coast. Braz J Oceanogr 60:323–330

Bagdonas E, Bukelskis E, Lazutka JR (2003) Frequency of micronucleated erythrocytes in wild fish from natural freshwater body. Ekologija 1:67–71

Barbee GC, Barich J, Duncan B, Bickham JW, Matson CW, Hintze CJ, Autenrieth RL, Zhou G-D, McDonald TJ, Cizmas L, Norton D, Donnelly KC (2008) In situ biomonitoring of PAH-contaminated sediments using juvenile coho salmon (*Oncorhynchus kisutch*). Ecotoxicol Environ Saf 71:454–464

Barbosa JS, Cabral TM, Ferreira DN, Agnez-Lima LF, Medeiros SRB (2010) Genotoxicity assessment in aquatic environment impacted by the presence of heavy metals. Ecotoxicol Environ Saf 73:320–325

Baršienė J, Lazutka J, Syvokiene J, Dedonyte V, Rybakovas A, Bagdonas E, Bjornstad A, Andersen OK (2004) Analysis of micronuclei in blue mussels and fish from the Baltic and North Seas. Environ Toxicol 19:365–371

Baršienė J, Lehtonen KK, Koehler A, Broeg K, VuorinenPJ LT, Pempkowiak J, Syvokiene J, Dedonyte V, Rybakovas A, Repecka R, Vuontisjärvi H, Kopecka J (2006) Biomarker responses in flounder (*Platichthys flesus*) and mussel (*Mytilus edulis*) in the Klaipeda-Butinge area (Baltic Sea). Mar Pollut Bull 53:422–436

Baršienė J, Rybakovas A, Lang T, Grygiel W, Andreikenaite L, Michailovas A (2012) Risk of environmental genotoxicity in the Baltic Sea over the period of 2009-2011 assessed by micronuclei frequencies in blood erythrocytes of flounder (*Platichthys flesus*), herring (*Clupea harengus*) and eelpout (*Zoarces viviparus*). Mar Environ Res 77:35–42

Baršienė J, Rybakovas A, Lang T, Andreikenaite L, Michailovas A (2013) Environmental genotoxicity and cytotoxicity levels in fish from the North Sea offshore region and Atlantic coastal waters. Mar Pollut Bull 68:106–116

Baršienė J, Butrimavičienė L, Rybakovas A, Grygiel W, Lang T, Michailovas A, Jackunas T (2014) Environmental genotoxicity and cytotoxicity in flounder (*Platichthys flesus*), herring (*Clupea harengus*) and Atlantic cod (*Gadus morhua*) from chemical munitions dumping zones in the southern Baltic Sea. Mar Environ Res 96:56–67

Baršienė J, Butrimavičienė L, Michailovas A, Grygiel W (2015) Assessing the environmental genotoxicity risk in the Baltic Sea: frequencies of nuclear buds in blood erythrocytes of three native fish species. Environ Monit Assess 187:4078–4089

Baršienė J, Butrimavičienė L, Grygiel W, Stunžėnas V, Valskienė R, Greiciūnaitė J, Stankevičiūtė M (2016) Environmental genotoxicity assessment along the transport routes of chemical munitions leading to the dumping areas in the Baltic Sea. Mar Pollut Bull 103:45–53

Başımoğlu Koca Y, Koca S, Yıldız S, Gürcü B, Osanç E, Tunçbaş O, Aksoy G (2005) Investigation of histopathological and cytogenetic effects on *Lepomis gibbosus* (Pisces: Perciformes) in the Çine stream (Aydın/Turkey) with determination of water pollution. Environ Toxicol 20:560–571

Batista NJC, Cavalcante AACM, Oliveira MG, Medeiros ECN, Machado JL, Evangelista SR, Dias JF, Santos CEI, Duarte A, Silva FR, Silva J (2016) Genotoxic and mutagenic evaluation of water samples from a river under the influence of different anthropogenic activities. Chemosphere 164:134–141

Baudou FG, Ossana NA, Castañé PM, Mastrángelo MM, González Núñez AA, Palacio MJ, Ferrari L (2019) Use of integrated biomarker indexes for assessing the impact of receiving waters on a native neotropical teleost fish. Sci Total Environ 650:1779–1786

Beaton ED, Gosselin I, Festarini A, Gagnaire B, Farrow F, Cavalié I, Shultz C, Kim SB, Walsh S, Chen HQ, Adam-Guillermin C, Stuart M (2019) Correlated responses for DNA damage, phagocytosis activity and lysosomal function revealed in a comparison between field and laboratory studies: fathead minnow exposed to tritium. Sci Total Environ 662:990–1002

Belfiore NM, Anderson SL (2001) Effects of contaminants on genetic patterns in aquatic organisms: a review. Mutat Res 489:97–122

Benincá C, Ramsdorf W, Vicari T, Ribeiro CAO, Almeida MI, Assis HCS, Cestari MM (2012) Chronic genetic damages in *Geophagus brasiliensis* exposed to anthropic impact in Estuarine Lakes at Santa Catarina Coast – Southern of Brazil. Environ Monit Assess 184:2045–2056

Bergamaschi B, Rodrigues MT, Silva JVS, Kluge M, Luz RB, Fleck JD, Bianchi E, Silva LB, Spilki FR (2015) Moving beyond classical markers of water quality: detection of enteric viruses and genotoxicity in water of the Sinos River. Braz J Biol 75(Suppl. 2):S63–S67

Bessa ML, Antunes SC, Pereira R, Gonçalves FJM, Nunes B (2016) Multibiomarker toxicity characterization of uranium mine drainages to the fish *Carassius auratus*. Environ. Sci Pollut Res 23:13355–13367

Beyer J, Petersen K, Song Y, Ruus A, Grung M, Bakke T, Tollefsen KE (2014) Environmental risk assessment of combined effects in aquatic ecotoxicology: a discussion paper. Mar Environ Res 96:81–91

Bianchi E, Goldoni A, Trintinaglia L, Lessing G, Silva CEM, Nascimento CA, Ziulkoski AL, Spilki FR, Silva LB (2015) Evaluation of genotoxicity and cytotoxicity of water samples from the Sinos River Basin, Southern Brazil. Braz J Biol 75(Suppl. 2):S68–S74

Bickham JW, Sandhu S, Hebert PDN, Chikhi L, Athwal R (2000) Effects of chemical contaminants on genetic diversity in natural populations: implications for biomonitoring and ecotoxicology. Mutat Res 463:33–51

Boettcher M, Grund S, Keiter S, Kosmehl T, Reifferscheid G, Seitz N, Rocha PS, Hollert H, Braunbeck T (2010) Comparison of *in vitro* and *in situ* genotoxicity in the Danube River by means of the comet assay and the micronucleus test. Mutat Res 700:11–17

Bogoni JA, Armiliato N, Araldi-Favassa CT, Techio VH (2014) Genotoxicity in *Astyanax bimaculatus* (Twospot Astyanax) exposed to the waters of Engano River (Brazil) as determined by micronucleus tests in erythrocytes. Arch Environ Contam Toxicol 66:441–449

Bolognesi C, Hayashi M (2011) Micronucleus assay in aquatic animals. Mutagenesis 26:205–213

Bolognesi C, Perrone E, Roggieri P, Pampanin DM, Sciutto A (2006a) Assessment of micronuclei induction in perypheral blood erythrocytes of fish exposed to xenobiotics under controlled conditions. Aquat Toxicol 78S:S93–S98

Bolognesi C, Perrone E, Roggieri P, Sciutto A (2006b) Bioindicators in monitoring long term genotoxic impact of oil spill: haven case study. Mar Environ Res 62:S287–S291

Bombail V, Dennis A, Gordon E, Batty J (2001) Application of the comet and micronucleus assay to butterfish (*Pholis gunnellus*) erythrocytes from the Firth of Forth, Scotland. Chemosphere 44:383–392

Bonnineau C, Moeller A, Barata C, Proia L, Sans-Piché F, Schmitt-Jansen M, Guasch H, Segner H (2012) Advances in the multibiomarker approach for risk assessment in aquatic ecosystems. In: Guasch H, Ginebreda A, Geiszinger A (eds) The handbook of environmental chemistry v.19 — emergency and priority pollutants in rivers. Springer, Berlin, pp 147–179

Brack W, Ait-Aissa S, Backhaus T, Dulio V, Escher BI, Faust M, Hilscherova K, Hollender J, Hollert H, Müller C, Munthe J, Posthuma L, Seiler T-B, Slobodnik J, Teodorovic I, Tindall AJ, de Aragão UG, Zhang X, Altenburger R (2019) Effect-based methods are key. The European Collaborative Project SOLUTIONS recommends integrating effect-based methods for diagnosis and monitoring of water quality. Environ Sci Eur 31:10

Braham RP, Blazer VS, Shaw CH, Mazik PM (2017) Micronuclei and other erythrocyte nuclear abnormalities in fishes from the Great Lakes Basin, USA. Environ Mol Mutagen 58:570–581

Bresler V, Bissinger V, Dizer AAH, Sturm A, Kratke R, Fishelson L, Hansen P-D (1999) Marine molluscs and fish as biomarkers of pollution stress in littoral regions of the Red Sea, Mediterranean Sea and North Sea. Helgol Mar Res 53:219–243

Bücker A, Conceição MB (2012) Genotoxicity evaluation of tilapia (*Oreochromis niloticus*) exposed to waters from two sites of Itajaí-Açu River (SC, Brazil). J Braz Soc Ecotoxicol 7:51–56

Bühler D, Marinowic DR, Barros MP, Silva LB (2014) Genetic damage induced by water pollutants in the freshwater fish *Hyphessobrycon luetkenii* (Characidae) in a reservoir of the Canela National Forest, Brazil. J Freshw Ecol 29:295–299

Buschini A, Martino A, Gustavino B, Monfrinotti M, Poli P, Rossi C, Santoro M, Dörr AJM, Rizzoni M (2004) Comet assay and micronucleus test in circulating erythrocytes of *Cyprinus carpio* specimens exposed in situ to lake waters treated with disinfectants for potabilization. Mutat Res 557:119–129

Butrimavičienė L, Baršienė J, Greiciūnaitė J, Stankevičiūtė M, Valskienė R (2018) Environmental genotoxicity and risk assessment in the Gulf of Riga (Baltic Sea) using fish, bivalves, and crustaceans. Environ Sci Pollut Res 25:24818–24828

Cajaraville MP, Hauser L, Carvalho G, Hylland K, Olabarrieta I, Lawrence AJ, Lowe D, Goksøyr A (2003) Genetic damage and the molecular/cellular response to pollution. In: Lawrence AJ, Hemingway KL (eds) Effects of pollution on fish: molecular effects and population responses. Blackwell Science, Oxford, pp 14–82

Campana MA, Panzeri AM, Escalante AH, Moreno VJ, Dulout FN (2001) Micronucleus test in fish from a pampasic pond (Argentina): an estimation of the presence of genotoxic compounds. J Environ Pathol Toxicol Oncol 20:325–331

Campos Júnior EO, Pereira BB, Morelli S, Pavanin EV, Pavanin LA (2014) Biological monitoring and B chromosome frequency in Bagre (*Rhamdia quelen*) in Southeast Brazil. Environ Toxicol Pharmacol 38:510–517

Campos Júnior EO, Pereira BB, Morelli S (2015) Monitoring genotoxicity potential in the Mumbuca. J Toxicol Environ Health A 78:1277–1287

Campos Júnior EO, Silva Oliveira RG, Pereira BB, Souto HN, Campos CF, Nepomuceno JC, Morelli S (2016) Assessment of genotoxic, mutagenic, and recombinogenic potential of water resources in the Paranaíba River Basin of Brazil: a case study. J Toxicol Environ Health A 79:1190–1200

Carrasco KR, Tilbury KL, Myers MS (1990) Assessment of the piscine micronucleus test as an in situ biological indicator of chemical contaminant effects. Can J Fish Aquat Sci 47:2123–2136

Carrola J, Santos N, Rocha MJ, Fontainhas-Fernandes A, Pardal MA, Monteiro RA, Rocha E (2014) Frequency of micronuclei and of other nuclear abnormalities in erythrocytes of the grey mullet from the Mondego, Douro and Ave estuaries – Portugal. Environ Sci Pollut Res 21:6057–6068

Carvalho-Neta RNF, Sousa DBP, Sobrinho ICM, Horton EY, Almeida ZS, Tchaicka L, Sousa AL (2015) Genotoxic and hematological parameters in *Colossoma macropomum* (Pisces, Serrasalmidae) as biomarkers for environmental impact assessment in a protected area in northeastern Brazil. Environ Sci Pollut Res 22:15994–16003

Castaño A, Bols N, Braunbeck T, Dierickx P, Halder M, Halder M, Kawahara K, Lee LE, Mothersill C, Pärt P, Repetto G, Sintes JR, Rufli H, Smith R, Wood C, Segner H (2003) The use of fish cells in ecotoxicology. The report and recommendations of ECVAM Workshop 47. Altern Lab Anim 31:317–351

Castilhos ZC, Almonsy N, Souto PS, Pereira da Silva LCC, Linde AR, Bidone ED (2004) Bioassessment of ecological risk of amazonian ichthyofauna to mercury. Bull Environ Contam Toxicol 72:671–679

Castro D, Mieiro CL, Coelho JP, Guilherme S, Marques A, Santos MA, Duarte AC, Pereira E, Pacheco M (2018) Addressing the impact of mercury estuarine contamination in the European eel (*Anguilla anguilla* L., 1758) – an early diagnosis in glass eel stage based on erythrocytic nuclear morphology. Mar Pollut Bull 127:733–742

Çavas T, Ergene-Gözükara S (2003) Micronuclei, nuclear lesions and interphase silver-stained nucleolar organizer regions (AgNORs) as cyto-genotoxicity indicators in *Oreochromis niloticus* exposed to textile mill effluent. Mutat Res 538:81–91

Çavas T, Ergene-Gözükara S (2005a) Micronucleus test in fish cells: a bioassay for in situ monitoring of genotoxic pollution in the marine environment. Environ Mol Mutagen 46:64–70

Çavas T, Ergene-Gözükara S (2005b) Induction of micronuclei and nuclear abnormalities in *Oreochromis niloticus* following exposure to petroleum refinery and chromium processing plant effluents. Aquat Toxicol 74:264–271

Chamberland K, Lindroth BA, Whitaker B (2002) Genotoxicity in Androscoggin river smallmouth bass. Northeast Nat 9:203–212

Cheng C-H, Yang F-F, Liao S-A, Miao Y-T, Ye C-X, Wang A-L, Tan J-W, Chen X-Y (2015) High temperature induces apoptosis and oxidative stress in pufferfish (*Takifugu obscurus*) blood cells. J Therm Biol 53:172–179

Cheong HSJ, Seth I, Joiner MC, Tucker JD (2013) Relationships among micronuclei, nucleoplasmic bridges and nuclear buds within individual cells in the cytokinesis-block micronucleus assay. Mutagenesis 28:433–440

Claxton LD, Houk VS, Hughes TJ (1998) Genotoxicity of industrial wastes and effluents. Mutat Res 410:237–243

Coimbra RSC, Mascarenhas MS, Saraiva VB, Santos CR, Lopes RM, Hauser-Davis RA, Oliveira VPS, Molisani MM, Almeida MG, Rezende CE, Carvalho CEV, Oliveira MM (2018) Metal loads and biomarker suite responses in a tropical carnivorous fish indicative of anthropogenic impacts in a Southeastern Brazilian lagoon. Environ Monit Assess 190:564

Colin N, Porte C, Fernandes D, Barata C, Padròs F, Carrassòn M, Monroy M, Cano-Rocabayera O, Sostoa A, Piña B, Maceda-Veiga A (2016) Ecological relevance of biomarkers in monitoring studies of macro-invertebrates and fish in Mediterranean rivers. Sci Total Environ 540:307–323

Corredor-Santamaria W, Gómez MS, Velasco-Santamaria YM (2016) Using genotoxic and haematological biomarkers as an evidence of environmental contamination in the Ocoa River native fish, Villavicencio-Meta, Colombia. Springerplus 5:351

Costa PM, Costa MH (2007) Genotoxicity assessment in fish peripheral blood: a method for a more efficient analysis of micronuclei. J Fish Biol 71(Suppl A):148–151

Costa PM, Lobo J, Caeiro S, Martins M, Ferreira AM, Caetano M, Vale C, DelValls TA, Costa MH (2008) Genotoxic damage in *Solea senegalensis* exposed to sediments from the Sado Estuary (Portugal): effects of metallic and organic contaminants. Mutat Res 654:29–37

Costa PM, Neuparth TS, Caeiro S, Lobo J, Martins M, Ferreira AM, Caetano M, Vale C, DelValls TA, Costa MH (2011) Assessment of the genotoxic potential of contaminated estuarine sediments in fish peripheral blood: laboratory versus *in situ* studies. Environ Res 111:25–36

Costa PM, Caeiro S, Vale C, DelValls TA, Costa MH (2012) Can the integration of multiple biomarkers and sediment geochemistry aid solving the complexity of sediment risk assessment? A case study with a benthic fish. Environ Pollut 161:107–120

Cotelle S, Ferard JF (1999) Comet assay in genetic ecotoxicology: a review. Environ Mol Mutagen 34:246–255

Cuevas N, Zorita I (2018) Baseline levels of environmental genotoxicity and potential confounding factors using common sole (*Solea solea*) as sentinel organism. Mar Environ Res 138:1–8

D'Costa A, Shyama SK, Praveen Kumar MK (2017) Bioaccumulation of trace metals and total petroleum and genotoxicity responses in an edible fish population as indicators of marine pollution. Ecotoxicol Environ Saf 142:22–28

Dalzochio T, Ressel Simões LA, de Souza MS, Rodrigues GZP, Petry IE, Andriguetti NB, Silva GJH, Gehlen G, da Silva LB (2017) Water quality parameters, biomarkers and metal bioaccumulation in native fish captured in the Ilha River, southern Brazil. Chemosphere 189:609–618

Dalzochio T, Rodrigues GZP, Simões LAR, Souza MS, Petry IE, Andriguetti NB, Silva GJH, Silva LB, Gehlen G (2018) In situ monitoring of the Sinos River, southern Brazil: water quality parameters, biomarkers, and metal bioaccumulation in fish. Environ Sci Pollut Res 25:9485–9500

Das RK, Nanda NK (1986) Induction of micronuclei in peripheral erythrocytes of fish *Heteropneustes fossilis* by mitomycin C and paper mill effluent. Mutat Res 175:67–71

Dashwood RH, Bailey GS (1998) Use of fish and fish transgenics in laboratory and field genotoxicology studies. Mutat Res 399:123–124

De Flora S, Bagnasco M, Zanacchi P (1991a) Genotoxic, carcinogenic, and teratogenic hazards in the marine environment, with special reference to the Mediterranean Sea. Mutat Res 258:285–320

De Flora S, Zanacchi P, Bagnasco M, Brunetti R, Majone F, Levis AG (1991b) Metabolic and genetic effects of marine pollution on aquatic organisms. In: Gledhill B, Mauro F (eds) Trends in biological dosimetry. Wiley-Liss, New York, pp 69–78

De Flora S, Viganò L, D'Agostini F, Camoirano A, Bagnasco M, Bennicelli C, Melodia F, Arillo A (1993) Multiple genotoxicity biomarkers in fish exposed *in situ* to polluted river water. Mutat Res 319:167–177

Deguchi Y, Toyoizumi T, Masuda S, Yasuhara A, Mohri S, Yamada M, Inoue Y, Kinae N (2007) Evaluation of mutagenic activities of leachates in landfill sites by micronucleus test and comet assay using goldfish. Mutat Res 627:178–185

Dertinger SD, Torous DK, Hayashi M, MacGregor JT (2011) Flow cytometric scoring of micronucleated erythrocytes: an efficient platform for assessing in vivo cytogenetic damage. Mutagenesis 26:139–145

Deutschmann B, Kolarevic S, Brack W, Kaisarevic S, Kostic J, Kracun-Kolarevic M, Liska I, Paunovic M, Seiler T-B, Shao Y, Sipos S, Slobodnik J, Teodorovic I, Vukovic-Gacic B, Hollert H (2016) Longitudinal profile of the genotoxic potential of the River Danube on erythrocytes of wild common bleak (*Alburnus alburnus*) assessed using the comet and micronucleus assay. Sci Total Environ 573:1441–1449

Dhawan A, Bajpayee M, Parmar D (2009) Comet assay: a reliable tool for the assessment of DNA damage in different models. Cell Biol Toxicol 25:5–32

Dixon DR, Pruski AM, Dixon LRJ, Jha AN (2002) Marine invertebrate eco-genotoxicology: a methodological overview. Mutagenesis 17:495–507

Dourado PLR, Rocha MP, Roveda LM, Raposo Junior JL, Cândido LS, Cardoso CAL, Morales MAM, Oliveira KMP, Grisolia AB (2017) Genotoxic and mutagenic effects of polluted surface water in the midwestern region of Brazil using animal and plant bioassays. Genet Mol Biol 40:123–133

Duarte IM, Dias MC, David JAO, Matsumoto ST (2012) A qualidade da água da Lagoa Jacuném (Espírito Santo, Brasil) em relação a aspectos genotóxicos e mutagênicos, mensurados respectivamente pelo ensaio do cometa e teste do micronúcleo em peixes da espécie *Oreochromis niloticus*. R Bras Bioci 10:211–219 (in Portuguese)

Edwards JW, Edyvane KS, Boxall VA, Hamann M, Soole KL (2001) Metal levels in seston and marine fish flesh near industrial and metropolitan centres in South Australia. Mar Pollut Bull 42:389–396

Erbe MCL, Ramsdorf WA, Vicari T, Cestari MM (2011) Toxicity evaluation of water samples collected near a hospital waste landfill through bioassays of genotoxicity piscine micronucleus test and comet assay in fish *Astyanax* and ecotoxicity *Vibrio fischeri* and *Daphnia magna*. Ecotoxicology 20:320–328

Ergene S, Çavas T, Çelik A, Köleli N, Aymak C (2007a) Evaluation of river water genotoxicity using the piscine micronucleus test. Environ Mol Mutagen 48:421–429

Ergene S, Çavas T, Çelik A, Köleli N, Kaya F, Karahan A (2007b) Monitoring of nuclear abnormalities in peripheral erythrocytes of three fish species from the Goksu Delta (Turkey): genotoxic damage in relation to water pollution. Ecotoxicology 16:385–391

Escher BI, Leusch FDL (2012) Bioanalytical tools in water quality assessment. IWA Publishing, London

Escher BI, Neal PA, Leusch FDL (2015) Effect-based trigger values for in vitro bioassays: reading across from existing water quality guideline values. Water Res 81:137–148

Escher BI, Aït-Aïssa S, Behnisch PA, Brack W, Brion F, Brouwer A, Buchinger S, Crawford SE, Du Pasquier D, Hamers T, Hettwer K, Hilscherová K, Hollert H, Kase R, Kienle C, Tindall AJ, Tuerk J, van der Oost R, Vermeirssen E, Neale PA (2018) Effect-based trigger values for in vitro and in vivo bioassays performed on surface water extracts supporting the environmental quality standards (EQS) of the European water framework directive. Sci Total Environ 628–629:748–765

European Community Marine Strategy Framework Directive (2008) Directive 2008/56/EC of the European Parliament and of the Council of 17 June 2008 establishing a framework for community action in the field of marine environmental policy (Marine Strategy Framework Directive). OJUnion, 1.164/19

European Union Water Framework Directive (2000) Directive 2000/60/EC of the European Parliament and of the Council of 23 October 2000 establishing a framework for community action in the field of water policy. OJEC 327:1–72

Falfushynska H, Horyn O, Fedoruk O, Khoma V, Rzymski P (2019) Difference in biochemical markers in the gibel carp (*Carassius auratus gibelio*) upstream and downstream of the hydro-power plant. Environ Pollut 255:113213

Farag MR, Alagawany M (2018) Erythrocytes as a biological model for screening of xenobiotics toxicity. Chem Biol Interact 279:73–83

Fasulo S, Marino S, Mauceri A, Maisano M, Giannetto A, D'Agata A, Parrino V, Minutoli R, De Domenico E (2010) A multibiomarker approach in *Coris julis* living in a natural environment. Ecotoxicol Environ Saf 73:1565–1573

Fenech M (2000) The *in vitro* micronucleus technique. Mutat Res 455:81–95

Fenech M, Chang WP, Kirsch-Volders M, Holland N, Bonassi S, Zeiger E, Human MicroNucleus Project (2003) HUMN project: detailed description of the scoring criteria for the cytokinesis-block micronucleus assay using isolated human lymphocyte cultures. Mutat Res 534:65–75

Fenech M, Kirsch-Volders M, Natarajan AT, Surralles J, Crott JW, Parry J, Norppa H, Eastmond DA, Tucker JD, Thomas P (2011) Molecular mechanisms of micronucleus, nucleoplasmic bridge and nuclear bud formation in mammalian and human cells. Mutagenesis 26:125–132

Fernandez WS, Dias JF, Ribeiro CAO, Azevedo JS (2011) Liver damages and nuclear abnormalities in erythrocytes of *Atherinella brasiliensis* (Actynopterigii, Atherinopsidade) from two beaches in southeast of Brazil. Braz J Med Biol Res 59:163–169

Feron VJ, Schwarz M, Hemminki K, Krewski D (1999) Long- and medium-term carcinogenicity studies in animals and short-term genotoxicity tests. In: Moolgavkar S, Krewski D, Zeise L, Cardis E, Moller H (eds) Quantitative estimation and prediction of human cancer risks. IARC scientific publications n. 131. IARC, Lyon, pp 103–129

Ferreira GR, Nepomuceno JC (2008) Poluição do Rio Santa Catarina no Municipio de Vazante (MG) detectada por meio do teste do micronùcleo em peixes. Perquirēre 5:277–284 (in Portuguese)

Floehr T, Scholz-Starke B, Xiao H, Koch J, Wu L, Hou J, Wolf A, Bergmann A, Bluhm K, Yuan X, Roß-Nickoll M, Schäffer A, Hollert H (2015) Yangtze Three Gorges Reservoir, China: a holistic assessment of organic pollution, mutagenic effects of sediments and genotoxic impacts on fish. J Environ Sci 38:63–82

Flores-Galván MA, Daesslé LW, Arellano-García E, Torres-Bugarín O, Macías-Zamora JV, Ruiz-Campos G (2020) Genotoxicity in fishes environmentally exposed to As, Se, Hg, Pb, Cr and toxaphene in the lower Colorado River basin, at Mexicali valley, Baja California, México. Ecotoxicology 29:493–502

Francisco CM, Bertolino SM, De Oliveira Júnior RJ, Morelli S, Pereira BB (2019) Genotoxicity assessment of polluted urban streams using a native fish *Astyanax altiparanae*. J Toxicol Environ Health A 82:514–523

Freire CA, Souza-Bastos LR, Chiesse J, Tincani FH, Piancini LDS, Randi MAF, Prodocimo V, Cestari MM, Assis HCS, Abilhoa V, Vitule JRS, Bastos LP, Oliveira-Ribeiro CA (2015) A multibiomarker evaluation of urban, industrial, and agricultural exposure of small characins in a large freshwater basin in southern Brazil. Environ Sci Pollut Res 22:13263–13277

Freire MM, Amorim LMF, Buch AC, Gonçalves AD, Sella SM, Cassella RJ, Moreira JC, Silva-Filho EV (2020) Polycyclic aromatic hydrocarbons in bays of the Rio de Janeiro state coast, SE – Brazil: effects on catfishes. Environ Res 181:108959

Frenzilli G, Falleni A, Scarcelli V, Del Barga I, Pellegrini S, Savarino G, Mariotti V, Benedetti M, Fattorini D, Regoli D, Nigro M (2008) Cellular responses in the cyprinid *Leuciscus cephalus* from a contaminated freshwater ecosystem. Aquat Toxicol 89:188–196

Frenzilli G, Nigro M, Lyons BP (2009) The comet assay for the evaluation of genotoxic impact in aquatic environments. Mutat Res 681:80–92

Furnus GNA, Caffetti JD, Garcìa EM, Benìtez MF, Pastori MC, Fenocchio AS (2014) Baseline micronuclei and nuclear abnormalities frequencies in native fishes from the Paranà River (Argentina). Braz J Biol 74:217–221

Fuzinatto CF, Flohr L, Melegari SP, Matias WG (2013) Induction of micronucleus of *Oreochromis niloticus* exposed to waters from the Cubatão do Sul river, Southern Brazil. Ecotoxicol Environ Saf 98:103–109

Gajski G, Žegura B, Ladeira C, Novak M, Sramkova M, Pourrut B, Del Bo' C, Milić M, Gutzkow KB, Costa S, Dusinska M, Brunborg G, Collins A (2019) The comet assay in animal models: from bugs to whales – (part 2 vertebrates). Mutat Res 781:130–164

Galindo TP, Moreira LM (2009) Evaluation of genotoxicity using the micronucleus assay and nuclear abnormalities in the tropical sea fish *Bathygobius soporator* (Valenciennes, 1837) (Teleostei, Gobiidae). Genet Mol Biol 32:394–398

Gallão M, Gnocchi KG, Carvalho LR, Silva BF, Barbosa AN, Chippari-Gomes AT (2019) The impact of sublethal concentrations of treated leachate on Nile tilápia (*Oreochromis niloticus*). Bull Environ Contam Toxicol 102:519–524

Ghisi NC, Oliveira EC (2013) Evidence of contamination in the main body of the public water supply of Campo Mourão-PR: an analysis from a bioindicator of environmental quality. SaBios-Rev Saúde Biol 8:13–22

Ghisi NC, Oliveira EC, Fávaro LF, Assis HCS, Prioli AJ (2014) In situ assessment of a neotropical fish to evaluate pollution in a river receiving agricultural and urban wastewater. Bull Environ Contam Toxicol 93:699–709

Ghisi NC, Oliveira EC, Mota TFM, Vanzetto GV, Roque AA, Godinho JP, Bettim FL, Assis HCS, Prioli AJ (2016) Integrated biomarker response in catfish *Hypostomus ancistroides* by multivariate analysis in the Pirapó River, southern Brazil. Chemosphere 161:69–79

Gomes LC, Chippari-Gomes AR, Miranda TO, Pereira TM, Merçon J, Davel VC, Barbosa BV, Pereira ACH, Frossard A, Ramos JPL (2019) Genotoxicity effects on *Geophagus brasiliensis* fish exposed to Doce River water after the environmental disaster in the city of Mariana, MG, Brazil. Braz J Biol 79:659–664

Gomes-Silva G, Pereira BB, Liu K, Chen B, Santos VSV, de Menezes GHT, Pires LP, Santos BMT, Oliveira DM, Machado PHA, de Oliveira Júnior RJ, de Oliveira AMM, Plath M (2020) Using native and invasive livebearing fishes (Poeciliidae, Teleostei) for the integrated biological assessment of pollution in urban streams. Sci Total Environ 698:134336

Gorbi S, Benedetti M, Virno Lamberti C, Pisanelli B, Moltedo G, Regoli F (2009) Biological effects of diethylene glycol (DEG) and produced waters (PWs) released from offshore activities: a multi-biomarker approach with the sea bass *Dicentrarchus labrax*. Environ Pollut 157:3166–3173

Gravato C, Santos MA (2002) Juvenile sea bass liver biotransformation and erythrocytic genotoxic responses to pulp mill contaminants. Ecotoxicol Environ Saf 53:104–112

Gravato C, Santos MA (2003) Dicentrarchus labrax biotransformation and genotoxicity responses after exposure to a secondary treated industrial/urban effluent. Ecotoxicol Environ Saf 55:300–306

Gravato C, Santos MA, Magalhães I (2000) Juvenile *Dicentrarchus labrax* L. biochemical and genotoxical responses after short-term exposure to ß-naphtoflavone and contaminated harbour waters. Fresenius Environ Bull 9:269–274

Grisolia CK, Starling FLRM (2001) Micronuclei monitoring of fishes from Lake Paranoá, under influence of sewage treatment plant discharges. Mutat Res 491:39–44

Grisolia CK, Oliveira ABB, Bonfim H, Klautau-Guimarães MN (2005) Genotoxicity evaluation of domestic sewage in a municipal wastewater treatment plant. Genet Mol Biol 28:334–338

Grisolia CK, Rivero CLG, Starling FLRM, da Silva ICR, Barbosa AC, Dorea JG (2009) Profile of micronucleus frequencies and DNA damage in different species of fish in a eutrophic tropical lake. Genet Mol Biol 32:138–143

Guilherme S, Válega M, Pereira ME, Santos MA, Pacheco M (2008) Erythrocytic nuclear abnormalities in wild and caged fish (*Liza aurata*) along an environmental mercury contamination gradient. Ecotoxicol Environ Saf 70:411–421

Gusso-Choueri PK, Choueri RB, Santos GS, Araújo GS, Cruz ACF, Stremel T, Campos SX, Cestari MM, Ribeiro CAO, Abessa DMS (2016) Assessing genotoxic effects in fish from a marine protected area influenced by former mining activities and other stressors. Mar Pollut Bull 104:229–239

Gutiérrez JM, Villar S, Plavan AA (2015) Micronucleus test in fishes as indicators of environmental quality in subestuaries of the Rìo de la Plata (Uruguay). Mar Pollut Bull 91:518–523

Hafez AM (2009) Mugil cephalus genome: a sensitive monitor for genotoxicity and cytotoxicity in aquatic environment. Aust J Basic Appl Sci 3:2176–2187

Hamann M, Boxall VA, Edyvane KS, Edwards JW, Soole KL (1999) Biochemical and cytogenetic biomarkers of pollutant exposure in marine fish (*Aldrichetta forsteri* Valenciennes and *Sillago schomburgkii* Peters) near industrial and metropolitan centres in South Australia. Biomarkers 4:290–302

Han Y, Li N, Oda Y, Ma M, Rao K, Wang Z, Jin W, Hong G, Li Z, Luo Y (2016) Evaluation of genotoxic effects of surface waters using a battery of bioassays indicating different mode of action. Ecotoxicol Environ Saf 133:448–456

Hansson T, Baršienė J, Tjärnlund U, Åkerman G, Linderoth M, Zebühr Y, Sternbeck J, Järnberg U, Balk L (2014) Cytological and biochemical biomarkers in adult female perch (*Perca fluviatilis*) in a chronically polluted gradient in the Stockholm recipient (Sweden). Mar Pollut Bull 81:27–40

Hariri M, Mirvaghefi A, Farahmand H, Taghavi L, Shahabinia A-R (2018) In situ assessment of Karaj River genotoxic impact with the alkaline comet assay and micronucleus test, on feral brown trout (*Salmo trutta fario*). Environ Toxicol Pharmacol 58:59–69

Hawkins WE, Overstreet RM, Walker WW (1988) Small fish models for identifying carcinogens in the aqueous environment. Water Res Bull 24:941–949

Hayashi M (2016) The micronucleus test – Most widely used in vivo genotoxicity test. Genes Environ 38:18

Hayashi M, Sofuni T, Ishidate M (1983) An application of acridine orange fluorescent staining to the micronucleus test. Mutat Res 120:241–247

Hayashi M, Ueda T, Uyeno K, Wada K, Kinae N, Saotome K, Tanaka N, Takai A, Sasaki YF, Asano N, Sofuni T, Ojima Y (1998) Development of genotoxicity assay systems that use aquatic organisms. Mutat Res 399:125–133

Hemachandra CK, Pathiratne A (2016) Combination of physico-chemical analysis, *Allium cepa* test system and *Oreochromis niloticus* erythrocyte based comet assay/nuclear abnormalities tests for cyto-genotoxicity assessments of treated effluents discharged from textile industries. Ecotoxicol Environ Saf 131:54–64

Hemachandra CK, Pathiratne A (2017) Cytogenotoxicity screening of source water, wastewater and treated water of drinking water treatment plants using two *in vivo* test systems: *Allium cepa* root based and Nile tilapia erythrocyte based tests. Water Res 108:320–329

Hooftman RN, de Raat WK (1982) Induction of nuclear anomalies (micronuclei) in the peripheral blood erythrocytes of the eastern mudminnow *Umbra pygmea* by ethyl methanesulphonate. Mutat Res 104:147–152

Hooftman RN, Vink GJ (1981) Cytogenetic effects on the eastern mudminnow *Umbra pygmea* exposed to ethyl methanesulphonate, benzo[a]pyrene, and river water. Ecotoxicol Environ Saf 5:261–269

Hook SE, Gallagher EP, Batley GE (2014) The role of biomarkers in the assessment of aquatic ecosystem health. Integr Environ Assess Manag 10:327–341

Hose JE (1994) Large-scale genotoxicity assessments in the marine environment. Environ Health Perspect 102(Suppl. 12):29–32

Hose JE, Cross JN, Smith SG, Diehl D (1987) Elevated circulating erythrocyte micronuclei in fishes from contaminated sites of Southern California. Mar Environ Res 22:167–176

Hoshina MM, de Franceschi de Angelis D, Marin-Morales MA (2008) Induction of micronucleus and nuclear alterations in fish (*Oreochromis niloticus*) by a petroleum refinery effluent. Mutat Res 656:44–48

Houk VS (1992) The genotoxicity of industrial wastes and effluents. A review. Mutat Res 277:91–138

Hughes JB, Hebert AT (1991) Erythrocyte micronuclei in winter flounder (*Pseudopleuronectes americanus*): results of field surveys during 1980-1988 from Virginia to Nova Scotia and in Long Island sound. Arch Environ Contam Toxicol 20:474–479

Hurtado-Alarcón JC, Solarte-David VA, López-Ortiz JB, Montoya-Herrera FL (2007) Prueba de micronúcleos en eritrocitos de sabaletas (*Brycon henni* E.) presentes en el río Porce y en el Embalse Porce II, Antioquia. Revista Electrón Ing Prod Acuícola 2:238–245 (in Spanish)

Hussain B, Sultana T, Sultana S, Masoud MS, Ahmed Z, Mahboob S (2018) Fish eco-genotoxicology: comet and micronucleus assay in fish erythrocytes as in situ biomarker of freshwater pollution. Saudi J Biol Sci 25:393–398

Hussain B, Fatima M, Al-Ghanim KA, Al-Misned F, Mahboob S (2020) Assessment of DNA integrity through MN bioassay of erythrocytes and histopathological changes in *Wallago attu* and *Cirrhinus mirigala* in response to freshwater pollution. Saudi J Biol Sci 27:251–260

Ilyinskikh NN, Ilyinskikh EN, Ilyinskikh IN (1998) Micronucleated erythrocytes frequency and radiocesium bioconcentration in pikes (*Esox lucius*) caught in the Tom River near the nuclear facilities of the Siberian Chemical Complex (Tomsk-7). Mutat Res 421:197–203

Ivanova L, Popovska-Percinic F, Slavevska-Stamenkovic V, Jordanova M, Rebok K (2016) Micronuclei and nuclear abnormalities in erythrocytes from barbel *Barbus Peloponnesius* revealing genotoxic pollution of the river Bregalnica. Maced Vet Rev 39:159–166

Jesus IS, Cestari MM, Bezerra MA, Affonso PRAM (2016) Genotoxicity effects in freshwater fish from a brazilian impacted river. Bull Environ Contam Toxicol 96:490–495

Jha AN (2004) Genotoxicological studies in aquatic organisms: an overview. Mutat Res 552:1–17

Jovanović J, Kolarević S, Milošković A, Radojković N, Simić V, Dojčinović B, Kračun-Kolarević M, Paunović M, Kostić J, Sunjog K, Timilijić J, Djordjević J, Gačić Z, Žegura B, Vuković-Gačić B (2018) Evaluation of genotoxic potential in the Velika Morava River Basin *in vitro* and *in situ*. Sci Total Environ 621:1289–1299

Kanev MO, Ozdemir K, Muranli FDG (2016) Genotoxic evaluation of the Ergene River, Turkey, on mosquito fish, *Gambussia affinis* (Baird and Girard, 1853) using the piscine micronucleus assay. Int J Aquat Biol 4:330–339

Katsumiti A, Valdez Domingos FX, Azevedo M, Silva MD, Damian RC, Almeida MIM, Silva de Assis HC, Cestari MM, Randi MAF, Oliveira Ribeiro CA, Freire CA (2009) An assessment of acute biomarker responses in the demersal catfish *Cathorops spixii* after the Vicuña Oil Spill in a harbour estuarine area in Southern Brazil. Environ Monit Assess 152:209–222

Kienzler A, Bopp SK, van der Linden S, Berggren E, Worth A (2016) Regulatory assessment of chemical mixtures: requirements, current approaches and future perspectives. Regul Toxicol Pharmacol 80:321–334

Kirschbaum AA, Seriani R, Pereira CDS, Assunção A, Abessa DMS, Rotundo MM, Ranzani-Paiva MJT (2009) Cytogenotoxicity biomarkers in fat snook *Centropomus parallelus* from Cananéia and São Vicente estuaries, SP, Brazil. Genet Mol Biol 32:151–154

Kligerman AD (1982) Fishes as biological detectors of the effects of genotoxic agents. In: Heddle JA (ed) Mutagenicity: new horizons in genetic toxicology. Academic, New York, pp 435–453

Kligerman AD, Bloom S (1975) A cytogenetics model for the study of chromosome aberrations in fishes. Mutat Res 31:334–335a

Kligerman AD, Bloom SE, Howell WM (1975) Umbra limi: a model for the study of chromosomes aberrations in fishes. Mutat Res 31:225–233

Klobučar GI, Stambuk A, Pavlica M, Peric MS, Hackenberger BK, Hylland K (2010) Genotoxicity monitoring of freshwater environments using caged carp (*Cyprinus carpio*). Ecotoxicology 19:77–84

Klymenko MO, Biedunkova OO (2016) Development stability and cytogenetic homeostasis of *Perca fluviatilis* (Percifomes, Percidae) in the rivers of Rivne Region. Vestn Zool 50:539–546

Klymenko MO, Biedunkova OO, Klymenko OM, Statnyk II (2018) Influence of river water quality on homeostasis characteristics of cypriniform and perciform fish. Biosyst Divers 26:16–23

Knasmüller S, Fenech M (2019) The micronucleus assay in toxicology. Issues in toxicology. The Royal Society of Chemistry, Cambridge

Koca S, Koca YB, Yildiz S, Gürcü B (2008) Genotoxic and histopathological effects of water pollution on two fish species, *Barbus capito pectoralis* and *Chondrostoma nasus* in the Büyük Menderes River, Turkey. Biol Trace Elem Res 122:276–291

Köhler A, Ellesat K (2008) Nuclear changes in blood, early liver anomalies and hepatocellular cancers in flounder (*Platichthys flesus* L.) as prognostic indicator for a higher cancer risk? Mar Environ Res 66:149–150

Kolak A, Treer T, Anicic I, Safner R (1999) Monitoring the genotoxicity of the river Sava by micronuclei in chub (*Leuciscus cephalus*). Cytobios 392:135–142

Kolarević S, Aborgiba M, Kračun-Kolarević M, Kostić J, Simonović P, Simić V, Milošković A, Reischer G, Farnleitner A, Gačić Z, Milačič R, Zuliani T, Vidmar J, Pergal M, Piria M, Paunović M, Vuković-Gačić B (2016) Evaluation of genotoxic pressure along the Sava river. PLoS One 11:e0162450

Kontaş S, Bostancı D (2020) Genotoxic effects of environmental pollutant heavy metals on *Alburnus chalcoides* (Pisces: Cyprinidae) inhabiting lower Melet River (Ordu, Turkey). Bull Environ Contam Toxicol 104:763–769

Kopecka J, Lehtonen KK, Barsiene J, Broeg K, Vuorinen PJ, Gercken J, Pempkowiak J (2006) Measurements of biomarker levels in flounder (*Platichthys flesus*) and blue mussel (*Mytilus trossulus*) from the Gulf of Gdansk (Southern Baltic). Mar Poll Bull 53:406–421

Kračun-Kolarević M, Kolarević S, Jovanović J, Marković V, Ilić M, Simonović P, Simić V, Gačić Z, Diamantini E, Stella E, Petrović M, Majone B, Bellin A, Paunović M, Vuković-Gačić B (2016) Evaluation of genotoxic potential throughout the upper and middle stretches of Adige river basin. Sci Total Environ 571:1383–1391

Kroon F, Streten C, Harries S (2017) A protocol for identifying suitable biomarkers to assess fish health: a systematic review. PLoS One 12:e0174762

Kushwaha B, Pandey S, Sharma S, Srivastava R, Kumar R, Nagpure NS, Dabas A, Srivastava SK (2012) In situ assessment of genotoxic and mutagenic potential of polluted river water in *Channa punctatus* and *Mystus vittatus*. Int Aquat Res 4:16

Lacerda D, Vergilio CS, Silva Souza T, Costa LHV, Rangel TP, Oliveira BCV, Almeida DQR, Pestana IA, Almeida MG, Rezende CE (2020) Comparative metal accumulation and toxicogenetic damage induction in three neotropical fish species with distinct foraging habits and feeding preferences. Ecotoxicol Environ Saf 195:110449

Lapuente J, Lourenço J, Mendo SA, Borràs M, Martins MG, Costa PM, Pacheco M (2015) The Comet assay and its applications in the field of ecotoxicology: a mature tool that continues to expand its perspectives. Front Genet 6:180

Lasheen MR, Abdel-Gawad FK, Alaneny AA, Abd El Bary HMH (2012) Fish as bioindicators in aquatic environmental pollution assessment: a case study in Abu-Rawash area, Egypt. World Appl Sci J 19:265–275

Lee RF, Steinert S (2003) Use of the single cell gel electrophoresis/comet assay for detecting DNA damage in aquatic (marine and freshwater) animals. Mutat Res 544:43–64

Lehtonen KK, Schiedek D, Köhler A, Lang T, Vuorinen PJ, Förlin L, Baršienė J, Pempkowiak J, Gercken J (2006) The BEEP project in the Baltic Sea: overview of results and outline for a regional biological effects monitoring strategy. Mar Poll Bull 53:523–537

Lemos CT, Rödel PM, Terra NR, Oliveira NCD, Erdtmann B (2007) River water genotoxicity evaluation using micronucleus assay in fish erythrocytes. Ecotoxicol Environ Saf 66:391–401

Lemos CT, Almeida Iranco F, Oliveira NCD, Souza GD, Fachel JMG (2008) Biomonitoring of genotoxicity using micronuclei assay in native population of *Astyanax jacuhiensis* (Characiformes: Characidae) at sites under petrochemical influence. Sci Total Environ 406:337–343

Leusch FDL, Snyder SA (2015) Bioanalytical tools: half a century of application for potable reuse. Environ Sci Water Res Technol 1:606–621

Lima Cardoso R, Carvalho-Neta RNF, Castro ACL, Ferreira CFC, Silva MHL, Jesus Azevedo JW, Sobrinho JRSC, Santos DMS (2018) Histological and genotoxic biomarkers in *Prochilodus lacustris* (Pisces, Prochilodontidae) for environmental assessment in a protected area in the Northeast of Brazil. Bull Environ Contam Toxicol 101:570–579

Lima LBD, Morais PB, Andrade RLT, Mattos LV, Moron SE (2018) Use of biomarkers to evaluate the ecological risk of xenobiotics associated with agriculture. Environ Pollut 237:611–624

Lima ARB, Torres RA, Jacobina UP, Pinheiro MAA, Adam ML (2019) Genomic damage in *Mugil curema* (Actinopterygii: Mugilidae) reveals the effects of intense urbanization on estuaries in northeastern Brazil. Mar Poll Bull 138:63–69

Lindberg HK, Wang X, Järventaus H, Falck GC-M, Norppa H, Fenech M (2007) Origin of nuclear buds and micronuclei in normal and folate-deprived human lymphocytes. Mutat Res 617:33–45

Linde-Arias AR, Inácio AF, Alburquerque C, Freire MM, Moreira JC (2008) Biomarkers in an invasive fish species, *Oreochromis niloticus*, to assess the effects of pollution in a highly degraded Brazilian River. Sci Total Environ 399:186–192

Liney K, Hagger JA, Tyler CR, Depledge MH, Galloway TS, Jobling S (2006) Health effects in fish of long-term exposure to effluents from wastewater treatment works. Environ Health Perspect 114(Suppl. 1):81–89

Llorente MT, Martos A, Castaño A (2002) Detection of cytogenetic alterations and blood cell changes in natural populations of carp. Ecotoxicology 11:27–34

Long ER, Buchman MF (1989) An evaluation of candidate measures of biological effects for the national status and trends program. NOAA Tech. Memo., NOS OMA 45

Longwell AC, Perry D, Hughes JB, Hebert A (1983) Frequencies of micronuclei in mature and immature erythrocytes of fish as an estimate of chromosome mutation rates-results of field surveys on windowpane flounder, winter flounder and Atlantic mackerel. International Council for the Exploration of the Sea, C.M. 1983/E:55, Marine Environmental Quality Committee, Ref. Demersal and Pelagic Fish Committees, 14 pp

Longwell AC, Chang S, Hebert A, Hughes JB, Perry D (1992) Pollution and developmental abnormalities in Atlantic fishes. Environ Biol Fish 35:1–21

Loro VL, Murussi C, Menezes C, Leitemperger J, Severo E, Guerra L, Costa M, Perazzo GX, Zanella R (2015) Spatial and temporal biomarkers responses of *Astyanax jacuhiensis* (Cope,

1894) (Characiformes: Characidae) from the middle Rio Uruguai, Brazil. Neotrop Ichthyol 13:569–578

Lugas'kova NV (2003) Species specificity of the cytogenetic stability of fish in a eutrophic water body. Russ J Ecol 34:210–214

Luzhna L, Kathiria P, Kovalchuk O (2013) Micronuclei in genotoxicity assessment: from genetics to epigenetics and beyond. Front Genet 4:131

Lyne TB, Bickham JW, Lamb T, Gibbons JW (1992) The application of bioassays in risk assessment of environmental pollution. Risk Anal 12:361–365

Maceda-Veiga A, Monroy M, Viscor G, De Sostoa A (2010) Changes in non-specific biomarkers in the Mediterranean barbel (*Barbus meridionalis*) exposed to sewage effluents in a Mediterranean stream (Catalonia, NE Spain). Aquat Toxicol 100:229–237

Maceda-Veiga A, Monroy M, Navarro E, Viscor G, De Sostoa A (2013) Metal concentrations and pathological responses of wild native fish exposed to sewage discharge in a Mediterranean river. Sci Total Environ 449:9–19

MacGregor JT, Wehr CM, Gould DH (1980) Clastogen-induced micronuclei in peripheral blood erythrocytes: the basis of an improved micronucleus test. Environ Mutagen 2:509–514

Mallick P, Khuda-Bukhsh AR (2003) Nuclear anomalies and blood protein variations in fish of the Hooghly-Matlah River system, India, as an indicator of genotoxicity in water. Bull Environ Contam Toxicol 70:1071–1082

Manna GK, Banerjee G, Gupta S (1985) Micronuclei test in the peripheral erythrocytes of the exotic fish, *Oreochromis massambica*. Nucleus 28:176–179

Maria VL, Correia AC, Santos MA (2003a) Genotoxic and hepatic biotransformation responses induced by the overflow of pulp mill and secondary-treated effluents on *Anguilla anguilla* L. Ecotoxicol Environ Saf 55:126–137

Maria VL, Correia AC, Santos MA (2003b) Genotoxic and biochemical responses in caged eel (*Anguilla anguilla* L.) after short-term exposure to harbour waters. Environ Int 29:923–929

Maria VL, Teles M, Pacheco M, Correia AC, Santos MA (2004) Biomarker responses in a polluted river: effects of pulp and paper mill contaminants on caged *Anguilla anguilla* L. Fresenius Environ Bull 13:317–325

Marlasca MJ, Sanpera C, Riva MC, Sala R, Crespo S (1998) Hepatic alterations and induction of micronuclei in rainbow trout (*Oncorhynchus mykiss*) exposed to a textile industry effluent. Histol Histopathol 13:703–712

Matsumoto ST, Mantovani MS, Malaguttii MIA, Dias AL, Fonseca IC, Marin-Morales MA (2006) Genotoxicity and mutagenicity of water contaminated with tannery effluents, as evaluated by the micronucleus test and comet assay using the fish *Oreochromis niloticus* and chromosome aberrations in onion root-tips. Genet Mol Biol 29:148–158

McGregor DB, Rice JM, Venitt S (1999) The use of short- and medium-term tests for carcinogens and data on genetic effects in carcinogenic hazard evaluation. IARC scientific publications, vol. 146. IARC, Lyon

Melo KM, Alves IR, Pieczarka JC, Oliveira David JA, Nagamachi CY, Grisolia CK (2013) Profile of micronucleus frequencies and nuclear abnormalities in different species of electric fishes (Gymnotiformes) from the Eastern Amazon. Genet Mol Biol 36:425–429

Metcalfe CD (1988) Induction of micronuclei and nuclear abnormalities in the erythrocytes of mudminnows (*Umbra limi*) and brown bullheads (*Ictalurus nebulosus*). Bull Environ Contam Toxicol 40:489–495

Meyer JN (2010) QPCR: a tool for analysis of mitochondrial and nuclear DNA damage in ecotoxicology. Ecotoxicology 19:804–811

Mijat V (2005) Micronucleus test in most recent ichthiology research. Ribarstvo 63:21–32 (in Serbian)

Minissi S, Ciccotti E, Rizzoni M (1996) Micronucleus test in erythrocytes of *Barbus plebejus* (Teleostei, Pisces) from two natural environments: a bioassay for the *in situ* detection of mutagens in freshwater. Mutat Res 367:245–251

Mitchelmore CL, Chipman JK (1998) DNA strand breakage in aquatic organisms and the potential value of the comet assay in environmental monitoring. Mutat Res 399:135–147

Mohmood I, Maria VL, Oliveira M, Ahmad I, Pacheco M, Santos MA (2008) Seasonal assessment of a contaminated coastal lagoon (Ria de Aveiro, Portugal) using *Dicentrarchus labrax* L. erythrocytic nuclear abnormalities. Fresenius Environ Bull 17:1924–1931

Monserrat JM, Martínez PE, Geracitano LA, Amado LL, Martins CM, Pinho GL, Chaves IS, Ferreira-Cravo M, Ventura-Lima J, Bianchini A (2007) Pollution biomarkers in estuarine animals: critical review and new perspectives. Comp Biochem Physiol C Toxicol Pharmacol 146:221–234

Montesano R, Bartsch R, Vainio H, Wilbourn J, Yamasaki H (1987) Long-term and short-term assays for carcinogens. A critical appraisal. IARC scientific publications, vol. 83. IARC, Lyon

Morais CR, Carvalho SM, Araujo GR, Souto HN, Bonetti AM, Morelli S, Campos Júnior EO (2016) Assessment of water quality and genotoxic impact by toxic metals in *Geophagus brasiliensis*. Chemosphere 152:328–334

Morina V, Aliko V, Gavazaj F, Kastrati D (2012) Use of blood parameters as biomarkers of contaminant exposure in fish specimens from Sitnica River, Kosovo. J Int Environ Appl Sci 7:971–977

Müller L (2009) In-vitro genotoxicity tests to detect carcinogenicity: a systematic review. Hum Exp Toxicol 28:131–133

Nagpure NS, Srivastava R, Kumar R, Dabas A, Kushwaha B, Kumar P (2015) Assessment of pollution of river Ganges by tannery effluents using genotoxicity biomarkers in murrel fish, *Channa punctatus* (Bloch). Indian J Exp Biol 53:476–483

Napierska D, Barsiene J, Mulkiewicz E, Podolska M, Rybakovas A (2009) Biomarker responses in flounder *Platichthys flesus* from the Polish coastal area of the Baltic Sea and applications in biomonitoring. Ecotoxicology 18:846–859

Obiakor MO, Ezeonyejiaku CD, Ezenwelu CO, Ugochukwu GC (2010) Aquatic genetic biomarkers of exposure and effect in catfish (*Clarias gariepinus*, Burchell, 1822). Am-Euras J Toxicol Sci 2:196–202

Obiakor MO, Okonkwo JC, Nnabude PC, Ezeonyejiaku CD (2012) Eco-genotoxicology: micronucleus assay in fish erythrocytes as *in situ* aquatic pollution biomarker. A review. J Anim Sci Adv 2:123–133

Obiakor MO, Okonkwo JC, Ezeonyejiaku CD (2014) Genotoxicity of freshwater ecosystem shows DNA damage in preponderant fish as validated by in vivo micronucleus induction in gill and kidney erythrocytes. Mutat Res 775–776:20–30

Odeigah PG, Osanyinpeju AO (1995) Genotoxic effects of two industrial effluents and ethyl methane sulfonate in *Clarias lazera*. Food Chem Toxicol 33:501–505

Ohe T, Watanabe T, Wakabayashi K (2004) Mutagens in surface waters: a review. Mutat Res 567:109–149

Oikari A (2006) Caging techniques for field exposure of fish to chemical contaminants. Aquat Toxicol 78:370–381

Ojima Y, Ueno K, Hayashi M (1976) A review of the chromosome number of fish. La Kromosomo 2:19–47

Okoye CN, Mac Donald-Jay N, Kamunde C (2019) Effects of bioenergetics, temperature and cadmium on liver mitochondria reactive oxygen species production and consumption. Aquat Toxicol 214:105264

Oliveira M, Ahmad I, Maria VL, Ferreira CSS, Serafim A, Bebianno MJ, Pacheco M, Santos MA (2010a) Evaluation of oxidative DNA lesions in plasma and nuclear abnormalities in erythrocytes of wild fish (*Liza aurata*) as an integrated approach to genotoxicity assessment. Mutat Res 703:83–89

Oliveira M, Maria VL, Ahmad I, Teles M, Serafim A, Bebianno MJ, Pacheco M, Santos MA (2010b) Golden grey mullet and sea bass oxidative DNA damage and clastogenic/aneugenic responses in a contaminated coastal lagoon. Ecotoxicol Environ Saf 73:1907–1913

Oliveira-Martins CR, Grisolia CK (2009) Toxicity and genotoxicity of wastewater from gasoline stations. Genet Mol Biol 32:853–856

Omar WA, Zaghloul KH, Abdel-Khalek AA, Abo-Hegab S (2012) Genotoxic effects of metal pollution in two fish species, *Oreochromis niloticus* and *Mugil cephalus*, from highly degraded aquatic habitats. Mutat Res 746:7–14

Osman AGM, Kloas W (2010) Water quality and heavy metal monitoring in water, sediments, and tissues of the African catfish *Clarias gariepinus* (Burchell, 1822) from the River Nile, Egypt. J Environ Prot 1:389–400

Osman AGM, Abd El Reheem AM, Moustafa MA, Mahmoud UM, Abuel-Fadl KY, Kloas W (2011) In situ evaluation of the genotoxic potential of the river Nile: I. Micronucleus and nuclear lesion tests of erythrocytes of *Oreochromis niloticus niloticus* (Linnaeus, 1758) and *Clarias gariepinus* (Burchell, 1822). Toxicol Environ Chem 93:1002–1017

Ossana NA, Eissa BL, Baudou FG, Castañé PM, Soloneski S, Ferrari L (2016) Multibiomarker response in ten spotted live-bearer fish *Cnesterodon decemmaculatus* (Jenyns, 1842) exposed to Reconquista river water. Ecotoxicol Environ Saf 133:73–81

Ossana NA, Baudou FG, Castañé PM, Tripoli L, Soloneski S, Ferrari L (2019) Histological, genotoxic, and biochemical effects on *Cnesterodon decemmaculatus* (Jenyns 1842) (Cyprinodontiformes, Poeciliidae): early response bioassays to assess the impact of receiving waters. J Toxicol 2019:4687685

Pacheco M, Santos MA (1996) Induction of micronuclei and nuclear abnormalities in the erythrocytes of *Anguilla anguilla* L. exposed either to cyclophosphamide or to bleached kraft pulp mill effluent. Fresenius Environ Bull 5:746–751

Pacheco M, Santos MA (1999) Biochemical and genotoxic responses of adult eel (*Anguilla anguilla* L.) to resin acids and pulp mill effluent: laboratory and field experiments. Ecotoxicol Environ Saf 42:81–93

Pacheco M, Santos MA (2001) Biotransformation, endocrine, and genetic responses of *Anguilla anguilla* L. to petroleum distillate products and environmentally contaminated waters. Ecotoxicol Environ Saf 49:64–75

Pacheco M, Santos MA, Teles M, Oliveira M, Rebelo JE, Pombo L (2005) Biotransformation and genotoxic biomarkers in mullet species (*Liza* sp.) from a contaminated coastal lagoon (Ria de Aveiro, Portugal). Environ Monit Assess 107:133–153

Palacio-Betancur I, Palacio-Baena JA, Camargo-Guerrero M (2009) Micronuclei test application to wild tropical ichthyic species common in two lentic environments of the low zones in Colombia. Actual Biol 31:67–77

Palm A, Krause T (1995) Micronucleated erythrocytes in fish from several Estonian waterbodies. Proc Estonian Acad Sci Biol 44:69–77

Palm A, Tuvikene A, Krause T (1992) Changes of hematological characteristics of the rainbow trout (*Oncorhynchus mykiss*) reared in the mixture of natural and oil-shale mine drainage water. Proc Estonian Acad Sci Biol 41:183–181

Pampanin DM, Brooks SJ, Grøsvik BE, Le Goff J, Meier S, Sydnes MO (2017) DNA adducts in marine fish as biological marker of genotoxicity in environmental monitoring: the way forward. Mar Environ Res 125:49–62

Pantaleão SM, Alcântara AV, Alves JPH, Spanó MA (2006) The piscine micronucleus test to assess the impact of pollution on the Japaratuba river in Brazil. Environ Mol Mutagen 47:219–224

Pavlica M, Stambuk A, Malovic L, Mladinic M, Klobucar GI (2011) DNA integrity of chub erythrocytes (*Squalius cephalus* L.) as an indicator of pollution-related genotoxicity in the river Sava. Environ Monit Assess 177:85–94

Pimenta VMSD, Silva JM, Nepomuceno JC, Pavanin LA (2013) In situ assessment of the Paraguay River water, in Brazilian Pantanal, by means of micronucleus assay with fish and chemical analysis. Bull Environ Contam Toxicol 90:427–433

Pinheiro-Sousa DB, Torres Junior AR, Silva D, Santos RL, Carvalho Neta RNF (2019) A screening test based on hematological and histological biomarkers to evaluate the environmental impacts in tambaqui (*Colossoma macropomum*) from a protected area in Maranhão, Brazilian Amazon. Chemosphere 214:445–451

Polard T, Jean S, Gauthier L, Laplanche C, Merlina G, Sánchez-Pérez JM, Pinelli E (2011a) Mutagenic impact on fish of runoff events in agricultural areas in south-west France. Aquat Toxicol 101:126–134

Polard T, Jean S, Merlina G, Laplanche C, Pinelli E, Gauthier L (2011b) Giemsa versus acridine orange staining in the fish micronucleus assay and validation for use in water quality monitoring. Ecotoxicol Environ Saf 74:144–149

Poongothai K, Shayin S, Usharani MV (1996) Induction of micronuclei in fish by polluted water and heavy metal. Cytobios 86:17–22

Porto JIR, Araujo CSO, Feldberg E (2005) Mutagenic effects of mercury pollution as revealed by micronucleus test on three Amazonian fish species. Environ Res 97:287–292

Powers DA (1989) Fish as model systems. Science 246:352–358

Prasse C, Stalter D, Schulte-Oehlmann U, Oehlmann J (2015) Spoilt for choice: a critical review on the chemical and biological assessment of current wastewater treatment technologies. Water Res 87:237–270

Praveena M, Sailaja MC, Rao KJ (2014) Clastogenic effects induced due to chromium intoxication in a freshwater fish, *Labeo rohita*. Int J Adv Sci Tech Res 4:647–656

Ramsdorf WA, Vicari T, Almeida MIM, Artoni RF, Cestari MM (2012) Handling of *Astyanax* sp. for biomonitoring in Cangüiri Farm within a fountainhead (Iraí River Environment Preservation Area) through the use of genetic biomarkers. Environ Monit Assess 184:5841–5849

Rebok K, Jordanova M, Slavevska-Stamenković V, Ivanova L, Kostov V, Stafilov T, Rocha E (2017) Frequencies of erythrocyte nuclear abnormalities and of leucocytes in the fish *Barbus peloponnesius* correlate with a pollution gradient in the River Bregalnica (Macedonia). Environ Sci Pollut Res 24:10493–10509

Ribeiro DL, Barcelos GRM, Arce LPG (2014) Genotoxic effects of water from São Francisco River, Brazil, in *Astyanax paranae*. Bull Environ Contam Toxicol 93:274–279

Risjani Y, Loppion G, Couteau J, Yunianta Y, Widowati I, Hermawati A, Minier C (2020) Genotoxicity in the rivers from the Brantas catchment (East Java, Indonesia): occurrence in sediments and effects in *Oreochromis niloticus* (Linnæus 1758). Environ Sci Pollut Res 27:21905–21913

Rocha PS, Luvizotto GL, Kosmehl T, Böttcher M, Storch V, Braunbeck T, Hollert H (2009) Sediment genotoxicity in the Tiete River (São Paulo, Brazil): *in vitro* comet assay versus *in situ* micronucleus assay studies. Ecotoxicol Environ Saf 72:1842–1848

Rocha CAM, Pessoa CMF, Rodrigues CAC, Pinheiro RHS, Costa ET, Guimarães AC, Burbano RR (2016) Investigation into the cytotoxicity and mutagenicity of the Marajó archipelago waters using *Plagioscion squamosissimus* (Perciformes: Sciaenidae) as a bioindicator. Ecotoxicol Environ Saf 132:111–115

Rocha MP, Dourado PLR, Cardoso CAL, Cândido LS, Pereira JG, Oliveira KMP, Grisolia AB (2018) Tools for monitoring aquatic environments to identify anthropic effects. Environ Monit Assess 190:61–73

Rodriguez-Cea A, Ayllon F, Garcia-Vazquez E (2003) Micronucleus test in freshwater fish species: an evaluation of its sensitivity for application in field surveys. Ecotoxicol Environ Saf 56:442–448

Ross SW, Dalton DA, Kramer S, Christensen BL (2001) Physiological (antioxidant) responses of estuarine fishes to variability in dissolved oxygen. Comp Biochem Physiol C Toxicol Pharmacol 130:289–303

Russo C, Rocco L, Morescalchi MA, Stingo V (2004) Assessment of environmental stress by the micronucleus test and the comet assay on the genome of teleost populations from two natural environments. Ecotoxicol Environ Saf 57:168–174

Rybakovas A, Barsiene J, Lang T (2009) Environmental genotoxicity and cytotoxicity in the offshore zones of the Baltic and the north seas. Mar Environ Res 68:246–256

Salamone M, Heddle J, Stuart E, Katz M (1980) Towards an improved micronucleus test. Studies on 3 model agents, mitomycin C, cyclophosphamide and dimethylbenzanthracene. Mutat Res 74:347–356

Saleh KAJ (2010) The intensity of pollutant genotoxicity in Lake Uluabat. Int J Biotechnol Biochem 6:625–631

Saleh KAJ, Alshehri MA (2011) The intensity of pollutant genotoxicity in Lake Uluabat: investigation of the peripheral erythrocytes of *Cyprinus carpio*. Afr J Biotechnol 10:16045–16050

Salvagni J, Ternus RZ, Fuentefria AM (2011) Assessment of the genotoxic impact of pesticides on farming communities in the countryside of Santa Catarina State, Brazil. Genet Mol Biol 34:122–126

Samanta P, Im H, Na J, Jung J (2018) Ecological risk assessment of a contaminated stream using multi-level integrated biomarker response in *Carassius auratus*. Environ Pollut 233:429–438

Sánchez-Galán S, Linde AR, Izquierdo JI, García-Vázquez E (1998) Micronuclei and fluctuating asymmetry in brown trout (*Salmo trutta*): complementary methods to biomonitor freshwater ecosystems. Mutat Res 412:219–225

Sánchez-Galán S, Linde AR, Ayllon F, García-Vázquez E (2001) Induction of micronuclei in eel (*Anguilla anguilla* L.) by heavy metals. Ecotoxicol Environ Saf 49:139–143

SanJuan-Reyes N, Gómez-Oliván LM, Galar-Martínez M, García-Medina S, Islas-Flores H, González-González ED, Cardoso-Vera JD, Jiménez-Vargas JM (2015) NSAID-manufacturing plant effluent induces geno- and cytotoxicity in common carp (*Cyprinus carpio*). Sci Total Environ 530–531:1–10

Schiedek D, Broeg K, Barsiene J, Lehtonen KK, Gercken J, Pfeifer S, Vuontisjärvi H, Vuorinen PJ, Dedonyte V, Koehler A, Balk L, Schneider R (2006) Biomarker responses as indication of contaminant effects in blue mussel (*Mytilus edulis*) and female eelpout (*Zoarces viviparus*) from the southwestern Baltic Sea. Mar Pollut Bull 53:387–405

Schlegel R, MacGregor JT (1982) The persistence of micronuclei in peripheral blood erythrocytes: detection of chronic chromosome breakage in mice. Mutat Res 104:367–369

Schlegel R, MacGregor JT (1983) A rapid screen for cumulative chromosome damage in mice. Accumulation of circulating micronucleated erythrocytes. Mutat Res 113:481–487

Schmid W (1975) The micronucleus test. Mutat Res 31:9–15

Schultz N, Norrgren L, Grawe J, Johannisson A, Medhage O (1993) Micronuclei frequency in circulating erythrocytes from rainbow trout (*Oncorhynchus mykiss*) subjected to radiation, an image analysis and flow cytometric study. Comp Biochem Physiol C 105:207–211

Seriani R, Ranzani-Paiva MJT, Silva-Souza AT, Napoleão SR (2011) Hematology, micronuclei and nuclear abnormalities in fishes from São Francisco river, Minas Gerais state, Brazil. Acta Sci Biol Sciences 33:107–112

Seriani R, Abessa DMS, Kirschbaum AA, Pereira CDS, Ranzani-Paiva MJT, Assunçao A, Silveira FL, Romano P, Mucci JLN (2012) Water toxicity and cyto-genotoxicity biomarkers in the fish *Oreochromis niloticus* (Cichlidae). J Braz Soc Ecotoxicol 7:67–72

Seriani R, Abessa DMS, Moreira LB, Cabrera JPG, Sanches JQ, Silva CLS, Amorim FA, Rivero DHRF, Silva FL, Fitorra LS, Carvalho-Oliveira R, Macchione M, Ranzani-Paiva MJT (2015) In vitro mucus transportability, cytogenotoxicity, and hematological changes as non-destructive physiological biomarkers in fish chronically exposed to metals. Ecotoxicol Environ Saf 112:162–168

Serrano-Garcìa L, Montero-Montoya R (2001) Micronuclei and chromatin buds are the result of related genotoxic events. Environ Mol Mutagen 38:38–45

Silva Souza T, Fontanetti CS (2006) Micronucleus test and observation of nuclear alterations in erythrocytes of Nile tilapia exposed to waters affected by refinery effluent. Mutat Res 605:87–93

Silva MD, Rossi SC, Ghisi NC, Ribeiro CAO, Cestari MM, Assis HCS (2014) Using multibiomarker approach as a tool to improve the management plan for a Private Reserve of Natural Heritage (RPPN). Bull Environ Contam Toxicol 92:602–608

Silva EB, Corrêa SAS, Abessa DMS, Silva BFX, Rivero DHRF, Seriani R (2018) Mucociliary transport, differential white blood cells, and cyto-genotoxicity in peripheral erythrocytes in fish from a polluted urban pond. Environ Sci Pollut Res 25:2683–2690

Smagin AI, Lugaskova NV, Menshikh TB (2005) Study of fish from the Techa river waste storage reservoir with micronuclei testing method. Izv Chelyabinskogo nauchnogo tsentra UrO RAN 1:105–109 (in Russian)

Smith IR (1990) Erythrocytic micronuclei in wild fish from Lakes Superior and Ontario that have pollution-associated neoplasia. J Great Lakes Res 16:139–142

Soldatov AA (1995) Peculiarities of organization and functioning of the fish red blood system. J Evol Biochem Physiol 41:272–281

Sommer S, Buraczewska I, Kruszewski M (2020) Micronucleus assay: the state of art, and future directions. Int J Mol Sci 21:1534

Sousa EL, Pereira NJ, Gomes JB, Santos MM, Santos DMS (2020) Ecotoxicological analyses of springs of a Brazilian Northeast Conservation Unit. Bull Environ Contam Toxicol 104:27–34

Souza IC, Duarte ID, Pimentel NQ, Rocha LD, Morozesk M, Bonomo MM, Azevedo VC, Pereira CDS, Monferrán MV, Milanez CRD, Matsumoto ST, Wunderlin DA, Fernandes MN (2013) Matching metal pollution with bioavailability, bioaccumulation and biomarkers response in fish (*Centropomus parallelus*) resident in neotropical estuaries. Environ Pollut 180:136–144

Spies RB, Stegeman JJ, Rice DW, Woodlin B, Thomas P, Hose JE, Cross JN, Prieto M (1990) Sublethal responses of *Platichtus stellatus* to organic contamination in San Francisco Bay with emphasis on reproduction. In: McCarthy JF, Shugart LR (eds) Biomarkers of environmental contamination. CRC Press, Boca Raton, pp 87–122

Sposito JCV, Francisco LFV, Crispim BA, Silva Dantas FG, Souza JP, Viana LF, Solórzano JCJ, Oliveira KMP, Barufatti A (2019) Influence of land use and cover on toxicogenetic potential of surface water from Central-West Brazilian rivers. Arch Environ Contam Toxicol 76:483–495

Šrut M, Stambuk A, Pavlica M, Klobucar GI (2010) Cage exposure of european sea bass (*Dicentrarchus labrax*) for *in situ* assessment of pollution-related genotoxicity. Arh Hig Rada Toksikol 61:29–36

Stahl RGJ (1991) The genetic toxicology of organic compounds in natural waters and waste waters. Ecotoxicol Environ Saf 22:94–125

Stegeman JJ, Lech JJ (1991) Cytochrome P-450 monooxygenase systems in aquatic species: carcinogen metabolism and biomarkers for carcinogen and pollutant exposure. Environ Health Perspect 90:101–109

Strunjak-Perovic I, Coz-Rakovac R, Topic Popovic N (2003) Micronucleus occurrence in diploid and triploid rainbow trout (*Oncorhynchus mykiss* Walbaum). Vet Med-Czech 48:215–219

Strunjak-Perovic I, Coz-Rakovac R, Topic Popovic N, Jadan M (2009a) Seasonality of nuclear abnormalities in gilthead sea bream Sparus aurata (L.) erythrocytes. Fish Physiol Biochem 35:287–291

Strunjak-Perovic I, Topic Popovic N, Coz-Rakovac R, Jadan M (2009b) Nuclear abnormalities of marine fish erythrocytes. J Fish Biol 74:2239–2249

Sula E, Aliko V, Pagano M, Faggio C (2019) Digital light microscopy as a tool in toxicological evaluation of fish erythrocyte morphological abnormalities. Microsc Res Tech 83:362–369

Summak S, Aydemir NC, Vatan O, Yılmaz D, Zorlu T, Bilaloglu R (2010) Evaluation of genotoxicity from Nilufer stream (Bursa/Turkey) water using piscine micronucleus test. Food Chem Toxicol 48:2443–2447

Suter GW (2006) Ecological risk assessment. CRC Press, Boca Raton

Tabrez S, Shakil S, Urooj M, Damanhouri GA, Abuzenadah AM, Ahmad M (2011) Genotoxicity testing and biomarker studies on surface waters: an overview of the techniques and their efficacies. J Environ Sci Health C Environ Carcinog Ecotoxicol Rev 29:250–275

Talapatra SN, Banerjee SK (2007) Detection of micronucleus and abnormal nucleus in erythrocytes from the gill and kidney of *Labeo bata* cultivated in sewage-fed fish farms. Food Chem Toxicol 45:210–215

Teles M, Pacheco M, Santos MA (2007) Endocrine and metabolic responses of *Anguilla anguilla* L. caged in a freshwater-wetland (Pateira de Fermentelos, Portugal). Sci Total Environ 372:562–570

Tomasello B, Copat C, Pulvirenti V, Ferrito V, Ferrante M, Renis M, Sciacca S, Tigano C (2012) Biochemical and bioaccumulation approaches for investigating marine pollution using Mediterranean rainbow wrasse, *Coris julis* (Linneaus 1798). Ecotoxicol Environ Saf 86:168–175

Torres-Bugarìn O, Zavala-Aguirre JL, Gòmez-RubioP B-OHR, Zuñiga-Gonzàlez G, Garcìa-Ulloa Gòmez M (2007) Potential fish species as genotoxicity biomarkers at lake "La Alberca", Michoacàn, Mexico. Hidrobiologica 17:75–81 (in Spanish)

Tsangaris C, Vergolyas M, Fountoulaki E, Goncharuk VV (2011a) Genotoxicity and oxidative stress biomarkers in *Carassius gibelio* as endpoints for toxicity testing of Ukrainian polluted river waters. Ecotoxicol Environ Saf 74:2240–2244

Tsangaris C, Vergolyas M, Fountoulaki E, Nizheradze K (2011b) Oxidative stress and genotoxicity biomarker responses in grey mullet (*Mugil cephalus*) from a polluted environment in Saronikos gulf, Greece. Arch Environ Contam Toxicol 61:482–490

Tuvikene A, Huuskonen S, Koponen K, Ritola O, Mauer Ü, Lindström-Seppä P (1999) Oil shale processing as a source of aquatic pollution: monitoring of the biologic effects in caged and feral freshwater fish. Environ Health Perspect 107:745–752

Udroiu I (2006) The micronucleus test in piscine erythrocytes. Aquat Toxicol 79:201–204

Ulupinar M, Okumus I (2002) Detection of mutagenic-carcinogenic pollutants in aquatic systems using cytogenetic methods in fish. Turk J Zool 26:141–148

Valskienė R, Baršienė J, Butrimavičienė L, Pažusienė J, Grygiel W, Stankevičiūtė M, Rybakovas A (2019) Induction of nuclear abnormalities in herring (*Clupea harengus membras*), flounder (*Platichthys flesus*), and Atlantic cod (*Gadus morhua*) collected from the southern part of the Gotland Basin – the Baltic Sea (2010-2017). Environ Sci Pollut Res 26:13366–13380

van der Burg B, van der Linden S, Man H-Y, Winter R, Jonker L, van Vugt-Lussenburg B, Brower A (2013) A panel of quantitative Calux® reporter gene assays for reliable high-throughput toxicity screening of chemicals and complex mixtures. In: Steinberg P (ed) High-throughput screening methods in toxicity testing. Wiley, pp 519–532

van der Linden SC, von Bergh AR, van Vught-Lussenburg BM, Jonker LR, Teunis M, Krul CA, van der Burg B (2014) Development of a panel of high-throughput reporter-gene assays to detect genotoxicity and oxidative stress. Mutat Res 760:23–32

van der Oost R, Beyer J, Vermeulen NP (2003) Fish bioaccumulation and biomarkers in environmental risk assessment: a review. Environ Toxicol Pharmacol 13:57–149

van der Oost R, Sileno G, Janse T, Nguyen MT, Besselink H, Brouwer A (2017a) SIMONI (smart integrated monitoring) as a novel bioanalytical strategy for water quality assessment: part II-field feasibility survey. Environ Toxicol Chem 36:2400–2416

van der Oost R, Sileno G, Suárez-Muñoz M, Nguyen MT, Besselink H, Brouwer A (2017b) SIMONI (smart integrated monitoring) as a novel bioanalytical strategy for water quality assessment: part I-model design and effect-based trigger values. Environ Toxicol Chem 36:2385–2399

Van Ngan P, Gomes V, Passos MJACR, Ussami KA, Campos DYF, Silva Rocha AJ, Pereira BA (2007) Biomonitoring of the genotoxic potential (micronucleus and erythrocyte nuclear abnormalities assay) of the Admiralty Bay water surrounding the Brazilian Antarctic Research Station "Comandante Ferraz", King George Island. Polar Biol 30:209–217

Venturoti GP, Boldrini-França J, Gomes AS, Chisté B, Gomes LC (2019a) Geophagus brasiliensis (Teleostei: Cichlidae) as an indicator of toxicity of ornamental stone processing wastes. Comp Biochem Physiol C Toxicol Pharmacol 226:108639

Venturoti GP, Boldrini-França J, Kiffer WP, Francisco AP, Gomes AS, Gomes LC (2019b) Toxic effects of ornamental stone processing waste effluents on *Geophagus brasiliensis* (Teleostei: Cichlidae). Environ Toxicol Pharmacol 72:103268

Viana LF, Súarez YR, Cardoso CAL, Amaral Crispim B, Grisolia AB, Lima-Junior SE (2017) Mutagenic and genotoxic effects and metal contaminations in fish of the Amambai River, upper Paraná River, Brazil. Environ Sci Pollut Res 24:27104–27112

Viana LF, Súarez YR, Cardoso CAL, Amaral Crispim B, Cavalcante DNC, Grisolia AB, Lima-Junior SE (2018a) The response of neotropical fish species (Brazil) on the water pollution: metal bioaccumulation and genotoxicity. Arch Environ Contam Toxicol 75:476–485

Viana LF, Súarez YR, Cardoso CAL, Solórzano JCJ, Amaral Crispim B, Grisolia AB, Lima-Junior SE (2018b) Erythrocyte nuclear abnormalities in *Astyanax lacustris* in response to landscape characteristics in two neotropical streams. Arch Environ Contam Toxicol 75:327–334

Vieira CED, Almeida MS, Galindo BA, Pereira L, Martinez CBR (2014) Integrated biomarker response index using a neotropical fish to assess the water quality in agricultural areas. Neotrop Ichthyol 12:153–164

Vieira CED, Costa PG, Lunardelli B, de Oliveira LF, da Costa CL, Risso WE, Primel EG, Meletti PC, Fillmann G, Martinez CBR (2016) Multiple biomarker responses in *Prochilodus lineatus* subjected to short-term *in situ* exposure to streams from agricultural areas in Southern Brazil. Sci Total Environ 542:44–56

Viganò L, Camoirano A, Izzotti A, D'Agostini F, Polesello S, Francisci C, De Flora S (2002) Mutagenicity of sediments along the Po River and genotoxicity biomarkers in fish from polluted areas. Mutat Res 515:125–134

Vosyliene MZ, Kazlauskiene N, Baršienė J, Andreikenaite L, Milukaite A, Taujanskis E (2010) Ecotoxicity and genotoxicity relating to fish in wastewaters discharged from the Vilnius treatment plant. Water Sci Technol 62:859–865

Wachtel CC, Oliveira EC, Maniglia TC, Smith-Johannsen A, Almeida Roque A, Castilhos Ghisi N (2019) Waterborn genotoxicity in Southern Brazil using *Astyanax bifasciatus* (Pisces: Teleostei). Bull Environ Contam Toxicol 102:59–65

Walia GK, Handa D, Kaur H, Kalotra R (2015) Ecotoxicological studies on fish, *Labeo rohita* exposed to tannery industry effluent by using micronucleus test. Nucleus 58:111–116

Wernersson A-S, Carere M, Maggi C, Tusil P, Soldan P, James A, Sanchez W, Dulio V, Broeg K, Reifferscheid G, Buchinger S, Maas H, Van Der Grinten E, O'Toole S, Ausili A, Manfra L, Marziali L, Polesello S, Lacchetti I, Mancini L, Lilja K, Linderoth M, Lundeberg T, Fjällborg B, Porsbring T, Larsson DGJ, Bengtsson-Palme J, Förlin L, Kienle C, Kunz P, Vermeirssen E, Werner I, Robinson CD, Lyons B, Katsiadaki I, Whalley C, den Haan K, Messiaen M, Clayton H, Lettieri T, Carvalho RN, Gawlik BM, Hollert H, Di Paolo C, Brack W, Kammann U, Kase R (2015) The European technical report on aquatic effect-based monitoring tools under the water framework directive. Environ Sci Eur 27:7

White PA, Rasmussen JB (1998) The genotoxic hazards of domestic wastes in surface waters. Mutat Res 410:223–236

Wirzinger G, Weltje L, Gercken J, Sordyl H (2007) Genotoxic damage in field-collected three-spined sticklebacks (*Gasterosteus aculeatus* L.): a suitable biomonitoring tool? Mutat Res 628:19–30

Witeska M (2013) Erythrocytes in teleost fishes: a review. Zool Ecol 23:275–281

Würgler FE, Kramers PGN (1992) Environmental effects of genotoxins (eco-genotoxicology). Mutagenesis 7:321–327

Zakhidov ST, Chebotareva YV, Savvaitova KA, Maksimov VA (1996) A cytogenetic study of hemopoietic cells in fish from water bodies of the Norilo-Pyasinsk water system (Taimyr). Biol Bull 23:5–9

Zapata-Restrepo LM, Orozco-Jiménez LY, Rueda-Cardona M, Echavarría SL, Mena-Moreno N, Palacio-Baena JA (2017) Evaluación genotóxica del agua del Río Grande (Antioquia, Colombia) mediante frecuencia de eritrocitos micronucleados de *Brycon henni* (Characiformes: Characidae). Rev Biol Trop 65:405–414 (in Spanish)

Zhang W, Liu W, Zhang J, Zhao H, Zhang Y, Quan X, Jin Y (2012) Characterisation of acute toxicity, genotoxicity and oxidative stress posed by textile effluent on zebrafish. J Environ Sci 24:2019–2027

Zhang J, Chen S, Zhang Y, Quan X, Zhao H, Zhang Y (2014) Reduction of acute toxicity and genotoxicity of dye effluent using Fenton-coagulation process. J Hazard Mater 274:198–204

Printed by Printforce, the Netherlands